Botanic Gardens and the World Conservation Strategy

Conference organised by

The International Union for Conservation of Nature
and Natural Resources
(IUCN)

Gobierno de las Canarias

Cabildo Insular de Gran Canaria

Sponsored by

The World Wide Fund For Nature
(WWF)

under the Patronage of

The Food and Agriculture Organisation
of the United Nations
(FAO)

in co-operation with

The United Nations Educational, Scientific
and Cultural Organization
(Unesco)

and with the support of

The United Nations Environment Programme
(UNEP)

Funding from the Plant Program of the
World Wildlife Fund/Conservation
Foundation (Washington, DC)
and from the US Fish and Wildlife Service
has enabled the editors to send copies
of this book to all botanic gardens
in Latin America

Botanic Gardens and the World Conservation Strategy

Proceedings of an International Conference
26–30 November 1985
held at
Las Palmas de Gran Canaria

edited by
D. BRAMWELL, O. HAMANN, V. HEYWOOD, H. SYNGE

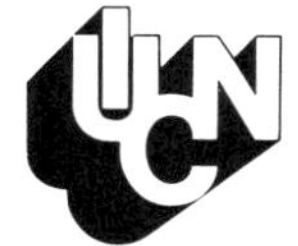

1987

PUBLISHED FOR IUCN BY ACADEMIC PRESS
London · Orlando · San Diego · New York · Austin · Boston · Sydney · Tokyo · Toronto
WITH THE SUPPORT OF R.J.R. NABISCO, INC.

ACADEMIC PRESS INC. (LONDON) LTD.
24/28 Oval Road,
London NW1 7DX

United States Edition published by
ACADEMIC PRESS INC.
Orlando, Florida 32887

British Library Cataloguing in Publication Data

Botanic Gardens and the
World Conservation Strategy.

1. Plant conservation   2. Botanical gardens

I. Bramwell, D.

639.9'9   QK86

ISBN 0-12-125462-3

Printed by St Edmundsbury Press,
Bury St Edmunds, England

IUCN

The International Union for Conservation of Nature and Natural Resources (IUCN) is a network of governments, non-governmental organisations (NGOs), scientists and other conservation experts, joined together to promote the protection and sustainable use of living resources.

Founded in 1948, IUCN has more than 500 member governments and NGOs in over 100 countries. Its six Commissions consist of more than 3000 experts on threatened species, protected areas, ecology, environmental planning, environmental policy, law and administration, and environmental education.

IUCN

- monitors the status of ecosystems and species throughout the world;
- plans conservation action, both at strategic level through the World Conservation Strategy and at the programme level through its programme of conservation for sustainable development;
- promotes such action by governments, inter-governmental bodies and non-governmental organisations;
- provides the assistance and advice necessary to achieve such action.

The IUCN Conservation Monitoring Centre is the division of the Union that provides a data service to IUCN and to the conservation and development communities. CMC is building an integrated and cross-referenced database on animals and plants of conservation concern, on wildlife trade, and on ecosystems (principally protected areas so far). The CMC produces a wide variety of specialist outputs and analyses, as well as major publications such as Red Data Books and Protected Area Directories. Other IUCN Centres focus on Environmental Law, and Conservation for Development. Collectively this information and expertise is used to assist IUCN members to prepare conservation programmes, and to provide information and advice to the conservation and development communities. Should you wish to consult these services please contact the:

IUCN Secretariat, World Conservation Centre
Avenue du Mont-Blanc, CH-1196 Gland, Switzerland.

Since 1984 IUCN and the World Wide Fund For Nature (WWF) have been implementing a Plants Conservation Programme, designed "to assert the fundamental importance of plants in all conservation activities".
The conference and this book that results is part of the Plants Conservation Programme.

The aim of this conference was to review the involvement of botanic gardens in implementing the World Conservation Strategy; in particular by considering their function as centres of information and education, and their capacity to carry out both *ex situ* and *in situ* conservation; by discussing how to make existing gardens more effective in plant conservation; and by agreeing on methods of international collaboration.

Organising Committee

Vice President, Autonomous Government
Director General for Environment, Autonomous Government
President of the Excmo. Cabildo Insular de Gran Canaria
President of the Environment Commission, Cabildo Insular de Gran Canaria
Representative of the Canarian Parliamentary Environment Committee
IUCN Regional Councillor
Director of the Jardín Botánico "Viera y Clavijo"
Representative of the Council of the Jardín Botánico "Viera y Clavijo"
The Head of Protocol of the Excmo. Cabildo Insular de Gran Canaria
Secretary Co-ordinator of the Organising Committee

Programme Committee

D. Bramwell (Canaries, Chairman)
C. Gómez-Campo (Spain)
O. Hamann (IUCN)
V.H. Heywood (Reading)
P. Raven (Missouri)
H. Synge (IUCN)

Contributors

AMBROSE, J.D. *University of Guelph Arboretum, Guelph, Ontario, N1G 2WI, Canada*

ASHTON, P.S. *The Arnold Arboretum of Harvard University, 22 Divinity Avenue, Cambridge, Massachusetts 02138, U.S.A.*

BEYER, R.I. *Royal Botanic Gardens, Kew, Richmond, Surrey TW9 3AB, U.K.*

BODEN, E.A. *Australian National Botanic Gardens, G.P.O Box 1777, Canberra, A.C.T. 2601, Australia.*

BODEN, R.W. *Australian National Botanic Gardens, G.P.O Box 1777, Canberra, A.C.T. 2601, Australia.*

BOLAÑOS, C. ARTILES *Cabildo Insular de Gran Canaria, Las Palmas, Islas Canarias, Spain.*

BRAMWELL, D. *Jardín Botánico Canario "Viera y Clavijo", Aptdo 14 de Tafira Alta, Las Palmas de Gran Canaria, Canary Islands, Spain.*

CREECH, J.L. *14 Legendary Road, Hendersonville, North Carolina 28739, USA.*

CRONK, Q. *Department of Botany, University of Cambridge, Downing Street, Cambridge, U.K.*

CULLEN, J. *Royal Botanic Garden, Edinburgh EH3 5LR, U.K.*

CURRAH, R.S. *Devonian Botanic Garden, University of Alberta, Edmonton, Alberta T6G 2E1, Canada.*

ELOFF, J.N. *National Botanic Gardens, Kirstenbosch, Claremont 7735, Republic of South Africa.*

ERN, H. *Botanischer Garten und Botanisches Museum, Königin-Luise-Strasse 6–8, D-1000 Berlin 33, F.R.G.*

FALK, D.A. *The Center for Plant Conservation, 125 The Arborway, Jamaica Plain, Massachusetts 02130, U.S.A.*

FORERO, E. *Missouri Botanical Garden, P.O. Box 299, St. Louis, Missouri 63166, U.S.A.*

GIVEN, D.R. *Botany Division, DSIR, Private Bag, Christchurch, New Zealand.*

GÓMEZ-CAMPO, C. *Escuela Técnica Superior de Ingenieros Agrónomos, Universidad Politécnica de Madrid, 28040 Madrid, Spain.*

GUNATILLEKE, C.V.S. *Department of Botany, University of Peradeniya, Peradeniya, Sri Lanka.*

GUNATILLEKE, I.A.U.N. *Department of Botany, University of Peradeniya, Peradeniya, Sri Lanka.*

HAMANN, O. *Botanisk Laboratorium, Køpenhauns Universitet, Guthersgade 140, 1123 Københaun K, Denmark.*

HAWKES, J.G. *c/o Department of Geological Sciences, The University of Birmingham, P.O. Box 363, Birmingham B15 2TT, U.K.*

HE SHAN-AN *Jiangsu Institute of Botany, Nanjing Botanical Garden, Mem. Sun Yat-Sen, Nanjing, Jiangsu, China.*

HERNÁNDEZ BERMEJO, J.E. *Jardín Botánico de Córdoba, Apdo 3048, 14071 Córdoba, Spain.*

HEYWOOD, V.H. *IUCN Conservation Monitoring Centre, 53 The Green, Kew, Richmond, Surrey TW9 3AA, U.K.*

JONES, D.T. *Department of Botany, University of Malaya, 59100 Kuala Lumpur, Malaysia.*

KHOSHOO, T.N. *Tata Energy Research Institute, 7 Jorbagh, New Delhi 110 003, India.*

LARSEN, K. *Botanical Institute, University of Aarhus, Nordlandsvej 68, DK-8240 Risskov, Denmark.*

LEAR, M. *Little Myrtle, The Fields, Mere, Wiltshire BA12 65A, U.K.*

MANGENOT, F. *Conservatoire et Jardins Botaniques de Nancy, 100 rue du Jardin Botanique, 54600 Villers-lès-Nancy, France.*

MACKINDER, D.C. *IUCN Conservation Monitoring Centre, 53 The Green, Kew, Richmond, Surrey TW9 3AA, U.K.*

MILLER, K.R. *International Union for Conservation of Nature and Natural Resources, Avenue du Mont-Blanc, CH-1196 Gland, Switzerland.*

MORLEY, B. *Botanic Gardens of Adelaide, N. Terrace, Adelaide, South Australia 5000, Australia.*

NAVARRO VALDIVIELSO, B. *Járdín Botánico Canario "Viera y Clavijo", Aptdo 14 de Tafira Alta, Las Palmas de Gran Canaria, Canary Islands, Spain.*

PARNELL, J. *School of Botany, Trinity College, Dublin, Ireland.*

PATTISON, G. *National Council for the Conservation of Plants and Gardens, R.H.S. Garden, Wisley, Ripley, Surrey GU23 6QB, U.K.*

RAVEN, P.H. *Missouri Botanical Garden, P.O. Box 299, St. Louis, Missouri 63166, U.S.A.*

REJDALI, MOH. *Institut Agronomique et Veterinaire Hassan II, B.P. 6202, Rabat, Morocco.*

SAJEVA, M. *Dipartimento di Scienze Botaniche dell'Universita di Palermo, Palermo, Via Archirafi 38, 90123 Italy.*

SÁNCHEZ-MEJORADAR, H. *Jardín Botánico, Instituto de Biología, Universidad Nacional Autónoma de México, Apto Postal 70-34, México 20 D.F., México.*

SASTRAPRADJA, S. *Puslitbang Bioteknologi, P.O. Box 323, Bogor, Indonesia.*

SASTRAPRADJA, D.S. *Puslitbang Bioteknologi, P.O. Box 323, Bogor, Indonesia.*

SEYMOUR, P.N.D. *Devonian Botanic Garden, University of Alberta, Edmonton, Alberta T6G 2E1, Canada.*

SHEN JIA-YU *Jiangsu Institute of Botany, Nanjing Botanical Garden, Mem. Sun Yat-Sen, Nanjing, Jiangsu, China.*

SIMMONS, J.B.E. *Royal Botanic Gardens, Kew, Richmond, Surrey TW9 3AB, U.K.*

SMRECIU, E.A. *Devonian Botanic Garden, University of Alberta, Edmonton, Alberta T6G 2E1, Canada.*

SOETISNA, U. *Puslitbang Bioteknologi, P.O. Box 323, Bogor, Indonesia.*

STRAHM, W. *c/o Forestry Quarters, Black River, Mauritius, Indian Ocean.*

SUMITHRAARACHCHI, B. *Botanic Gardens, Peradeniya, Sri Lanka.*

SYNGE, H. *49 Kelvedon Close, Kingston-upon-Thames, Surrey KT2 5LF, U.K.*

TAO JIN-CHUAN *Jiangsu Institute of Botany, Nanjing Botanical Garden, Mem. Sun Yat-Sen, Nanjing, Jiangsu, China.*

THIBODEAU, F.R. *The Center for Plant Conservation, 125 The Arborway, Jamaica Plain, Massachusetts 02130, U.S.A.*

VALCK, P. *Conservatoire et Jardins Botaniques de Nancy, 100 rue du Jardin Botanique, 54600 Villers-lès-Nancy, France.*

WANG MING-JUN *Jiangsu Institute of Botany, Nanjing Botanical Garden, Mem. Sun Yat-Sen, Nanjing, Jiangsu, China.*

WILLIAMS, J.T. *International Board for Plant Genetic Resources, FAO, Via delle Terme di Caracalla, 00100 Rome, Italy.*

WYSE JACKSON, P. *Botanic Gardens Conservation Secretariat, 53 The Green, Kew, Richmond, Surrey TW9 3AA U.K.*

XU ZAIFU *Yunnan Institute of Tropical Botany, Academia Sinica, P.O. Box 302, Xishuang Banna, Yunnan, China.*

YANG ZHI-BING *Jiangsu Institute of Botany, Nanjing Botanical Garden, Mem. Sun Yat-Sen, Nanjing, Jiangsu, China.*

ZHONG SHI-XIAN *Jiangsu Institute of Botany, Nanjing Botanical Garden, Mem. Sun Yat-Sen, Nanjing, Jiangsu, China.*

Foreword

KENTON R. MILLER

Director General
International Union for Conservation of Nature and Natural Resources (IUCN)
Gland, Switzerland

Between 26 and 30 November 1985, about 175 botanists and botanic garden managers from 39 countries participated in the Conference on Botanic Gardens and the World Conservation Strategy held in Las Palmas, Gran Canaria, to review the involvement of botanic gardens in implementing the World Conservation Strategy and to discuss how to make gardens more effective in plant conservation. The Conference, organised jointly by IUCN, the Gobierno de Canarias and the Cabildo Insular de Gran Canaria, provided a full discussion on the ways and means to mobilise the botanic gardens of the world and on how they can contribute much more strongly to the conservation of plant diversity.

On behalf of IUCN, I am delighted to introduce the Proceedings of the conference which include not only the many excellent papers presented but also the outputs of particular concern to IUCN, namely the "Declaration of Gran Canaria" and the Conference Recommendations.

IUCN is the world's largest and most experienced conservation alliance, in action since 1948, devoted to the protection and sustainable use of the Earth's living natural resources. The concerns are global in scope, and the operating programme of the Union is based on the concepts outlined in the World Conservation Strategy and the World Charter for Nature.

The World Conservation Strategy (WCS), prepared by IUCN, with the advice, cooperation and financial assistance of the United Nations Environment Programme (UNEP) and the World Wildlife Fund (WWF), and in cooperation with Unesco and FAO, provides a philosophical approach and framework linking conservation with development. It defines conservation as "the management of human use of the biosphere so that it may yield the greatest sustainable benefit to present generations while maintaining its potential to meet the needs and aspirations of future generations".

The concerns of the World Conservation Strategy have been widely accepted in principle by governments, development agencies and conservation organisations. The question is no longer whether conservation is a necesary part of social and economic development, but rather *how* conservation can be achieved.

This was the theme of the IUCN General Assembly in Madrid (November 1984), which adopted a 3-year programme of work for the Union. The programme included an increased commitment to the conservation of plant life following the encouraging start made with the IUCN/WWF Plants Conservation Programme, outlined by Ole Hamann in the present volume.

As a specific project under the IUCN/WWF Plants Conservation Programme, a "Botanic Gardens Conservation Strategy" is being prepared. Indeed, the first draft, prepared principally by Professor Vernon H. Heywood, was the main business document to be discussed at the Conference in Las Palmas. The aim of the document is to give a rationale for the various ways in which a botanic garden can have an involvement in conservation and to detail what the conservation movement believes each garden can and should contribute.

Following the Conference's discussion of the draft "Botanic Gardens Conservation Strategy", work on the document is being carried forward, taking into account the deliberations of the delegates and the present proceedings, with a view to publishing the final document in the near future.

We in IUCN believe that the botanic gardens of the world could, and should, develop into a major new global force for conservation. To achieve this ambitious goal, botanic gardens will surely have to link together into a functioning network. IUCN, which is an international organisation comprising networks of scientists, of conservation NGOs and of governments and governmental agencies, is pleased to assist botanic gardens to develop such a network.

Already a start was made by those gardens linked to IUCN's Conservation Monitoring Centre through the Botanic Gardens Conservation Co-ordinating Body. This programme is now being developed further, applying the capacity of the Conservation Monitoring Centre to monitor and analyse plant data worldwide for those who need it.

With generous start-up grants from industries and foundations in the U.S.A., IUCN established on 1 January 1987 a Botanic Gardens Conservation Secretariat within its Conservation Monitoring Centre. The mechanism has therefore been created through which cooperation between the botanic gardens of the world and the international conservation movement can be facilitated.

Thus, further developments in establishing the conceptual and institutional framework for the conservation work of botanic gardens have followed the Conference in Las Palmas, allowing for the joining of efforts to conserve the endangered plant life on earth. Indeed, the enthusiasm displayed by the botanic gardens of the world for conservation as the leading edge of a botanic garden's work bodes well for the future.

In conclusion, the Conference in Las Palmas represented a major step forward towards the implementation of the objectives of the World Conservation Strategy and IUCN is looking forward to an enlarged and fruitful collaboration with the community of botanic gardens in conserving plant diversity.

Preface

CARMELO ARTILES BOLAÑOS

President, Cabildo Insular de Gran Canaria

Botanic gardens are an ancient institution which, like every creation of man, have evolved through the years adapting to the demands made upon them by each culture and by each generation.

The inspiring philosophy behind the concept of the botanic garden has changed but not in its substance nor in its relevance to man. Now botanic gardens face a challenge: to integrate their activities within the conservation movement, which today inspires the most progressive attitudes towards the global natural resources in man's care.

From my own point of view I am proud of the fact that the Island Council of Gran Canaria has been amongst the pioneers of conservation. In the early 1950s the then Council President and his colleagues accepted the proposals made by Enrique Sventenius to conserve the Canarian flora by establishing a special botanic garden. Since then we have accumulated over 30 years experience in this field and have seen the interest which the ''Viera y Clavijo'' Botanic Garden, popularly known as the *Jardin Canario*, has generated in conservation in the islands.

For this reason it is a great pleasure for me as President of the Island Council to write the preface to this volume, which reports the results of the International Symposium, where the botanic gardens of the world freely decided to meet, very appropriately in Gran Canaria, to work out their future rôle in the conservation of the world's flora. I am sure that the intense working sessions reached the results that the effort involved deserved.

The Island Council of Gran Canaria is conscious not only of the extraordinary scientific importance of the endemic Canarian flora and the island's vegetation but also of their significance as a potential natural resource. Especially at this time when the developed nations are having to pursue high technology to maintain their position on the world stage, we find ourselves involved in one of them, biotechnology, which although still at an embryonic stage indicates that because we still know little about the useful properties that our flora can provide, we must do all we can to conserve it in case valuable resources are lost.

The Canarian flora has a very high proportion of species in immediate danger of extinction, about 200 species in all, which is a similar figure to that for the whole of continental Europe. The flora undoubtedly needs urgent and adequate protection to safeguard its genetic potential, today unknown, but tomorrow perhaps of strategic importance for the needs, not only of the Canary Islands but of all mankind. So naturally conservation and biotechnology will have to unite towards a common objective — to serve mankind in a more rational and balanced way.

It is, therefore, extremely important that we take advantage of the existence of the large, international network of botanic gardens and the immense wealth of knowledge held within their collections and in the experience of their personnel. I believe that this book will go a long way towards creating a suitable environment for this to happen. I welcome the proposal to set up an IUCN Botanic Gardens Conservation Secretariat to promote and coordinate conservation in them and I wish it every success in its very important mission of helping to save the world's natural resources. In particular, I welcome *The Declaration of Gran Canaria,* which without a doubt will be a positive contribution to conservation, and will echo in the institutions, organisations and governments all over the world who are involved in the greatest challenge of our time — conservation.

Contents

Part 1
Botanic Gardens and the World Conservation Strategy

Part 2
Botanic Gardens and the Community

Part 3
Research and Rescue

Part 4
Botanic Gardens for Sustainable Development

Part 5
Botanic Gardens — A World Network?

Part 6
Short Communications

Introduction

HUGH SYNGE

United Kingdom

This book forms the proceedings of IUCN's conference on "Botanic Gardens and the World Conservation Strategy", 26–30 November 1985, and is an important part of the IUCN/WWF Plants Conservation Programme. Started in 1984, this programme is a major venture to "assert the fundamental importance of plants in all conservation activities". It was developed from a sense that plants were being neglected in conservation work and that the strategic and conceptual basis for the conservation of plant life was lacking.

The programme, now consisting of 49 projects and activities with a budget of several million dollars, is described in detail in Ole Hamann's paper. Essentially, the programme has six specific objectives, one of which is to strengthen the rôle of botanic gardens in conservation. IUCN felt that botanic gardens could and should become the specialist agencies for plant conservation and that gardens could be much more effective if they could work together more closely as a network with shared goals.

Within that objective, in 1982–3 IUCN started designing a number of projects with the specific aims of:

Drawing botanic gardens closer into the conservation network;
Co-ordinating them as the *ex situ* network for threatened plants;
Promoting the use of botanic gardens to tell the public about plants and their conservation.

IUCN took these rather broad aims to the newly formed IUCN/WWF Plant Advisory Group, which met for the first time at St Louis in December 1984 under the Chairmanship of Dr Peter Raven. Faced with a list of desirable projects, the Group strongly advised IUCN and WWF to focus on three. These were:

The conference

It was felt that by far the best way to discuss the fundamental issues concerned was by calling a conference on how botanic gardens can contribute to implementing the World Conservation Strategy. This would provide the forum to receive comments on the draft Botanic Gardens Conservation Strategy (see below) and to address the fundamental issues on how botanic gardens can undertake a larger rôle in conservation and collaborate more effectively together. Only when this had been done would it be sensible for IUCN to decide how best it could promote these aims.

The Botanic Gardens Conservation Strategy

Partly resulting from the conference, IUCN would prepare a Strategy modelled along the lines of the World Conservation Strategy but specifically for botanic gardens. It would outline precisely what a botanic garden's rôle in conservation should be and give garden directors the facts and arguments that will assist them in pressing their case for funds and facilities. A draft of the strategy was the main business document of the conference and a summary of it is given in the next section of this book.

Developing the monitoring programme

It was clear that whatever the conference would decide on how botanic gardens could work more closely together, considerable developments were due for the IUCN Botanic Gardens Conservation Co-ordinating Body (described below). To be truly effective, it had to move beyond simple monitoring of plant collections towards offering help and advice in the co-ordination of *ex situ* conservation for all threatened plant species.

IUCN saw the conference as defining the details of the programme, providing intellectual background to it, and acting as IUCN's spearhead into the botanic garden community, ensuring that the programme represented the views of botanic garden managers and answered their needs as far as possible.

IUCN was most fortunate in receiving an invitation from the authorities in the Canary Islands to host the conference. This was readily accepted and the meeting was jointly organised by IUCN, the Gobierno de las Canarias and the Cabildo Insular de Gran Canaria. Its success was greatly due to Dr David Bramwell, Director of the Jardín Botánico Canario "Viera y Clavijo" on the hills above Las Palmas. This garden, as Bramwell's contribution shows, has been very successful in effecting plant conservation in the Canary Islands, in particular in the reserve system on the island of Gran Canaria, and, as many delegates commented, forms an excellent model for other gardens to follow. Staff of the garden, ably led by Sr Alphonso Luezas, carried out most of the detailed work of local organization for the conference.

Sponsorship was received from WWF, patronage from FAO, co-operation from Unesco and support from UNEP. In particular, WWF gave a generous grant which covered much of the costs of preparing the scientific programme and the attendance costs of some of both the speakers and non-speakers. Unesco, UNEP and IBPGR also paid for a number of delegates to attend who would not otherwise have been able to come. The organisers of the meeting are most grateful to these institutions for this much valued support and to the Canarian hosts for the splendid hospitality they lavished upon the delegates. Appreciation is also extended to those who have enabled the preparation of the proceedings book, in particular to R.J.R. Nabisco Inc., U.S.A., for financial support to keep the price at a reasonable level.

The first section of the meeting, as an introduction to the theme, covered general papers on botanic gardens and the World Conservation Strategy. V.H. Heywood explains how the functions of botanic gardens have changed over the years and reviews the present network of botanic gardens, based upon his own unrivalled experience and upon the detailed questionnaire he circulated to botanic gardens as part of the preparation of the draft Botanic Gardens Conservation Strategy. P.H. Raven gives a review of the scope of the plant conservation problem worldwide and outlines the reasoning behind IUCN's startling prediction that 60,000 species of flowering plants — nearly 1 in 4 of the world's total — could become extinct by 2050 if present trends continue. This grave situation is the reasoning behind the IUCN Plants Programme, of which this conference and theme is one part, described by Ole Hamann, IUCN Plants Programme Officer, in the next contribution. The overview that this gives of IUCN's and WWF's activities to define a conceptual and strategic basis for plant conservation may be of interest and use to botanic gardens in designing their own conservation programmes.

The next section, on botanic gardens and the community, covers a theme of great importance. Some botanic gardens today are brilliant exponents and enthusiasts for plants in their local communities, but most, sadly, are not. In this section we provide accounts from gardens that are recognised to be world leaders in this field. Some have long-established programmes, such as the botanic gardens of South Africa and Australia. Others are more recent, such as the work described at the Jardins Botaniques de Nancy, the Jardín Botánico ''Viera y Clavijo'' and the newly created Jardín Botánico de Córdoba. Community involvement is not just a good way to promote and achieve conservation but also to ensure a garden's survival by building a groundswell of local interest and support.

The third session was entitled Research and Rescue. It concentrated on what is usually seen as the botanic gardens' most obvious task in conservation — rescuing endangered species and ensuring their survival. Gardens fully recognise that conservation in the habitat is the best option and many of the papers attempt to define the crucial relationship between *ex situ* and *in situ* conservation. D.R. Given and P.S. Ashton, in two keynote contributions, explain from a scientific perspective some of the drawbacks of living plant collections for *ex situ* conservation. Turning this into practice, D. Bramwell outlines the impressive rescue programme of the Jardín Botánico "Viera y Clavijo" and C.V.S. and I.A.U.N. Gunatilleke and B. Sumithraarachchi give a tropical perspective from their impressive work on the critically endangered lowland forest plants of Sri Lanka.

For long-term conservation, the emphasis should be on seed-banks wherever possible: J.G. Hawkes and C. Gómez-Campo both provide substantial papers full of practical details on seed banking. Professor Hawkes concentrates on the scientific aspects of collecting the seeds and on ensuring that the seeds survive in the bank, based on his experience with the crop seed-bank network; Professor Gómez-Campo emphasises more the ease by which a small seed-bank can be created in a botanic garden, especially for the routine preparation of a regular seed list. There is clearly some difference of emphasis between these papers on the extent to which a seed bank for long-term conservation requires a substantial long-term scientific commitment, and developing a consensus on this most important point is clearly an important priority for IUCN and the botanic garden community. This is especially pressing in view of IBPGR's offer of collaboration with botanic gardens, proposed in the paper of J.T. Williams and J.L. Creech, and the recommendation on the importance of seed banking agreed by the conference.

The fourth session was on how botanic gardens can contribute to sustainable development, which is at the heart of the World Conservation Strategy. Of particular interest will be the papers by T.N. Khoshoo on the National Botanical Research Institute, Lucknow, and by Xu Zaifu on the Xishuangbanna Tropical Botanical Garden in Yunnan. Both are accounts of most impressive programmes to promote conservation and development, programmes that are at the heart of the work of those gardens. Promoting a better and more sustainable use of plants is not only a proper thing to do in its own right, to improve living standards, but is also the way to reduce the impacts on many wild plant species, presently endangered by habitat destruction but for which reserves cannot be created. Both gardens make good models for others to follow. There are similar messages in the papers by E. Forero on Latin American botanic gardens and by S. and D.S. Sastrapradja and U. Soetisna on Indonesian botanic gardens.

The fifth session discussed international collaboration and considered how gardens could work more effectively together. The first paper, by J.B.E. Simmons and R.I. Beyer of Kew, outlined how gardens could collaborate in training personnel and could help each other through provision of technical advice. This was written from their own experience at Kew, where they receive numerous requests for help from gardens all over the world. Working out ways of establishing a clearing house for requests of this type and maybe even of making twinning relationships between botanic gardens, was one of the themes discussed by many delegates and was reflected in the recommendations.

Whereas twinning and technical aid normally occur across international boundaries, frequently between north and south, there is a need for very extensive and close collaboration between gardens within a country. This collaboration should clearly be in addition to whatever international co-ordination there may be, such as in the programmes of IUCN and the International Association of Botanic Gardens (IABG). The draft Botanic Gardens Conservation Strategy outlines a model set of objectives for a national co-ordinating group; these are very different to those at international level, and both are needed. Perhaps the best model for a national collaborative programme is that of the Center for Plant Conservation in the U.S.A., and its impressive programme is outlined in detail by its directors, F.R. Thibodeau and D.A. Falk. A similar programme is being created in Canada, as described by P.N.D. Seymour and colleagues at the Devonian Botanic Garden. The International Association of Botanic Gardens (IABG) are also active in setting up regional networks, as explained by K. Larsen, B. Morley and H. Ern in their contribution.

Two evening sessions concentrated on technical aspects of international collaboration. One covered the proposals by IUCN for an International Transfer Format for the exchange of computerised information on botanic garden plant records. This is outlined in detail in the paper by J. Cullen, M. Lear and colleagues, which was distributed before the meeting. Essentially, the paper provides scientific and technical data on the various elements of information ("fields") that every botanic garden record database would need to include and proposes an international format for transfer of such information from one garden to another. The advantage of this system, once adopted, is that not only can gardens exchange data between each other but also that monitoring of the global plant collection becomes possible through access to the individual garden databases. In particular this would greatly facilitate and improve the present manual monitoring done by IUCN of the plants held in 270 leading botanic gardens around the world. The explanation and final definition of the ITF is nearly complete and soon will be available as a booklet.

Another evening session covered the work of the Botanic Gardens Conservation Co-ordinating Body (generally known as the "Body") and considered a plan for an extension of its work. The results of this session are incorporated in the workplan for the new structure, described below. The conference was in general very supportive of the work done by the "Body", but wanted it to take a more proactive, more interventionist rôle moving beyond the simple monitoring of the presence of threatened species in botanic garden collections. 88% of those attending the session believed that a readily transferable system of plant records was a "good idea", indicating their support for this wider rôle. Also delegates made it clear in a series of straw polls that their gardens would be prepared to pay a realistic subscription to cover the costs involved; a good number of representatives of gardens declared that their gardens would each be prepared to pay a rate of around US$ 500 per year, an increase from the present subscription of $40. The session concluded with unanimous backing for the "Body" to expand its work and for the botanic garden community to provide the money to do this.

It was felt, however, that the "Body" needed to cover more botanic gardens before it can declare which species are conserved *ex situ* and that data on the *sources* of the plants was also needed. F.R. Thibodeau of the Center for Plant Conservation proposed that the "Body" should cover technical advice on individual species, in particular on how to grow them, acting as a clearing house for information. Information from other parts of IUCN Conservation Monitoring Centre, in particular on which species were safe in protected areas, would help assign priorities for *ex situ* conservation (M. Lear).

On the final day there were sessions on the Botanic Gardens Conservation Strategy and on the Resolutions. The draft Strategy was greatly welcomed and was accepted in principle, as demonstrated in the Recommendations. The discussion provided many suggestions of great interest and value, and IUCN is taking them into full account in the completion of the document. It is clear from the discussions, both formal and informal, that the Strategy will be very useful to gardens. IUCN is most anxious to complete it as soon as possible and then to make it widely available. An intermediate summary of its main contents and conclusions are given on p. xxxvii–xxxix.

The formal conclusions of the conference centred around a set of recommendations (pp. 365-373), aimed at the botanic garden community and its supporting bodies, and the Declaration of Gran Canaria (p. 363), which was written as a public statement to affirm the commitment of botanic gardens to working together in the defence of the plant kingdom, united behind the Botanic Gardens Conservation Strategy.

There were naturally many discussions about what structures would be needed to ensure follow-up to the conference and to start building the network of gardens. In the recommendations, gardens formally invited IUCN "to assist ... (them) ... in setting up a secretariat to co-ordinate and promote the implementation of the Strategy, incorporating the activities of the IUCN Botanic Gardens Conservation Co-ordinating Body, with the aim of making this Secretariat self-supporting". As Kenton Miller says in the Foreword, IUCN is pleased to accede to this request. From 1 January 1987, the Botanic Gardens Conservation Co-ordinating Body is wound up and the IUCN Botanic Gardens Conservation Secretariat created to take its place with a much more substantial programme, as requested by the conference.

Before describing the work done to prepare for the new organisation and outlining its programme, it may be helpful to give a short review of the work of the Botanic Gardens Conservation Co-ordinating Body. This is an abbreviated and updated version of the account I wrote in the Abstracts booklet for the conference.

The Botanic Gardens Conservation Co-ordinating Body

Following two conferences at Kew in 1975 and 1978, IUCN became closely involved in efforts to conserve threatened plants *ex situ* in botanic garden collections. This work was a natural consequence of studies to prepare threatened plant lists for various regions. By 1972, IUCN's then Threatened Plants Committee Secretariat had prepared a list of threatened plants for Europe, and during the mid 1970s, national institutions prepared lists for Australia, New Zealand, United States, South Africa and elsewhere. Today lists have been prepared for most of the developed countries.

It seemed sensible, therefore, to start by finding which threatened plants were already in cultivation. Between 1975 and 1978, IUCN undertook a small project to do this for plants on the European threatened plant list. The results of this survey were submitted to the second of the Kew conferences, in 1978. As a result IUCN was invited to set up a small structure to carry this work forward, with the eventual aim of linking botanic gardens in their conservation activities and keeping them in touch with conservation around the world. To begin with, the only large part of this programme was its name — the Botanic Gardens Conservation Co-ordinating Body — but it was always intended that it should be small so that it would not pose any great financial burden on the gardens that paid to subscribe for its services.

Some 90 botanic gardens subscribed from the beginning. By the end of 1986 there were 154 subscribing institutions, to which one can add the 116 gardens in the USSR, which contributed data through the Moscow Main Botanic Garden but could not participate financially.

The key to success has been a small but permanent Secretariat, run as part of IUCN's Threatened Plants Unit at Kew: as Liaison Officer, Winifred Worth carried out the Body's Programme from 1980 to 1986. Most generously, Ian Beyer, Deputy Curator of the Royal Botanic Gardens, Kew, has acted as Treasurer during these years and ensured that the finances have been on a sound basis; subscription for the year 1984–85 was set at £25 or US$ 40, which has paid only a small percentage of the costs.

The programme carried out two aims: keeping gardens in touch with international conservation issues, and monitoring *ex situ* conservation of threatened plants. It has not involved itself in wider issues such as liaison and training, advice on creation and direction of botanic gardens, etc.

From the beginning, subscribers received the *Threatened Plants Newsletter,* which is issued about twice a year. Since the beginning of 1985, they also received the quarterly *IUCN Bulletin*, available in French as well as English.

Several times a year, each garden received a list of threatened plants for a region or plant group of special interest. (These lists are derived from IUCN's database on threatened plants, maintained by the Threatened Plants Unit who are in touch with botanists and conservationists all over the world.) Gardens are asked to mark off which plants they have in cultivation and return the questionnaire to IUCN at Kew. The results are put on computer. So far most of the known threatened plants have been screened in this way.

From the data two types of report were issued:

> "Botanic Garden Reports" showing which threatened species in a group or region (e.g. Cycads or India) are in cultivation. Alongside each plant name are the plant codes for the gardens that cultivate it. 18 were published, listed in Fig. 1.
>
> "Garden Printouts" which are sent to each subscriber once a year. These list all conservation plants recorded by that garden, showing for each the full distribution, degree of threat worldwide, the source of the acquisition, and, most important, the number of *other gardens* who have records of growing that plant. An example is given in Fig.2.

Of course, these reports and printouts are simply the beginning of a long-term venture and do not achieve conservation in themselves. They do, however, provide garden managers with the information from which they can develop their living collections not in isolation but as part of an international conservation network. Asterisks on the printouts draw attention to plants in no other gardens, and in less than 6 gardens, so these can be propagated and distributed as a priority.

By the end of 1986 the database contained 22,232 records, one for each occurrence of a threatened plant in a botanic garden. Of the c. 18,000 known threatened species, 4946 are in botanic gardens, but the true total must be considerably higher. Although some plants were in nearly 100 gardens, very many are in too few gardens for safe-keeping: 36% of taxa are in only one garden (two-asterisk plants) and 77% in 5 gardens or less (one-asterisk plants).

Fig. 1. Lists circulated and reports issued by the Botanic Gardens Conservation Co-ordinating Body

Group or region	List	Report
Europe	July 1978	July 1979 (No. 1)
Madagascan succulents	December 1979	October 1980 (No. 2)
Southern Africa	June 1980	June 1982 (No. 5)
Macaronesia	October 1980	July 1981 (No. 4)
Cycads	December 1980	December 1980 (No. 3)
South Atlantic Is.	May 1981	February 1983 (No. 7)
Tree Ferns	September 1981	February 1983 (No. 6)
New Zealand	December 1981	November 1983 (No. 8)
Australia	June 1982	November 1983 (No. 9)
W. Indian Ocean	August 1982	March 1984 (No. 10)
Hawaii	November 1982	May 1985 (No. 14)
Galapagos/J. Fernandez Is.	February 1983	March 1984 (No. 11)
Europe (update)	March 1983	August 1984 (No. 12) (full list)
Mexican Cacti	July 1983	January 1985 (No. 13)
India	November 1983	October 1985 (No. 16)
Euphorbia and Aloe of Africa	March 1984	August 1985 (No. 15) (including Madagascan succulents update)
North Africa	August 1984	August 1986 (No. 17)
Middle America	January 1985	August 1986 (No. 18)
Cuba	August 1985	
U.S.A.	June 1986	
U.S.S.R.	August 1986	

Fig. 2. Sample garden printout

PLANTES CONSERVEES AU CONSERVATOIRE BOTANIQUE DE STANGELARC'H, BREST (BRES)

5 Novembre 1986 — 3

		Categorie mondiale	Source CWUD	Donnees dans d'autres jardins C	W	U	D	Total
TAXACEAE								
* *Torreya taxifolia* Arn.	Floride (E), Georgie (EU) (E)	E	C	1	1	2		4
ZAMIACEAE								
Zamia floridana A.DC.	Floride (V), Georgie (EU) (I)	V	U	12	7	12		27
ANGIOSPERMAE								
ACANTHACEAE								
* *Ruellia insignis* Balf.f.	Socotra	nt	C		1			1
AGAVACEAE								
* *Agave arizonica* H.S.Gentry & J.Weber	Arizona	E	C	2	1			3
Furcraea bedinghausii K.Koch	Mexique	R	C	5	1	2	1	8
AIZOACEAE								
** *Delosperma napiforme* (N.E.Br.) Schwantes	La Reunion	E	W D					0
Neohenricia sibbettii L.Bolus	E. Lib d'Orange	V	C	6	3	4	3	13
AMARYLLIDACEAE								
Crinum mauritianum Lodd.	I. Maurice	E	W D	2	6	1	2	8

Legende C: source cultivee connue; W: site naturel connu; U: source inconnue; D: plantes disponibles.
Ex: eteint; E: menace; V: vulnerable; R: rare; I: Indetermine; K: insuffisamment connu; nt: non menace.

As the work continued during the early 1980s, the feeling grew that a more substantial programme was required; on the one hand there was a need to use the data gathered to avoid excessive duplication and to ensure that all the most relevant species were conserved, and on the other a perception that IUCN should work with botanic gardens on policy issues, on education, on research, and so on. To be truly effective, the "Body" had to move beyond simple monitoring towards offering help and advice in the co-ordination of living collections. The need for a larger, more substantial programme, is its epitaph.

Progress since the conference

In January 1986, IUCN decided to accede to the request of the conference to assist in the creation of a more substantial programme. The "Body" would be maintained as a low-key operation during 1986, while the plans for the Secretariat were made. Fortunately, IUCN soon received generous funding for its detailed plan. Grants have now been promised from four donors:

R.J.R. Nabisco, Inc., for the preparatory work of the Secretariat and for continuing the work of the 'Body' during 1986;

An anonymous donor to complete the Botanic Gardens Conservation Strategy;

The Jessie Smith Noyes Foundation, for development of the Secretariat, 1986–1988.

The W. Alton Jones Foundation Inc., for development of the Secretariat, 1987–88.

By the end of 1986 the preparatory work had been done and the "Body" was wound up. The Botanic Gardens Conservation Secretariat started work on 1 January 1987. It is housed with the IUCN Conservation Monitoring Centre (CMC) at its offices at Kew and run in close association with the Threatened Plants Unit (TPU). This joint operation is under the direction of Professor V.H. Heywood, since 1 January 1987 Associate Director for Plant Conservation of the CMC.

In the initial period, the IUCN/WWF Plant Advisory Group, under the chairmanship of Dr Peter Raven, will advise on the programme. The Secretariat will actively recruit as many botanic gardens as possible to become members.

Its prospectus, issued in March 1987, defined its programme as follows:

1. To promote the implementation of the Botanic Gardens Conservation Strategy

The aim is to help gardens, and networks of gardens, to carry out specific tasks under the Strategy to conserve plants. Some co-ordination is essential to ensure work is not duplicated and that important tasks are not left out. The Secretariat will also help gardens to use the Strategy as a tool for their own development.

The following activities are planned for the initial period:

(a) Missions to selected gardens to explore problems of implementing the Strategy and to formulate plans. The Secretariat will not employ a large staff itself, but will seek to find highly qualified consultants from member gardens to undertake these missions.

(b) National or regional workshops to develop coherent botanic garden programmes. These may either be independent meetings or can be sessions in other, related conferences. As an example of the latter, IUCN is co-sponsoring and helping to organise a conference in Rabat, Morocco, in 1987, on conservation of the plant genetic resources of North Africa; the rôle of botanic gardens will be a major theme. It is also supporting a meeting of East European botanic gardens being held in Sofia, Bulgaria and is co-sponsoring a Symposium being held in connection with the opening of the Jardín Botánico de Córdoba, Spain, both in 1987. Regional conferences are being planned for botanic gardens in tropical Africa and the West Mediterranean littoral.

(c) Preparation of guidelines for national, regional and international networking. The Strategy envisages a large rôle for national networks and outlines a clear division of labour between national linking bodies and the international Secretariat. The Secretariat will also develop liaison with the regional chapters of the International Association of Botanic Gardens (IABG).

Specific products will include:

An expanded World Directory of Botanic Gardens in association with IABG; it is intended that this will be available not just as a book but in a computer-readable form, with software to run it, so that gardens can use it for analysis and for mail shots.

A study and analysis of *in situ* reserves run by botanic gardens, with management guidelines to be prepared jointly with IUCN's Commission on National Parks and Protected Areas.

2. *To monitor and co-ordinate ex situ collections of conservation-worthy plants*

The intention is to develop the present monitoring of collections into a programme that will not just monitor but also help co-ordinate *ex situ* collections worldwide to ensure as wide a range of conservation-worthy plants as possible are maintained in cultivation. Simply, IUCN wants to help the gardens develop their collections into one international network where excessive duplication is avoided, although with a sufficient geographical spread to act as a safeguard, and with all the relevant species covered.

As before, each garden in the scheme will be expected to contribute data on its holdings of conservation-worthy plants, preferably in electronic form using the newly designed International Transfer Format (ITF). For gardens on computer, this will mean converting their file into ITF and sending a copy on a diskette or magnetic tape to IUCN each year. Physical compatibility is not required. Gardens will be encouraged to computerise their records in their own institutions, using PCs where appropriate. To encourage better recording and to assist computerisation, IUCN is planning a Manual on Botanic Garden Plant Records. Each garden needs to have its own effective data management system, compatible with that of other gardens and with IUCN.

Gardens *not* on computer will be asked to select those regions for which they have plants of known wild source origin. They will then be asked to annotate lists of threatened plants for these regions with their holdings; they will also receive a list of commonly cultivated, threatened species, namely those already recorded in ten gardens or more. In this way, the Secretariat will aim to have a reasonably up-to-date information base for gardens not yet on computer.

In return, all gardens will receive:

Annual printouts of their collection in relation to the collections of other members. Gardens on computer will be offered the option of also receiving a tape or diskette with the data included, so they can add it to their own datafiles.

General reports on the status of plants in cultivation;

Answers to individual enquiries, e.g. Who is growing this plant and from where did they acquire it? Which group of plants is least recorded in cultivation? Where is the best place to see Madagascan plants in Europe?

Online access to the CMC database is possible and can be negotiated at an extra price, but at present the hardware and telecommunications costs for each remote user are not cheap. IUCN is exploring options of reducing this cost through connection to international networks, following the opening of the first remote terminal on the CMC system in September 1986 at the U.S. National Park Service, Washington, D.C. Undoubtedly costs will come down over time and eventually one could envisage most, if not all, gardens having online access to the database; it is being designed and built with this in mind. This would be in addition to the gardens' own databases, whether on PCs or on larger computers.

IUCN is considering the feasibility of creating a database of experience gained by member gardens in cultivating and propagating individual threatened species. This would be set up in such a way that gardens could contribute to it directly. It would have a strong bibliographic element, drawing upon TPU's existing computerised bibliography of c. 9000 papers and reports relevant to plant conservation.

An important long-term goal will be to secure effective long-term storage in seed-banks for all threatened species that do not have recalcitrant seeds. It is hoped to do this in close collaboration with the International Board for Plant Genetic Resources (IBPGR), who co-ordinate the existing network of seed banks for crop plants. The Royal Botanic Gardens, Kew, have offered extensive help from their own seed bank at Wakehurst Place.

3. *To develop a programme for liaison and training*

One of the most frequent requests received from botanic gardens is for help with training of technical staff and for advice on various aspects of botanic garden operation in respect of conservation activities. Almost as frequent are appeals for general support and encouragement. IUCN feels that this can be most effectively achieved by fostering links at national level and at regional level, and by international co-operation between gardens in the developing world and in the developed countries. The Secretariat plans, therefore, to act as a clearing house for this work and will aim to put those who request assistance in touch with those who might be able to help. It may be appropriate in some cases for a "twinning" arrangement to be made; in other cases a multilateral arrangement involving several gardens may be more suitable. The various ways in which cooperation between botanic gardens can be effected will be the subject of a detailed study and it is planned to prepare guidelines and mechanisms to achieve these goals.

4. *To arrange the Botanic Gardens Conservation Congress every 3 years*

The intention is that the next meeting, in 1988, will be held in a tropical botanic garden. As with the Las Palmas meeting, efforts will be made to find sponsors who will contribute to the attendance costs of delegates who could not otherwise come. The proceedings will be published. The main theme of the 1988 Congress will be Tropical Botanic Gardens — Conservation and Development.

So, by early 1987, the preparation has been done, the ''Body'' has been buried and the new structure is in place. The revised Strategy is nearing completion and a database has been compiled on the 1400 botanic gardens of the world from which the new World Directory will be made. The IUCN database on threatened plants now contains virtually all recorded threatened plants in the world and is ready to be harnessed. A Transfer Format for computerised botanic garden plant records is almost complete. The time has come for the botanic gardens' new initiative to begin.

The IUCN Botanic Gardens Conservation Strategy: a summary

by IUCN

There are about 1400 botanic gardens and arboreta in the world, visited by over 100 million members of the public each year. They are therefore ideally placed to convince the public of the importance of plant conservation, and play a leading part in achieving conservation of plant life — in the wild, in cultivation and in gene banks.

The Botanic Gardens Conservation Strategy is intended to stimulate a far greater involvement by botanic gardens in implementing the World Conservation Strategy than they have shown hitherto. It provides a rationale for the involvement of botanic gardens in conservation and gives policy guidance on how this can be achieved. It is designed principally for those who work in them, and for those who could make better use of them.

The Strategy outlines the contribution that botanic gardens can make to achieving what the World Conservation Strategy identifies as the three main objectives of living resource conservation:

(a) To maintain essential ecological processes and life-support systems;

(b) To preserve genetic diversity;

(c) To ensure that the utilisation of species and ecosystems is sustainable.

The World Conservation Strategy outlines the reasons why these objectives must be achieved as a matter of urgency and the obstacles to their achievement.

The main rôle of botanic gardens in this process will be to contribute to the preservation of plant genetic diversity and to help ensure the sustainable utilisation of plant species and the ecosystems in which they occur. This is urgent because:

(a) As many as 60,000 plant species may be in danger of extinction during the next 30–40 years (out of the world's 250,000), most of which is due to the destruction of their habitats, principally in the tropics;

(b) Many others such as rattans, quality hardwoods and medicinal plants are taken from the wild to an extent that cannot be sustained and already some of their plant populations are becoming exhausted;

(c) People presently depend on as few as 20 plant species for over 85% of their food, plants that are suffering a decline in their genetic diversity;

(d) Many other plants have never been examined for useful products and their potential as commercial crops has not been explored. Thousands of species have not yet been given a name or described scientifically, and we are, therefore, ignorant of the value that they may have for mankind.

The Strategy first provides an analysis of the present situation in botanic gardens, based partly on the answers to a questionnaire sent to all the botanic gardens of the world. This shows the great and very desirable diversity between the types of botanic garden. It also shows a great imbalance between where most of the plants occur — two-thirds are in the tropics — and where botanic gardens tend to be situated — most are in the temperate realm, c. 400 in Europe alone. Many botanic gardens are already committed to conservation, both *in situ* and *ex situ*. Others, however, have not been able to meet this new challenge and some are moribund or faltering. More crucial, the botanic garden community by virtue of its history and diversity has not yet organised itself into a network that will be an effective force for plant conservation worldwide.

The Strategy:

(a) Recommends that each individual garden clarifies its commitment to conservation in a Mission Statement and adopts more professional standards of management to achieve its Mission;

(b) Provides the basis for a more coherent accessions policy that takes account of conservation needs and of what plants are held in *other* botanic gardens;

(c) Outlines ways to improve the documentation of plant records and the verification of plant holdings, including computerisation to improve management of the collection and to facilitate exchange of data between institutions;

(d) Explores the relationships between wild and managed conservation for botanic gardens' efforts; for in-the-wild conservation, outlines the rôle of the garden in habitat evaluation, rare species monitoring, 'habitat gardening' and in managing protected areas; for managed conservation, proposes strict rules and procedures for the establishment of reserve collections, gene banks and other germplasm, and suggests methods of sampling populations to maintain adequate genetic variation;

(e) Emphasises the facilities which botanic gardens can bring to educating their estimated 100 million visitors each year;

(f) Recommends that each garden provides a service to its local community as a resource and information service;

(g) Provides a framework for training of personnel, with emphasis on conservation.

The Strategy is also concerned with how botanic gardens can best work with each other and with other organisations in achieving conservation. The Strategy outlines:

Who will participate in its operation;

Which are the priority regions where botanic gardens can best work with each other and with other organisations in achieving conservation;

Which are the priority species that most need conservation.

It proposes that most emphasis be put on collaboration between gardens at a *national level*, or where appropriate, regional level. The Strategy proposes the structure for a national group and outlines a set of objectives.

The Strategy also recommends collaboration at *international* level, though this should be less formal and less structured. The Strategy outlines a programme of 'North-South' technical liaison between botanic gardens and calls for a global programme of monitoring and co-ordination of cultivated collections. It suggests the creation of a high-level Secretariat, small but with full-time staff, to oversee the implementation of the Strategy.

1

Botanic Gardens and the World Conservation Strategy

The changing rôle of the botanic garden

V.H. HEYWOOD

IUCN Conservation Monitoring Centre, Kew, U.K.

Summary

Botanic gardens, in one form or another, have been a significant instrument in man's scientific and cultural development over the centuries. So prestigious is the name "Botanic Garden" that it has been applied to many parks and similar sites that have no real claim to the title, however widely it may be defined. Individual gardens have played an important rôle in economic development and commercial advance in many parts of the world. Tropical botanic gardens, in particular, have been considered a major tool of colonial expansion in the 18th and 19th centuries, and were responsible for the introduction and transfer of germplasm from one part of the world to another, thereby laying the basis for the agricultural pattern that persists to the present day.

The original botanic gardens of Europe were medicinal foundations, intended to provide living specimens and supply drugs to students of medicine. Subsequently, as their collections expanded with material brought from various parts of the world that were being opened up by exploration, the main rôle of botanic gardens became the scientific study of plant diversity for its own sake, combined with the development of horticultural skills.

The expansion of the living collections in botanic gardens in Europe and other temperate countries was often stimulated by generalised pretensions of comprehensiveness rather than in response to any clearly articulated scientific policy. Other botanic gardens developed collections of their local, national or regional flora, thus providing a more or less comprehensive source of material for scientific study by the staff associated with the garden.

Botanic Gardens and the
World Conservation Strategy

ISBN: 0-12-125462-3

In recent years there has been a reversal of rôles. No longer are the larger Western botanic gardens dominant in terms of their exotic collections, rather it is their herbaria and laboratories which justify their reputation. Today, in accepting a major conservation rôle, it is the botanic gardens in those countries with rich floras that have a clearly defined rôle to play. There is, therefore, a *crise d'identité* in many botanic gardens in the West. It is now being recognised that their research policies, as regards the herbarium collections and laboratories, are separate and often distinct from those that motivated the accumulation of the living collections, which are often neglected or ignored by the scientific staff. These gardens too have a rôle to play in conservation but this rôle will increasingly involve cooperation with gardens in tropical and subtropical countries with richer floras.

It is ironic that just as the conservation mission of botanic gardens is becoming widely accepted, some of the major tropical gardens, which have in the past occupied a brilliant rôle in plant introduction and domestication, have now been reduced to little more than public parks, their scientific research programmes having been largely stripped away.

Resumen

El cambiante papel del jardín botánico

Los jardines botánicos, de una manera u otra, han sido un significativo instrumento en el desarrollo cultural y científico del hombre a lo largo de los siglos. Tan prestigiosa es la denominación de ''Jardín Botánico'' que ha sido aplicada a muchos parques y lugares similares, sin que ellos correspondan realmente a ese titulo; no obstante, por extensión, pueden ser asi definidos. Los jardines individuales han jugado un importante papel en el desarrollo económico y avance comercial de muchos lugares del mundo. Los jardines botánicos tropicales en particular, han sido considerados como una importante herramienta en la expansión colonial de los siglos XVIII y XIX; ellos fueron responsables de la introducción y transferencia de germoplasma de una a otra parte del mundo, suministrando de este modo, las bases para modelos agrícolas que persisten en la actualidad.

Los primeros jardines botánicos de Europa fueron fundaciones de carácter médico, encaminadas a suministrar especimenes vivos y fármacos a los estudiantes de medicina. Paralelamente, a medida que las colecciones se iban ampliando con el material traido de aquellas partes del mundo que se iban explorando, el principal papel de los Jardines Botánicos fué el estudio científico de la diversidad vegetal en sí misma, combinado con el desarrollo de prácticas hortícolas.

La ampliación de las colecciones de plantas vivas en los jardines botánicos europeos y de otros países del área templada, fué estimulada a menudo por la pretensión generalizada de hacerlas exhaustivas, más que como respuesta a una politica cientifica claramente articulada. Otros jardines botánicos desarrollaron colecciones de sus floras regionales, nacionales o locales, proveyendo así una fuente de material más o menos amplia, para los estudios del personal científico asociado al Jardín.

En los últimos años se ha producido una inversión de papeles. Ya no dominan los grandes jardines botánicos occidentales en términos de sus exóticas colecciones; más bien son sus laboratorios y herbarios los que ganan su prestigio. Aceptando el importante papel de la conservación, hoy los jardines botánicos de los países con ricas floras tienen un papel que jugar claramente definido. Existe desde luego, una "crisis de identidad" en muchos de los jardines botánicos del occidente. Ahora se ha reconocido que su politica de investigación — en cuanto a herbarios y laboratorios se refiere — está separada y, a menudo, diferenciada de aquellas que motivaron la acumulación de las colecciones vivas, las cuales son con frecuencia despreciadas o ignoradas por el personal cientifico. Estos jardines también tienen un papel que jugar en la conservación, pero tal papel se incrementará cooperando con jardines de países tropicales y subtropicales con floras más ricas.

Resulta irónico que, justo cuando la misión conservacionista de los jardines botánicos está llegando a ser ampliamente aceptada, algunos de los principales jardines tropicales, que en el pasado jugaron un brillante papel en la introducción y domesticación de plantas, han quedado reducidos ahora a poco más que parques públicos, con sus programas de investigación científica ampliamente recortados.

Introduction

Botanic gardens, in the sense of collections of living plants grown for some educational, economic, medicinal, or scientific purpose, have played a significant rôle in many cultures and civilizations over the ages. Their contribution to cultural development, to economic progress and commercial expansion has been of very great significance, and this is often forgotten today. Tropical botanic gardens in particular were often created by governments as instruments of colonial expansion and commercial development, playing a major part in establishing the patterns of agriculture in several parts of the world, most notably in south-east Asia. Although it is only in recent years that the importance of the conservation of biodiversity, in the form of germplasm, has been explicitly recognised both by agricultural agencies and by governments, many botanic gardens in the tropics, and to a lesser extent in temperate countries, have been engaged in such operations for at least the past two centuries. As we shall see, from a scientific point of view, this process had serious limitations and was not conducted with the rigour that we would expect today, but it was nonetheless an impressive achievement.

In practice different kinds of botanic gardens have undertaken a series of distinct rôles during their history. In this paper I shall outline the major episodes and their characteristics, and endeavour to suggest in what new directions it would be profitable to proceed.

Botanic gardens: no simple definition

The question of what constitutes a botanic garden and how one defines one is complex and the reader is referred to the discussion in the IUCN Botanic Gardens Conservation Strategy for a discussion of the issues involved (IUCN 1985). In brief, botanic gardens can occupy many different rôles and there is no one single characteristic by which to define them. In general it may be said that the plant collections held in them are managed in a scientific way and have a particular purpose other than simply pleasure and amenity, however valid those particular aims may be. From this it follows that the collections are normally labelled and that they are available for study by *bona fide* students and, by extension, open to the public at large. There are, however, exceptions to any of the criteria one may propose and no one single model that can be pointed to as ideal or typical. In fact there have been several different models over the centuries.

The early European Medicinal Gardens

In the western tradition the first botanic gardens to be founded were medicinal gardens whose primary rôle was to provide material and instruction for students in the faculties of medicine in Italy, France, Switzerland, the Netherlands and other western countries. Certainly earlier gardens which could claim to be "botanic" were created by previous civilisations in other parts of the world such as China, pre-Hispanic Mexico and the Arab world, but we know little of them or their detailed functioning, and they do not seem to have greatly influenced the development of the later European foundations (Stafleu 1969).

The earliest medicinal gardens were established in Italy in the 16th century. The first was at Pisa, and was founded in 1543 by Luca Ghini who held the post of *lectura simplicium* in the university and was accordingly the first to lecture in what today we would call pharmacognosy (Atkinson 1955). It was followed closely by Padua (1545), Florence (1545) and Bologna (1547). Then came Zurich (1560), Leiden (1577), Leipzig (1579), Paris (1597), Montpellier (1598), Oxford (1621), Uppsala (1655), Edinburgh

(1670), Berlin (1679) and Amsterdam (1682). All of these botanic gardens exist to the present day, most of them in their original locations, but it is sad to report that Amsterdam is currently under threat of closure.

The functions of these medicinal or physic gardens, sometimes called gardens of simples (as in the Giardino dei Semplici in Florence, which still bears this name), was initially quite restricted. The properties of many of the plants cultivated were known from the descriptions in the herbals recorded over the centuries since the publication of the classic treatise of Dioscorides, *De Materia Medica*, published in the 1st century AD and subsequently revised for about 1,500 years. Other plants, whose properties, medicinal values or other attributes were little known, were introduced into these gardens and it became a legitimate field of study for such plants to be investigated. Thereafter it became normal practice to grow increasing numbers of different plant species in botanic gardens for scientific study, and thus it is from these original medicinal foundations that many of today's major European botanic gardens have arisen, changing their functions as they developed. The medicinal rôle has now been revived in some cases and specialist medicinal plant botanic gardens have been created in recent decades of this century.

The classic European model

While many European botanic gardens, as we have noted, expanded their collections and became centres for the study of plant diversity, the focus of research soon moved away from the living collections themselves and focussed increasingly on taxonomic studies in the herbaria that were founded in the institutions associated with the gardens. This was a reflection of the accelerating exploration of different parts of the world by naturalists, particularly in the 18th and 19th centuries, leading to the amassing of large collections of plant and animal material that had, on the one hand to be stored in newly created museums and herbaria, and on the other hand to be identified. The need for the handling and identification of such large amounts of material led to the establishment of the professional taxonomist.

Not surprisingly, with the dominance of botanical science by taxonomy in the 18th and much of the 19th century, the term botanic garden came to be more identified with the herbarium, library (and later laboratories) housed there and with the programme of taxonomic research undertaken than with the garden itself, i.e. the living collections on which often little research was undertaken. This identification of botanic gardens with taxonomy persists to the present day, especially in those cases where the herbarium collections are of international significance. An inevitable consequence of this was the frequent appointment of a taxonomist as a director and in some instances the neglect of other aspects of the garden.

This classic European model was also followed in other parts of the world such as North America and Australia, and an analogous situation developed in some of the principal European tropical gardens, as described in the next section.

The colonial tropical botanic garden

It was I.H. Burkill, in one of his chapters on the history of botany in India, who called botanic gardens the botanists' first caravanserais. Certainly the early tropical botanic gardens were more motivated by considerations of trade and commerce than by science for its own sake. This was especially true of those that were established by the British in the colonial territories, but a similar situation occurred in the other colonial powers in the 18th and 19th centuries. The case of the celebrated Calcutta Botanic Garden illustrates the point well. It was originally intended to establish what today we would call a commercial nursery rather than a botanic garden. Its founder, Colonel Kyd, in his submission to the Bengal government, was not at all clear as to what to call the garden he proposed and talked about it in terms of a horticultural nursery. Yet as Burkill (1953) notes, he wanted it to be something more than a garden, and so called it a botanic garden although no botany was intended to be practised there and he explicitly wrote that there would be none! We have Roxburgh to thank for introducing botany into the Calcutta Garden.

It was in 1787 that the Calcutta Botanic Garden was founded, during a period of great prosperity for the developing city through its rôle as a trading centre of the East India Company. The Company was concerned at the shortage of supplies of oak for its shipyard at Deptford on the Thames; one of the original aims of the new garden at Calcutta was to carry out trials to see if teak could be grown in the vicinity of Calcutta. So much enthusiasm was there for the idea of a garden that the city set aside 310 acres for the purpose. In the words of Kyd, the garden was to be created "not for the purpose of collecting rare plants as things of curiosity or furnishing articles of luxury, but for establishing a stock for disseminating such articles as may prove beneficial to the inhabitants as well as the natives of Great Britain, and which ultimately may tend to the extension of the national commerce and riches" (Kyd quoted in Gager 1938).

Calcutta was not, however, the first botanic garden to be founded in the tropics. That distinction goes to the Pamplemousses garden in Mauritius

(Ile de France), which was founded by Governor Labourdonnais in 1735, initially to provide fresh fruit and vegetables for the settlement and for ships calling at the port. He also introduced many spice plants as well as cassava for feeding the slaves. Later developments, including the establishment of a trials garden and the introduction of many economically useful plants, some of which were then introduced into Réunion, Madagascar and Cayenne, led to the ensemble of gardens known as the Parc de Pamplemousses and today as the Royal Botanic Gardens Pamplemousses.

This was the pattern for most of the botanic gardens created in India, in south-east Asia and in the Caribbean, mainly by the British and the Dutch. This was a period of intense commercial rivalry; a botanic garden in Singapore, for example, was started in 1822 by Stamford Raffles of the British East India Company when the British were trying to break the monopoly of the Dutch in the spice trade. (It was subsequently abandoned in about 1846 — Purseglove 1959.)

The rôle of these early tropical botanic gardens is well documented (Heywood 1983; Holttum 1970; Purseglove 1959). It does us no harm, however, to be reminded of the salient facts. Many crop plants were introduced by or through these gardens — often in association with European botanic gardens such as Amsterdam and Kew — and included cloves, chocolate, cinchona, tea, coffee, oil palm and breadfruit. The emphasis was very much on economic development and several of these gardens were created specifically to act as nurseries or propagation centres for the receipt of germplasm of commercial crops. Notable amongst these were the Royal Botanic Gardens at Peradeniya, founded in 1821, which wielded considerable influence on the development of agriculture in Ceylon and indeed far beyond. It was the director of this garden, Thwaites, who in response to a request from the U.K. government, took action to rescue the local economy after the failure of the coffee plantations caused by disease. It was decided to introduce *Cinchona* as a crop and for this purpose a new station or mountain botanic garden, that of Hagkala, was created. From 1873 to 1876 the almost incredible numbers of 3,486,000 seedlings of *Cinchona* were made available for local planters. Later a lowland garden, that of Heneratgoda, was established to receive and grow on seedlings of Para rubber (*Hevea*) from the Royal Gardens at Kew after their introduction from South America. The original rubber tree which was the first to be successfully tapped still stands today in the garden.

A third botanic garden involved in the rubber saga was Singapore, originally founded in 1859 by an Agri-Horticultural Society. After a chequered history the garden was taken over by Cantley. Among the economic species he introduced into the garden was Para rubber, a plant normally more associated with his successor H.N. Ridley. Ridley showed how Para rubber trees could be effectively tapped on a regular and frequent system and this made possible the development of the highly successful rubber plantation industry on which the prosperity of Malaya and later other parts of south-east Asia became based.

The combined efforts of these tropical botanic gardens in transporting what today we would call germplasm across the continents were largely responsible for the major agricultural patterns of the tropics in several parts of the world. Unfortunately in their very success often lay the seeds of their future decline. The increasing involvement of these tropical gardens in agriculture, horticulture and forestry led inevitably to the creation of separate institutes, departments or schools of agriculture or, as in the Caribbean, the foundation of Agricultural Experimental Stations. This deprived the gardens which fathered these new institutions of one of their major justifications for existence. In Peradeniya, for example, after Willis's retirement the Garden was made the basis of a Department of Agriculture for Ceylon, and much to the disappointment of Willis most of the botanical work of the garden ceased apart from the mycological work of T. Petch. On the other hand the ornamental horticultural work of the garden continued to develop and the Curator, H.F. Macmillan, published a very substantial work on tropical gardening which is still regarded as a standard work to this day.

Of course not all the activity of these tropical botanic gardens was economically orientated. As Holttum (1970) and Purseglove (1959) remind us, they were also major centres for the scientific study of tropical botany, not just floristics and taxonomy, and their publications are perhaps not widely enough known by botanists in temperate regions. Willis himself wrote that "it would be difficult to exaggerate the value of travel in other countries to the working botanist, especially if his work lie in the departments of systematic botany, geographical distribution, ecology, morphology or economic botany, whilst to the physiological, or anatomical worker there are also innumerable problems that can only be solved by research in tropical countries" — a point that many of our physiological and biochemical colleagues would do well to heed today.

Civic and Municipal Botanic Gardens

A large number of what might be termed civic or municipal botanic gardens were founded in the 19th and 20th centuries — both in Europe, in France, the United Kingdom and Italy, for example, and in British Commonwealth countries such as Australia. In the 20th century, such gardens were a particular feature in the U.S.A.

Most of these gardens did not develop major scientific facilities or programmes in taxonomy or allied fields as found in the mainstream gardens mentioned above.

There were however notable exceptions, such as Missouri Botanical Garden (Henry Shaw's garden), which was founded in 1859 and was the first botanic garden in the United States. From the start it was a scientific institution. Another exception was the Palmengarten in Frankfurt, Germany, founded in 1869 and today with one of the world's leading orchid and succulent collections.

What most of these civic botanic gardens did develop was the horticultural aspect — they were botanic gardens in the sense of building up and maintaining collections of usually well-labelled plants, and exchanging seeds with other botanic gardens throughout the world. They did this through that unique and characteristic mechanism that had its origin in the 18th century, the Index Seminum. Their mission, if at all explicit, was somewhat limited and in the absence of scientific as opposed to horticultural leadership, stored up the seeds of disaster for the future. There was soon to be little of a "botanic" nature in many such gardens other than in name. Similarly in the United States and in Australia the term botanic garden was sometimes applied to gardens which had no real claim to the title at all.

The position in North America is interesting in that a large number of botanic gardens were created in the 20th century that do not fit easily into any of the usual models. They are often private foundations and community-orientated and do not belong to the mainstream international tradition. Seldom do they offer an Index Seminum and are not part of the world network of seed exchange. Again their emphasis is educational and horticultural and they usually have a strong programme of public education, often with a vigorous society of friends or supporters in the local community.

Special kinds of botanic gardens

In addition to the main kinds of botanic gardens described above, others of a more specialised nature have been developed. Examples are the Agro-Botanical Gardens, especially those in Europe, which specialise in the cultivation of and research on agricultural or horticultural plant species and their allies, often as parts of University or state Agriculture Institutes, Departments or Experiment Stations. Similar Agro-Horticultural Botanic Gardens were founded in India in the 19th century such as those at Bombay, Madras, Ootacamund and Alipore.

Germplasm collection gardens and Experimental Station gardens have been created in Java. These are not usually open to the public. A good example is that at Paseh in West Java, which is arranged like a botanic garden and includes amongst its functions, *ex situ* conservation of fruit trees, seed stock gardens and recreation.

Several botanic gardens specialising in tropical economic plants have been established in China in recent years. An example is the Hainan Botanical Garden of Tropical Economic Plants, whose goal is to introduce and collect tropical and subtropical economic plants of the world so as to enrich China's national plant resources. Others include the South China Botanical Garden at Guangzhou, the Xishuangbanna Botanical Garden of Tropical Plants and the Xiamen Botanic Garden.

Mention has already been made of medicinal plant gardens. Modern versions of these are found in various parts of the world. Their main aim as the name implies is the collection and cultivation of medicinal plants and the dissemination of knowledge about them to the public, usually in association with research laboratories or associations. Examples include the Drug Plant Gardens, Seattle (U.S.A), and in Japan the five Experiment Stations of Medicinal Plants maintained by the National Institute of Hygienic Sciences.

Not surprisingly a number of specialist orchid gardens have been created in various countries, especially although not exclusively in the tropics. These specialise in the cultivation and propagation of orchids, often with commercial exploitation a major consideration. Examples include the Orchid Research and Development Centre at Tipi (Arunachal Pradesh, India), which maintains an orchid sanctuary for *in situ* conservation at Sessa and two *ex situ* substations for cultivation of germplasm; and also in India the National Orchidarium and Experimental Garden of the Botanical Survey of India Eastern Circle at Shillong. The Wheeler Orchid Collection and Species Bank at Bell State University, Indiana (U.S.A) maintains 3,000 orchid species in cultivation, including 150 which are rare or endangered in nature as well as several which are probably now extinct in the wild.

There are also specialist gardens for the cultivation of many other groups of plants such as bamboos, Rhododendrons, Grevilleas, palms, native plants and so on. Emphasis on the cultivation of native plants has been pioneered by gardens such as the Rancho Santa Ana Garden at Claremont, California and this theme has been taken up by several gardens which are primarily interested in conservation. Examples are the Conservatoires of Brest, Porquerolles and Nancy in France, which form an Association of Conservatoires of Plant Species in association with the Muséum National d'Histoire Naturelle, Paris, charged with the cultivation and preservation of endangered species of particular regions; the Jardín Botánico Viera y

Clavijo, Tafira Alta, Gran Canaria which specialises in the cultivation, propagation and protection of endemic species of the Macaronesian area (Bramwell 1981); the National Botanic Gardens, Lae (Papua New Guinea) which concentrates on the cultivation and preservation of native plants; and the Vumba Botanical Reserve and Botanical Garden, Zimbabwe whose aim is to preserve, protect, cultivate and propagate indigenous plants of Zimbabwe, especially those that are endangered.

The number and distribution of botanic gardens and arboreta

It is surprisingly difficult to ascertain with any accuracy just how many botanic gardens exist in the world at any one time. The latest estimates that can be given derive from research that is in progress in connection with the preparation of the IUCN/WWF Botanic Gardens Conservation Strategy. This indicates that there are currently around 1,400 botanic gardens and arboreta although the figure does change almost by the week. This figure is much higher than previous recent estimates, although it is worth noting that as many as 1,600 botanic gardens were believed to have existed in Europe alone at the end of the 18th century. The actual figures depend, of course, on how botanic gardens and arboreta are defined and there are numerous marginal cases. Some form of categorisation as suggested in the Strategy will have to be employed.

The current distribution of botanic gardens and arboreta in the world is very uneven. By far the largest number is found today in the North Temperate zone, especially in Europe, the U.S.S.R. and North America (see Table 1). It is strikingly evident that this distribution is based on historical factors and bears no relationship to the floristic richness of the areas concerned. The number of botanic gardens and arboreta in the tropics is difficult to assess precisely, but is probably much larger than generally realised, largely due to the creation of a considerable number in recent years. During the past 10 years alone over 100 new botanic gardens have been opened or planned, many of them in tropical or subtropical countries. Altogether there are about 230 tropical botanic gardens. The greatest concentration is in south and south-east Asia — with an impressive network of 36 in India alone, followed by about 60 in Central and South America, some 30 in tropical Africa, 30 in the region of Oceania (including Hawaii with 20), 17 in the Caribbean, and a small number in Australia, China and the United States (in addition, of course, to the considerable numbers of gardens in temperate zones in these countries).

In tropical Africa and in Latin America the patterns of development and the parts played by botanic gardens are quite different from those we have described for south-east Asia and for the West Indies. With very few

Table 1
Plants and gardens

	Plants	*Gardens*
Europe	11,000	459
North America	21,000	262
Central and South America	90,000	59
Africa (with Madagascar)	30,000	46
Australia	25,000	41
U.S.S.R.	21,200	156

exceptions most are 20th century foundations. In South America only Rio de Janeiro Botanic Garden dates from the 19th century (1808) and in Central America an early botanic garden was founded in Mexico by Charles III of Spain's scientific expedition there in 1781 (Lozoya 1984); otherwise all other botanic gardens in Latin America are recent, many of them in the last decade or two. The colonial history and the timing of the Spanish and Portuguese occupation combined against the creation of the kinds of botanic gardens founded in the colonial territories of the British and Dutch for example. What the Spanish did favour was the creation of acclimatisation gardens such as that of La Orotava in the Canary Island of Tenerife, founded in 1788 by orders of the crown, to permit the introduction into Spain after due acclimatisation of the countless new species of plants being discovered in the Spanish territories in America and Asia. This garden exists to the present day, unlike another Spanish acclimatisation garden at Puerto de Santa Maria on the Spanish mainland, which received the material from La Orotava before it was sent on to the royal gardens at Aranjuez outside Madrid.

The situation is similar in tropical Africa where, with the exception of the Royal Botanic Gardens, Pamplemousses, Mauritius, all botanic gardens are relatively modern apart from three founded towards the end of the 19th century — Entebbe, Zumba and Limbe. By this time most of the agricultural crops had already been introduced into cultivation as the result of earlier initiatives in other parts of the tropics. The African and Latin American gardens did not become involved in major plant introductions in a way comparable with what happened in south-east Asia. Possibly it was as a consequence of this lack of involvement in major agricultural initiatives that most of these botanic gardens did not become important research centres and many of them are today in serious decline.

What is not often realised is the fluctuation in the number of gardens over the centuries and even over the years. Not only have some tropical botanic gardens found themselves under threat but a similar fate has overtaken many temperate botanic gardens and arboreta, including some quite celebrated ones such as the historic Hortus Botanicus of Amsterdam which faces closure or transformation into a public park.

The beginnings of the decline

In the 18th and 19th centuries both the numbers and the richness of the collections developed vary considerably, as we have seen. Their public and scientific rôles often went their separate ways. The public, visiting them in increasing numbers, viewed what was put on show uncritically on the whole, with the gardens making few concessions to public education other than the provision of labels with minimal information (which were probably aimed more at students than at the general public). The educational rôle of botanic gardens was slow to develop and is very much a phenomenon of the last few decades.

The expansion of the collections themselves was stimulated by generalised pretensions of building up comprehensive samples of plant diversity for general scientific study (who actually studied the collections is another matter!), rather than in response to any clearly articulated policy. Often, too, any collections policy was determined by the individual interests of those in day-to-day charge of the gardens, often not the director, and with changing interests and fashions it proved difficult to dispose of material acquired in response to earlier interests.

Some botanic gardens focussed their attention on the plants of their local or national flora and provided material in cultivation that could be used by staff associated with the garden for various kinds of scientific research. This was, however, the exception, especially in European botanic gardens.

In the west, during the period of industrial expansion, many botanic gardens were able to expand and build very substantial and expensive glasshouses and conservatories, derived from the original orangeries of earlier centuries. Even tropical botanic gardens felt constrained to provide what for them must have been somewhat superfluous constructions, and today the public now expects a botanic garden to contain facilities that were often built precisely for display purposes.

The scientific rôle of botanic gardens underwent a major change in the 19th and 20th centuries, but this was in respect of the associated facilities such as herbaria, libraries and laboratories that had been built up in their associated institutions or within the gardens themselves (Raven 1981), as a consequence of the floristic exploration of many parts of the world. This exploration had gathered momentum as a result of the expeditions that many of the gardens or their parent institutions had sponsored. As a consequence there was a shift away from the formerly dominant living collections to the herbarium collections, and staff were employed to work on these, leading to the series of Floras and other publications that emanated from herbaria. In many cases today, it is the herbaria and similar facilities by which the scientific reputation of a botanic garden is judged, not the living collections, which are often neglected or even ignored by the scientific staff. Thus many botanic gardens are suffering from a *crise d'identité* and one has in fact changed its title from botanic garden to botanical research institute.

The decline of many tropical botanic gardens that we noted above comes at a time when new and important rôles have been proposed for them and we must hope that in the coming years there will be a revival in their fortunes. After the loss of a major economic rôle tropical gardens that had associated herbaria focussed their attention increasingly on taxonomic and floristic studies. The gardens themselves have become little more than beautifully maintained public parks and even in some cases transferred to parks administrations. Some are viewed as ripe for urban development. It is unfortunate that the distinction between botanic gardens and public parks is not generally understood by the public and often not, it would seem, by politicians. Even the taxonomic activity in some tropical gardens is becoming difficult to maintain. One example of this is Singapore where, at a recent symposium on botanic gardens in the tropics, held at Penang, Dr Wee noted laconically that "currently taxonomic research is in abeyance at Singapore botanic garden".

Botanic gardens and conservation of plant genetic resources

In recent years an increasing number of botanic gardens are beginning to look to conservation as one of their major goals and indeed justifications. A considerable literature has developed on this field and is reported elsewhere in this book.

We have already referred to the major contribution that tropical botanic gardens made to the assessment and introduction of the principal economic crops grown today. Viewed today this would be seen to have had serious limitation. For one thing the range of crops actually introduced successfully was quite small and only represented a tiny part of the potential. And what is more, the species were often introduced from other continents or at least from distant lands while local species of economic potential were neglected or ignored. Furthermore little if any attention was paid to the problems of sampling with the result that all too often major crop species comprise only a very narrow range of the genotypes, if not just one, to be found in the wild genepool.

Today a great deal of attention and expertise has been focussed on genetic conservation of plant species and populations, and a considerable body of expertise has been developed by population geneticists and by organisations such as FAO and IBPGR. The latter bodies have concentrated on the 100 or so high priority species of economic plants, initially the staple food crops, so as to provide breeders in the future with adequate stocks. And while IBPGR has recently expanded its interests, the responsibility for the genetic conservation of the genepools of the thousands of wild species that are endangered, are of local economic potential such as medicinals and aromatics, or are of ecological importance or of scientific interest, rests with the botanical community. It is here that botanic gardens must play a major part.

In its Botanic Gardens Conservation Strategy IUCN will be giving this question high priority and will be preparing a series of guidelines and procedures. Other bodies such as the Center for Plant Conservation at the Arnold Arboretum and the American Association of Botanical Gardens and Arboreta are already, to some degree, engaged in such activities. There is, however, an enormous amount to be done and cooperation is essential both within the botanic garden community, including some form of multilateral association, and with FAO and IBPGR in both *in situ* and *ex situ* conservation and propagation under quarantine conditions.

Herbaria and botanic gardens have contrived a situation whereby they are not regarded as the most obvious centres for plant resource conservation research and activity. On the contrary they should be in the forefront of this and botanic gardens should be the flagships of our international botanical efforts in the service of science and in the service of man.

References

Atkinson, R.G. (1955). Herbal and botanic gardens. *Pharmaceutical Journal* March 5.

Bramwell, D. (1981). Conservation-orientated research in local botanic gardens. *Bot. Jahrb.* **102**, 125–132.

Brockway, L.H. (1979). 'Science and Colonial Expansion: the rôle of the British Royal Botanic Gardens'. Academic Press, London and New York.

Burkhill, I.H. (1953). Chapters on the history of botany in India — 1. From the beginning to the middle of Wallich's service. *Journal Bombay National History Society* **51**, 846–878.

Gager, C.S. (1938). Botanic gardens of the world. *Brooklyn Botanic Garden Record* **27**, 151–406.

Heywood, V.H. (1983). Botanic gardens and taxonomy — their economic rôle. *Bull. Bot. Survey India* **25**, 134–147.

Holttum, R.E. (1970). The historical significance of botanic gardens in S.E. Asia. *Taxon* **19**(5), 707–714.

IUCN (1985). 'Botanic Gardens Conservation Strategy: First Draft Proposal by Vernon H. Heywood'. WWF and IUCN, Gland.
Lozoya, X. (1984). 'Plantas y Luces e México'. Ediciones del Serbal, Barcelona.
Powell, D. (1973). 'The Voyage of the Plant Nursery, H.M.S. Providence, 1791–1793'. The Institute of Jamaica.
Purseglove, J.W. (1959). History and functions of botanic gardens with special reference to Singapore. *Gardeners' Bulletin, Singapore* **17**(2), 125–154.
Raven, P.H. (1981). Research in botanical gardens. *Bot. Jahrb.* **102**, 53–72.
Stafleu, F.A. (1969). Botanical gardens before 1818. *Boissiera* **14**, 31–46.

The scope of the plant conservation problem world-wide

PETER H. RAVEN

Missouri Botanical Garden, St Louis, U.S.A. and Chairman, IUCN/WWF Plant Advisory Group

Summary

Of the roughly 250,000 kinds of vascular plants, probably no more than 1,000 have become extinct during the past century, but the pace of extinction is increasing rapidly at present and will almost certainly continue to do so over the next few decades. About a third of the world's vascular plants occur in temperate regions; here threatened and endangered plant species are in general well understood and well monitored. Conservatively, about 10% of the species in temperate regions might be classified as threatened or endangered, the greatest number of such species being in the Cape region of southern Africa.

The greatest threat to plant species, however. exists in the tropics. About 170,000 species occur in the tropics and subtropics. Especially notable is the fact that more than 40,000 plant species — a quarter of the world's total diversity — occurs in the three northern Andean countries of Colombia, Equador, and Peru combined, countries in which the plants, animals and microorganisms are the most poorly known assemblage on earth.

The U.N. Food and Agriculture Organisation (FAO) of the United Nations has estimated that tropical closed forests amounted to 56% of their potential extent, and were being clear-cut at the rate of 1·1% per year. Although this rate of clear-cutting would allow the forests to survive for no more than 90 years, the actual picture is much worse, because the FAO did not make future projections and did not take into account population growth; the deterioration of the forests brought about by shifting cultivation; or the uneven distribution of forest

Botanic Gardens and the World Conservation Strategy

ISBN: 0-12-125462-3

destruction throughout the world. Furthermore, the seasonal forests of the tropics and subtropics have been and are being subjected to far greater pressures than are the closed forests on which the FAO concentrated its efforts.

Island biogeography theory indicates that when an area is reduced to 10% of its former size, half of the species will be lost. Using this theory, one can calculate that probably at least 7,500 plant species and 150,000 species of animals and microorganisms are already extinct or on the verge of extinction in regions such as Madagascar, lowland western Ecuador and the Atlantic forests of Brazil. In contrast the forests of the western Brazilian Amazon and the interior of the Guyanas, together with those of the Zaire Basin in Central Africa, will probably last the longest in a relatively undisturbed condition, but one may calculate that all other tropical and subtropical forests will be destroyed or at least severely disturbed during the next 30 years. These facts indicate that more than 60,000 species of plants are at risk of extinction in the tropics and subtropics. An extinction event of this magnitude has not occurred since the end of the Cretaceous Period, some 65 million years ago! In order to ameliorate this situation, we must find and learn about these plants and move quickly and effectively to save as many of them as possible, emphasising those of particular scientific or economic importance, either actual or potential.

Resumen extendido

Envergadura del problema de la conservación de plantas en todo el mundo

Existen aproximadamente 265,000 especies de plantas en el mundo, de las cuales alrededor de 16,000 son briofitas; 13,000 son helechos, licofitas, psilofitas y esfenofitas; 550 son coníferas; 100 son cicadales; 1 es ginkofita; 70 son gnetofitas y 235,000 son fanerógamas. Aquí vamos a limitar nuestro debate sobre las aproximadamente 250,000 plantas vasculares existentes. De éstas, probablemente no más de 1,000 se hayan extinguido durante el pasado siglo, pero el ritmo de extinción se ha incrementado en la actualidad con rapidez y casi con certeza, continuará haciéndolo así en las próximas décadas. Pasaremos revista aquí a las razones existentes para esta predicción.

Alrededor de un tercio de las plantas vasculares del mundo — sobre unas 80,000 especies — se dan en las regiones templadas. Esta cifra incluye alrededor de 10,500 especies en Europa; unas 20,000 en los Estados Unidos y Canada; y quizás 25,000 en las zonas templadas de China. En las regiones templadas del mundo las especies de plantas amenazadas o en peligro son bien conocidas en general, se hallan bajo seguimiento y, en consecuencia, están en menor peligro de extinción que aquellas de los países tropicales y subtropicales. Desde el punto de vista de la conservación, alrededor del 10% de las especies de las regiones templadas pueden clasificarse como amenazadas o en peligro, con el mayor número de tales especies en el Sur de Africa. Claramente, hay más especies de plantas en riesgo en la región del Cabo en Sudáfrica, que en cualquier otra parte del mundo con un área comparable, pero merecen con toda claridad una atención especial los abundantes endemismos muy locales de las regiones con clima mediterráneo en general e islas templadas.

Sin embargo, la mayor amenaza para las especies vegetales está en los trópicos y subtrópicos, incluyendo quizás 85,000 en Iberoamérica, 35,000 en Africa tropical y subtropical y, al menos 50,000 en el Asia tropical y subtropical. Es especialmente de resaltar el hecho de que más de 40,000 especies de plantas — la cuarta parte del total de la diversidad mundial — crecen en los tres países andinos septentrionales: Colombia, Ecuador y Perú combinados, en un área de alrededor de 2.7 millones de kilómetros cuadrados. Es decir, estos tres países juntos, tienen aproximadamente el tamaño de la cuarta parte de Europa, la cual alberga no más del 5% de la diversidad biológica mundial. Puesto que tantas especies aparecen en estos países, ello debe exigir una alta prioridad en la acción de conservación; el hecho de que sus vegetales, animales y microorganismos constituyan la multitud peor conocida de la Tierra, hace que la tarea de su conocimiento y preservación sea formidablemente laboriosa.

En 1980 la Organización de las Naciones Unidas para la Alimentación y la Agricultura (FAO), estimaba que las selvas ecuatoriales contaban con el 56% de extensión potencial y que estaban siendo clareadas a razón del 1.1% anual. Aunque esta razón de clareamiento no permitiría a estas selvas sobrevivir por más de 90 años, en realidad el panorama actual es mucho peor porque la FAO no hace proyecciones de futuro y no toma en cuenta el crecimiento de la población; el deterioro de las selvas se produce por la práctica de los cultivos rotativos, o por la desigual distribución de la destrucción de los bosques en el mundo. Además, los bosques estacionales de los trópicos y subtrópicos han sido y están siendo objeto de mayores presiones que las selvas ecuatoriales sobre las que la FAO concentra sus esfuerzos.

La teoría de la biogeografía insular indica que cuando un área se reduce al 10% de su tamaño original, desaparecen la mitad de las especies. Considerando que la vegetación natural de Madagascar, las tierras bajas occidentales del Ecuador y los bosques atlánticos del Brasil se han reducido a menos del 10% de su extensión hace 50 años y que, en una estimación conservadora, cada una de estas áreas tendría originalmente alrededor de 5,000 especies de plantas endémicas (y por extrapolación también conservadora, quizás 100,000 especies de organismos de todas las clases), puede calcularse que, probablemente al menos 7,500 especies de plantas y 150,000 de animales y microorganismos se han extinguido ya o están al borde de la extinción, solamente en los citados países.

Las selvas del Amazonas occidental brasileño y el interior de las Guayanas, junto con las de la meseta del Zaire en Africa central, serán las que probablemente perdurarán por más tiempo en unas condiciones más o menos inalteradas, pero puede calcularse que todas las demás selvas tropicales o subtropicales serán destruídas o al menos, severamente perturbadas durante los próximos 30 años, que será un tiempo durante el cual se duplicarán las poblaciones de Africa, Iberoamérica y Asia meridional.

Quizás 25,000 especies de plantas se encuentren en las selvas de Sudamérica que perdurarán más y quizás, unas 15,000 especies adicionales se hallan en las selvas ecuatoriales del Zaire. Así, 40,000 especies — o alrededor de la cuarta parte de las que albergan los trópicos van a perdurar probablemente en estas regiones más allá de mediados del próximo siglo. Unas 130,000 especies vegetales más, sin embargo, se encuentran en los trópicos, pero no es en estas regiones, únicamente en áreas donde la vegetación será aniquilada en las próximas décadas inmediatas. Asumiendo liberalmente que la mitad de estas especies pueden ser como malas hierbas o persistir en pequeñas bolsas de vegetación, llegamos a la horrorosa conclusión que más de 60,000 especies de plantas — casi un cuarto del total de la diversidad mundial — están en riesgo de extinguirse en los trópicos y subtrópicos en el periodo de tiempo de nuestras vidas y la de nuestros hijos. Un acontecimiento de extinción de tal magnitud no ha ocurrido desde el final del período Cretácico, hace unos 65 millones de años!

Para mejorar esta situación, deberíamos averiguar y aprender sobre estas plantas, así como movernos rápida y efectivamente para salvar a tantas de ellas como sea posible, haciendo hincapié en aquellas con un particular interés científico o importancia económica, bien sea actual o potencial.

In assessing the scope of the plant conservation problem worldwide, it is necessary first to review the numbers of species of plants in the world; next, to consider the factors that in broad terms are threatening them; and, finally, to provide an assessment of the degree of threat and the numbers of species that are endangered in different parts of the world. I shall deal with each of these topics in turn.

There are approximately 265,000 species of plants in the world, of which about 16,000 are bryophytes; 13,000 are ferns, lycophytes, psilophytes, and sphenophytes; 550 are conifers; 100 are cycads; 1 a ginkgophyte; 70 gametophytes; and about 235,000 are flowering plants. In this paper, I shall confine my discussion primarily to the approximately 250,000 kinds of vascular plants. Of these, probably no more than 1,000 have become extinct during the past century, but the pace of extinction is increasing rapidly at present and will almost certainly continue to do so over the next few decades. Here we review the reasons for this prediction.

In general, it is the very rapid growth of the human population and the increasing expectations of human beings that are endangering the world's plants. The human population of the world in the year 1950 was approximately 2·5 billion, but today it has climbed to more than 5 billion people. In the developed world, here defined as Europe, the Soviet Union, Japan, the United States, Canada, Australia, and New Zealand, a population that amounted to some 830 million people in 1950 has climbed to about 1·2 billion at the present time: and in the rest of the world, a population of approximately 1·7 billion people in 1950 has climbed to about 3·8 billion now. Excluding China and focussing on countries that are partly or wholly tropical, a total of 1·1 billion people in 1950 has now grown to approximately 2·7 billion people — well over a doubling. In the next 35-year period another doubling in the number of people in the tropics and subtropics is expected, with the consequent enormous pressures on the environment that these increases will bring. Even though most demographers calculate a relatively stable world population by somewhere in the second half of the next century, the period of very rapid growth that we are now experiencing, together with rising human expectations, is putting a very great pressure on all aspects of the global environment.

In tropical countries, the problem is especially acute, since very rapid growth is coupled with a standard of living that is generally about a tenth that in developed countries, and for which there are legitimate expectations of improvement. Unfortunately, improvements mean more stress on the environment, more development of natural habitats, and more threats of extinction. In order to approach the problem at a more specific level, it is best to treat it regionally.

About a third of the world's vascular plants, some 80,000 species occur in temperate regions. According to statistics presented in "Plants in Danger: What Do We Know?" (Davis *et al.*, 1986) about 10% of these are endangered, threatened, rare or extinct. The figures include 1,927 rare and threatened species, 117 endangered species and 20 extinct species, among the estimated 11,300 species in Europe; comparable figures of 2,122, 107 and 39 for an estimated 20,500 species in southern Africa (Gibbs Russell *et al.*, 1984) and about 2,150 threatened, endangered and extinct species among the estimated 20,000 in the United States and Canada. It would be assumed that there may be something like 2,500 species in these categories among the estimated 25,000 species of plants in China, and that the total for all the temperate regions of the world of threatened, endangered, and extinct species of plants might amount to about 8,000 species — approximately a tenth of the total.

Of particular significance is the flora of the Cape Floristic Province of South Africa, where about 8,578 species occur in an area of only 90,000 square kilometres — two-thirds the size of England — and an estimated 5,850 of these are endemic, often very narrowly so. This region is therefore the most critical area of all the temperate regions of the world in terms of numbers of endangered and threatened plants species. Other regions with mediterranean climates, such as California and Western Australia, clearly warrant general attention also. Species found on many different temperate islands are likewise of great importance in view of the high levels of endemism in these regions.

The overall pattern is clear, however, and the point has been made above that the threatened, endangered and rare species of temperate regions are, in general, much better known and better monitored than those of the tropical and subtropical regions of the world. With world totals probably approximating 8,000 of 80,000 species threatened, several hundred endangered, and another several hundred extinct, the numbers are obviously significant. The threat to plant species in temperate regions should by no means be considered a trivial one, and continued monitoring in the development of schemes to preserve the plants of these areas is highly desirable.

Turning now to the approximately 170,000 species of vascular plants that occur in the tropics and subtropics, we find the greatest problems of plant conservation. Of these estimated 170,000 species, perhaps 85,000 occur in Latin America, 35,000 in tropical and subtropical Africa, and at least 50,000 in tropical and subtropical Asia. Especially notable among these statistics is the fact that more than 40,000 plant species — about a quarter of the total tropical diversity — occur in the three northern Andean countries of Colombia, Ecuador and Peru combined, in an area of about 2·7 million square kilometres. These three countries together are about a quarter of the size of the continent of Europe, but are home to about one-sixth of the world's total biological diversity — roughly more than three times as much as is found in all of Europe. Since so many species of plants, as well as animals and microorganisms, occur in Colombia, Ecuador and Peru, these areas must command a high priority for conservation action. The fact that their plants, animals and microorganisms are the most poorly known assemblage on earth makes the task of learning about them and preserving them an especially formidable one, particularly in view of the rapid changes that are characteristic of these countries, which will be outlined below.

Since the documentation of particular species is often inadequate as a measure of the extent to which organisms are threatened, endangered, or extinct in the tropics, it is necessary to think in terms of broad vegetation types. Richest in terms of species representation are the tropical closed forests, or rain forests, which are estimated by the Food and Agriculture Organisation (FAO) of the United Nations to occupy an area of approximately 6·7 million square kilometres in 1980 (FAO/UNEP, 1981).

The FAO estimated that this constituted some 56% of the original area, ranging regionally from two-thirds of the original area (4·55 million square kilometres) in Latin America to a third (just over 1 million square kilometres) in Asia. Again according to FAO estimates, these rain forests were being clear-cut in the late 1970s at the rate of about 1·1% per year.

Clear-cutting at this rate would amount to the removal of all tropical evergreen forests — the richest biological communities on earth — by the second half of the next century, within approximately 90 years, but there are compelling reasons for estimating the effects on organisms as much more serious than even these impressive statistics would suggest. The three most important reasons follow.

First of all, the FAO estimates, made in the late 1970s, did not take into account growth in human population; they were instantaneous estimates of the rate of clear-cutting of tropical closed forests at that time. Since the populations of both tropical South America and southern Asia are estimated to double in the next 29 years, and that of tropical Africa to double in less than 25 years, one can be certain that human pressures on these forests will actually accelerate enormously over the next three decades, with concommitant, much more rapid losses in the forest than the FAO figures would suggest. There appears to be no reason for assuming that a slackening of the pressure on the forest will accompany these very rapid regional rates of growth in the tropics of the world.

Second, the FAO took into account only clear-cutting — the literal removal of tropical closed forests — and did not consider the effects of shifting agriculture, firewood gathering, and other forms of intensive forest exploitation that are actually having a much more extensive effect, cumulatively, on forest destruction than simple figures based on clear-cutting would suggest. Of the total of 2·7 billion people estimated to live in the tropics and subtropics in 1986, the World Bank has estimated that approximately 1 billion, or about 40%, are living in absolute poverty. A substantial proportion of these people, about half of whom are actually malnourished by United Nations standards, are engaged in shifting cultivation in tropical forests. When one flies over tropical forests at relatively low elevations, one may see the effects of their activities everywhere. Reserves and sections of forests that may appear undisturbed from the outside are actually, on closer inspection, found to be badly damaged and filled with small settlements. Tropical organisms are characteristically very narrow in their ecological preferences, and the kinds of disturbance that effect them lead much more rapidly to the extinction of organisms than do comparable alterations in temperate forests. For these reasons, the widespread practice of shifting cultivation in tropical forests, and even more widespread gathering of fuel wood and other commodities in tropical forests, are leading to their decimation and the extinction of species much more rapidly than is the case in temperate regions.

Third and finally, the destruction of tropical forests is by no means uniform throughout the world; all of the evergreen closed forests will not disappear simultaneously. Rather, three large blocks of tropical closed forests — those in the western and north-western Brazilian Amazon, those in the interior of the Guyanas in South America, and those in the central Zaire (Congo) Basin in Africa — are expected to last longer than the other blocks of tropical forests, owing to their relatively low population densities. However, even these three relatively extensive blocks of forest will probably have been eliminated by about the middle of the next century. What is significant for the argument here, however, is the fact that all *other* tropical closed forests — those in Mexico and Central America, the West Indies, Andean South America, the southern and eastern fringes of the Amazon Basin and all other tropical and subtropical forests of South America, all forests in Africa outside of the Zaire Basin, and *all* of the tropical forests of Asia — will have been eliminated or severely damaged during the next three decades. This kind of elimination will clearly have an enormous and extensive effect on the extinction of organisms throughout the world, and forms the basis for the projections offered in this paper.

Lest it be thought that the tropical evergreen forests are being destroyed unusually rapidly, one should point out that all of the other forests of the tropics — the seasonal and dry forests — are much more severely damaged than are the tropical closed forests at the present time. Throughout the tropical and subtropical world, the picture is the same: human exploitation of seasonal forests has proceeded much more rapidly and completely than that of tropical closed forests, which in general are far less congenial for human occupation. For example, of the very extensive tropical dry forest that occurred along the western coast of Mexico and Central America five centuries ago, less than 2% approximates its original condition, according to estimates by Daniel H. Janzen of the University of Pennsylvania.

What happens when an area of forest is drastically reduced in extent? The synthetic theory of island biography developed by Robert H. MacArthur and E.O. Wilson has established a relationship between species number and area that holds relatively constant in biological communities throughout the world, whether these are on islands or on continental areas. The relationship is a logarithmic one and indicates that the multiplication of an area by a factor of approximately 10 times is necessary to double its number of species. If we turn the relationship around and consider what happens when an area is reduced to a tenth of its original size, we can predict that half of the species that occurred in the area originally will either become extinct immediately or will enter a process of extinction in the fragments of biological communities that persist. In areas such as Madagascar, the pacific forests of Ecuador, or the atlantic forests of Brazil, the natural vegetation has already been reduced to less than 10% of its original extent. I would estimate that each of these three areas, for example, had something approaching 10,000 species of native plants, of which conservatively a half, on average, were endemic. In these three areas taken together then, one may estimate that some 15,000 species of plants are threatened, endangered or extinct at the present time, and that a total approximating 7,000 to 8,000 of these would be endemic, and thus, threatened, endangered or extinct globally as a result of the local situations outlined. By extrapolation, one could estimate that a *minimum* of several hundred thousand species of endemic animals and microorganisms are also threatened, endangered or extinct in these areas. Even though many of the plants that are enumerated in the three-area total of 15,000 species endangered may still be persisting as smaller, scattered populations, such populations are clearly in danger of extinction during the next few years based on the principles of chance alone, as described by the island biogeography hypothesis.

Turning these figures around, and applying them on a worldwide basis, one may say that perhaps 25,000 plant species occur in the two large forest blocks in South America that will last the longest, those in western and north-western Brazilian Amazon and those in the interior of the Guyanas, and that perhaps an additional 15,000 species occur in the closed forests of Zaire. On the basis of these figures, it may be estimated that about 40,000 species, a quarter of the total found in the world's tropics, occur in forests that may last as long as the middle of the 21st century. This leaves an additional 130,000 species of plants that occur in the tropics and subtropics of the world, but not within forest blocks expected to last beyond the next two or three decades! Some of the plants in this estimated total of 130,000 species are weedy; others occur in desert or other dry areas that are not as heavily impacted by human activities as others; and others may simply persist in small and isolated populations, in part because of their own biological peculiarities. Nevertheless, the reduction of vegetation in the tropics and subtropics that is the exclusive home to some 130,000 species of plants during the course of the next 2 to 3 decades clearly will bring many of these plants to a threatened or endangered status or literally reduce them to extinction during this period of time. It seems fully reasonable in terms of what is known about the dynamics of small populations to assume that more than 60,000 species of plants, fully a quarter of the world's total diversity, are at risk of extinction in the tropics and subtropics during our lives and those of our children; if anything, these estimates may well be conservative. They would indicate, however, that the extinction rate for plants alone over the next 30 years may approximate 2,000 species per year, and this extinction may be occurring at the rate of a species or two a day now and may increase tenfold by the end of the 30-year period we are considering. If we extend our consideration to animals and microorganisms as well, well over a million species are at risk within the same period of time. And if recent estimates that the number of species of organisms that actually occur in the tropics greatly exceeds the conservative estimates of 4–5 million, then many more species may be at risk, by some estimates as many as 6 million species or more. For organisms as a whole, therefore, the average rate of extinction over the next 30-year period could be estimated conservatively as 30,000 species a year or something approximating 100 species a day!

We should not let our lack of knowledge of tropical organisms allow us to minimize the importance of these figures or to pretend that they are in some degree unreal. We would do so only at our peril and while we may argue about the exact details involved, the problem is clearly one of enormous proportions and drastic consequences for the future of humanity. We base our entire existence on our ability to utilise different kinds of organisms, which alone have the ability to convert the energy that reaches the earth from the sun into a form in which it is accessible for our maintenance and that of the other organisms on earth. Calculations that as many as a fifth of the world's plants may be suitable as sources of food and much higher proportions may be useful for their biochemical or other properties would suggest that, in addition to the aesthetic and scientific dimensions of the losses that we collectively are now experiencing, such losses may have immediate practical consequences also. Their extent suggests that we are losing species at the present time at something approaching 1,000 times the background rate characteristic of the past millions of years. It is difficult to be certain exactly what was the rate of extinction of plants and other organisms at the close of the Cretaceous Period, some 65 million years ago, but the loss may have approximated 50 or 60% of the world's total, according to estimates made by David Raup (pers. comm.). Following that period of time, no rate of extinction remotely comparable to that we are experiencing now has occurred for at least 65 million years.

We are living at a time, therefore, of utmost threat to the world's biota. In order to ameliorate this situation and secure the aesthetic, scientific and economic benefits of these species for our children and grandchildren, we must find out as much as possible about the plants, animals and microorganisms that share this planet with us now, and use this information to move quickly and effectively to save as many of them as possible. In carrying out such schemes, the scientific or economic importance of particular organisms, either actual or potential, should be kept in mind, and logical schemes should be devised that allow the application of such information directly to conservation strategies.

References

Davis, S.D., Droop, S.J.M., Gregerson, P., Henson, L., Leon, C.J., Lamlein Villa-Lobos, J., Synge, H., and Zantovska, J. (1986). "Plants in Danger: What do we know?" IUCN, Cambridge, U.K.

FAO/UNEP (1981). "Tropical Forest Resources Assessment Project (in the Framework of the Global Environment Monitoring System — GEMS)". UN 32/6.1301-78-04. Technical Reports nos. 1-3, Food and Agriculture Organization of the United Nations, Rome. (Comprises 3 separate reports: "Los Recursos Forestales de la América Tropical" (Forest Resources of Tropical America; in Spanish); 2 — "Forest Resources of Tropical Africa." 108, 586 pp. (in English and French); 3 — "Forest Resources of Tropical Asia," 475 pp. (in English and French).)

Gibbs Russell, G.E. *et al.* (1984). List of species of Southern African plants. *Mem. Bot. Surv. S. Afr.* **48**. 144 pp.

The IUCN/WWF Plants Conservation Programme in action

OLE HAMANN

Botanisk Laboratorium, Køpenhauns Universitet, Denmark (formerly Plants Programme Officer, IUCN, Switzerland)

Summary

The overall objective of the IUCN/WWF Plants Conservation Programme is to promote, formulate and support efforts to conserve plants, particularly plant genetic resources and plants of actual or potential economic value in developing countries in the tropics, thereby contributing to the preservation of a sustainable environment for socioeconomic development and progress.

To ensure that the harvest of living natural resources can be sustained, the general objectives of the World Conservation Strategy have to be met. On a more specific level, the following operational objectives are identified in the programme: firstly to build the public awareness that will help to stimulate action by pressure on governments and industries, by citizen action and by encouraging donations. Secondly, to strengthen IUCN's and WWF's ability to promote plant conservation more effectively among their constituencies and beyond. Thirdly, to conserve sufficient diversity within species to ensure that all their potential is available for the future. Fourthly, to emphasise the conservation of wild plants of importance, particularly for rural people in developing countries. Fifthly, to strengthen the rôle of botanic gardens in conservation, and finally to identify, develop and implement on-the-ground model projects in priority countries.

To meet these objectives, the Plants Conservation Programme comprises a set of activities and projects designed to address plant conservation from several directions. These activities, of which

Botanic Gardens and the World Conservation Strategy

ISBN: 0-12-125462-3

examples are given and discussed, range from publicising and distributing information explaining the rationale for plant conservation to long-term strategic activities such as preparing a Botanic Gardens Conservation Strategy, and to supporting the establishment and management of plant-rich sites such as tropical forests in National Parks.

In conclusion, a year and a half after the launching of the Programme, an impact is starting to show, this impact is evident from real on-the-ground conservation of endangered plants and ecosystems and from a catalytic effect on other organisations concerned with conservation of natural resources.

Resumen

El Programa para la Conservación de Plantas de la UICN/WWF en acción

El objectivo primordial del Programa de Conservación de Plantas es promover, formular y apoyar los esfuerzos en la conservación de vegetales y plantas de valor económico actual o potencial en los países en vías de desarrollo de los trópicos, contribuyendo así a sostener el medio ambiente para el progreso y desarrollo socio-económico.

Los objetivos generales de la Estrategia Mundial para la Conservación deben ser encontrar el modo de asegurar que puedan ser sostenidos los productos de los recursos naturales vivos. Los siguientes objetivos operativos están identificados en el programa y son: concienciar a la opinión pública, así como estimular la acción presionado a los Gobiernos y a las industrias por medio de la acción directa ciundadana y por donativos; fortalecer la actitud de UICN/WWF para promover la conservación de plantas con más efectividad entre sus elementos constituyentes e incluso más allá; cubrir una diversidad suficiente dentro de cada especie que asegure que todo su potencial genético es obtenible en el futuro; poner énfasis en la conservación de plantas silvestres de importancia, particularmente para gente rural en los países en via de desarrollo; fortalecer el papel de los jardines botánicos en la conservación e identificar, desarrollar, e instrumentar modelos de proyectos sobre el terreno en países prioritarios.

El Programa de Conservación de Plantas se diseña para abordar estos objectivos. Es una serie de actividades y proyectos que se dirigen a la conservación de plantas desde varias direcciones. Las actividades, de las cuales se dan ejemplos y se debaten, van desde la publicación y distribución de información en la lógica para la conservación de plantas hasta actividades estratégicas a largo plazo, tales como preparar una estrategia de conservación en jardines botánicos, y apoyar el establecimiento y gestión de lugares ricos en plantas tales como la selva tropical en parques nacionales.

Un año y medio después de haber lanzado el Programa está a punto de aparecer. Esto se muestra por medio de la conservación de plantas amenazadas especificas y ecosistemas en el campo, y también por un efecto catalítico en otras organizaciones implicadas con la conservación de recursos naturales.

Introduction

The World Wide Fund For Nature (WWF) is a private, international conservation foundation with 23 Affiliate Organisations throughout the world. Its scope is the conservation of nature and the natural environment in all its forms: fauna, flora, landscape, soils, water, air and other natural resources.

WWF mobilises on a world-wide basis the strongest possible moral and financial support for safeguarding the living world and seeks to convert such support into action based on scientific priorities. WWF ensures that its programme has a sound scientific basis by close collaboration with IUCN.

Since its founding in 1961, WWF has channelled over U.S.$95 million into more than 4,200 projects in some 130 countries. These projects have saved many animals and plants from extinction and helped to conserve natural areas all over the world. WWF has served as a catalyst for conservation action, brought its influence to bear on critical conservation situations, and provided a link between conservation needs, the scientific resources necessary to meet them, and the governments and other authorities concerned.

The International Union for Conservation of Nature and Natural Resources (IUCN), founded in 1948, is the leading independent international organisation concerned with conservation. It is a network of governments, non-governmental organisations, scientists and other specialists dedicated to the conservation and sustainable use of living resources.

The unique rôle of IUCN is based on its 537 member organisations in 116 countries. The membership includes 58 states, 123 government agencies and virtually all the major national and international non-governmental conservation organisations. Over 1,500 experts support the work of IUCN's 6 commissions — on ecology; environmental education; environmental planning; environmental policy, law and administration; national parks and protected areas; and the survival of species.

The IUCN Secretariat conducts or facilitates IUCN's major functions, which are: monitoring biological data and the status of resources requiring conservation, developing plans (such as the World Conservation Strategy) for dealing with conservation problems, supporting action arising from these plans by governments or other appropriate organisations, and finding ways and means to implement them. The IUCN Secretariat provides the major input into the programming of WWF's conservation activities around the world. IUCN also serves as the Secretariat for the Ramsar Convention (Convention on Wetlands of International Importance, Especially as Waterfowl Habitat).

The Plants Conservation Programme

The World Conservation Strategy (WCS) defines conservation as "the management of human use of the biosphere so that it may yield the greatest sustainable benefit to present generations while maintaining its potential to meet the needs and aspirations of future generations" (IUCN/UNEP/WWF, 1980). The major objectives of conservation reflected in the World Conservation Strategy are:

(1) to maintain essential ecological processes and life-support systems,
(2) to preserve genetic diversity, and
(3) to ensure that any utilisation of species and ecosystems is sustainable.

The gravest challenge facing mankind today is the irreversible reduction of the planet's capacity to support people. Some major causes for this, of concern to conservation and development alike, are based on the excessive and increasing demands placed on the world's natural resources. Millions of people in developing countries are compelled through short-term needs to destroy those resources necessary to free them from poverty in the long term. Huge areas of natural habitats are being lost, especially in tropical countries and on islands, resulting in loss of genetic diversity, of species and of ecosystems, and in a breakdown of ecological processes. In addition, deforestation and poor land management cause the loss of millions of tons of soil every year. This continuing depletion of natural resources will have adverse consequences reaching far beyond those countries where the resources are located.

Because of the accelerating world-wide deterioration of natural habitats, entire ecosystems may disappear and as many as 60,000 plant species may become extinct before the middle of the next century, according to the IUCN/WWF Plant Advisory Group. The loss of biological diversity and genetic resources, specifically plant genetic resources, is a threat to our own life-support system. Should tens of thousands of plants disappear forever, it would amount to a fundamental and permanent change in the character of life on earth. IUCN recognises the problems involving the utilisation of natural resources by man, but asserts that such usage can be managed in a manner compatible with the conservation of these resources. Accordingly, IUCN aims at achieving the practical implementation of the objectives of the World Conservation Strategy.

To meet the threats to plant life, IUCN and WWF launched the Plants Conservation Programme in March 1984 to promote plant conservation more effectively around the world (IUCN and WWF 1984). This major programme is designed to address plant conservation from several directions, with the emphasis on achieving conservation of plant diversity, of plant genetic resources and of plants of actual or potential economic value (particularly in the tropics). The programme offers a unique opportunity to bridge the present gap between scientists, protected area managers and conservationists on the one side and foresters and agronomists on the other, in the common cause of saving plant life.

Objectives of the Plants Conservation Programme

The overall objective is to promote, support and formulate efforts to conserve plant genetic resources and plants of actual or potential economic value, particularly in developing countries in the tropics, thereby contributing to the preservation of a sustainable environment for socioeconomic development and progress. To ensure that the harvest of living natural resources can be sustained, the three general objectives of the World Conservation Strategy (mentioned above) have to be met. On a more specific level, the programme consists of a set of projects and activities to meet the following operational objectives:

(1) *Spreading the message*
To build public awareness to help stimulate action by pressure on governments and industries, by citizen action, and by encouraging donations.

(2) *Building the capacity to conserve*
To enable IUCN and WWF to promote plant conservation more effectively among their members, the world's conservation organisations, and governments.

(3) *Conserving plant genetic resources*
To conserve sufficient diversity within species to ensure that all their potential is available for use in the future.

(4) *Conserving wild plants of economic value*
To emphasise the conservation of wild plants of importance, particularly for rural people in developing countries.

(5) *Botanic gardens — a vital link*
To strengthen the rôle of botanic gardens in conservation.

(6) *Promoting plant conservation in selected countries*
To identify, develop and implement on-the-ground model projects in priority countries.

Spreading the message

An international awareness campaign on plant conservation was launched in Spring 1984 at the Royal Botanic Gardens, Kew, in the U.K., by HRH Prince Philip. For this campaign, a range of information and educational materials was produced. These included information leaflets; a booklet describing the Conservation Programme (IUCN and WFF 1984); press kits with fact sheets and feature articles; Plant Pacs, comprising selected articles on crops and relatives, medicinal plants, threatened plants and botanic gardens; a popular book on plant conservation *The Green Inheritance* by Anthony Huxley; an audio-visual presentation *Saving the Plants that Save Us*; and an advertisement, developed by Ogilvy & Mather, which was given free space in newspapers and magazines such as *Time, Newsweek, International Herald Tribune* and *South*.

A number of the 23 national WWF organisations launched their own Plants Awareness Campaigns (and Programmes), developing their own materials or adapting those of IUCN/WWF-International to their local needs. For example, WWF-Spain adapted the WWF/IUCN audiovisual presentation "Saving The Plants That Save Us" for the Spanish public, including reference to particular Spanish endangered plants. The main focus of WWF-Spain's activities on plant conservation are on education and awareness and on specific field projects. In South Africa, a very extensive plants conservation programme is being implemented, starting with a superb exhibition in Cape Town, *Flora 83*, that displayed nearly 5,000 different species from throughout the country, and was followed by numerous activities. Especially noteworthy is the creation of four new nature reserves to conserve some of the unique plant life of South Africa, such as the flora of Cape Fynbos and the Highweld. In Sweden, a very comprehensive Plants Conservation Campaign was initiated. The core is the famous Project Linnaeus, the aim of which is to ensure the conservation of as much as possible of the Swedish flora; around this core, a range of awareness activities and practical plant conservation projects is being carried out, aimed at both national and international plant conservation.

In addition some IUCN/WWF projects on awareness and education are being implemented in selected countries. In India, for example, a major project is underway to achieve a nationwide spread of plant conservation awareness among selected youth audiences, such as educational institutes of engineering, technology, agriculture, civil service, etc. and all the major secondary educational institutes including public and central government schools. As an integral part of the activities, two field projects are being developed to demonstrate socioeconomic and developmental benefits of reforestation and of conservation of existing climax forest. This project will be used to develop further links with government institutions.

These awareness activities outline the basic concepts of plant conservation. The arguments emphasise the value of plants to people, in particular the value of medicinal plants, wild relatives of crops, plants for energy and industry, and the value of vegetation, e.g. for water catchment and ecosystem stability. Thus, the rationale for plant conservation is presented in numerous ways.

Building the capacity to conserve

A main tactic of the programme is to persuade other organisations to conserve plants, and to help provide them with the resources to do the job, rather than for IUCN and WWF to attempt everything themselves. Those in a position to conserve plants include governments, international organisations, conservation bodies, industry, citizen groups as well as individual activists. All may have great influence and ability, but many lack the experience and information needed. They would benefit from a more coordinated approach which would identify the top priority tasks.

In order to provide the broad creative and strategic thinking needed, IUCN and WWF have established a Plant Advisory Group to help provide new ideas on plant conservation, to give guidance on long-term issues, and to set the information and knowledge of the professional botanists into the framework of conservation concern. The Plant Advisory Group is a small committee of leading botanists from around the world. It is chaired by Professor Peter Raven, Director of the Missouri Botanical Garden in St Louis, U.S.A. The group met for the first time in December 1984; during 2 days they provided very substantial guidance to IUCN and WWF on the science behind the Plants Conservation Programme and suggested a number of activities which will be of far-reaching significance. For example, the group initiated the process leading to this Conference on

Botanic Gardens and the World Conservation Strategy, and the members were instrumental in designing the conference programme. The draft *Botanic Gardens Conservation Strategy* has been prepared by Professor Vernon H. Heywood, with the assistance of Dr Peter S. Ashton, both members of the Plant Advisory Group.

The group is also involved in the preparation and publication of a state-of-the-art book on plant conservation: "Plant Conservation: Principles and Practice". The author will be Dr David Given of New Zealand, and the Plant Advisory Group will serve as the editorial committee.

Conserving plant genetic resources

In order to feed, clothe, house and provide fuel for the human population of the next century, a considerable increase beyond present farm and forest productivity will be required. This can only be achieved by increasing present yields or by bringing new land, mostly marginal land, under cultivation. In both cases, many of the adaptive qualities that will be required will need to be drawn from the genes of wild species. Many industrial and medical products are of natural origin or have been synthesised on the basis of natural models. As our knowledge of wild species increases, so will the number of marketable products derived from them. Furthermore, advances in biological sciences depend heavily on the preservation of wild species as the basic material from which new discoveries of biological processes and substances will be made. The genes of wild species will be the essential raw material or blueprints for future applications through genetic engineering.

Genetic resources conservation is a relatively new topic for IUCN, and the first activity has been to establish collaborative relationships with those organisations heavily involved in this area. In particular IUCN is now working closely with FAO on *in situ* conservation of plant genetic resources. FAO contracted (and funded) IUCN to prepare part of the background documentation for the First Session of the FAO Commission on Plant Genetic Resources, Rome, 11–15 March 1985. An IUCN delegation attended this intergovernmental meeting and presented the IUCN position on *in situ* conservation, on monitoring and on international data bank systems. With endorsement from the FAO Commission, the IUCN activities are followed up through the Ecosystem Conservation Group, comprising FAO, UNEP, UNESCO and IUCN. A special working group, including also IBPGR, has been established to work towards an Action Plan for *in situ* conservation of plant genetic resources.

As new initiatives following the Ecosystem Conservation Group meeting,

IUCN is preparing, with IBPGR, the publication of an information leaflet on *in situ* conservation of crop relatives, and with all the members of the Ecosystem Conservation Group, plus IBPGR, under the leadership of FAO, a booklet on *in situ* conservation of plant genetic resources.

IUCN has also established closer links with IBPGR on *in situ* conservation of wild relatives of important tropical crops. A joint project aiming at conserving wild species of Mango (*Mangifera*), an economically and socially important plant in tropical Asia, is being carried out on the island of Borneo, with funding from IBPGR and WWF. An extension to peninsular Malaysia is being prepared with the WWF-Malaysia. The first step in conserving wild *Mangifera* is to determine the extent to which the genetic diversity of the genus *Mangifera* is adequately conserved in the existing protected areas of Borneo (and peninsular Malaysia). The ultimate aim is to enable the needs for *in situ* wild crop relatives conservation to be integrated into the design and management of the protected area system. The first phase of field work has been concluded in East Kalimantan (the Indonesian part of Borneo). Where previously 10 species of *Mangifera* had been recorded, the field survey showed about 20 species to exist in East Kalimantan, two of which may be new to science. The preliminary results indicate that the genus *Mangifera* is not adequately conserved in existing protected areas, but further field surveys are needed, also in Sabah and Sarawak (the Malaysian part of Borneo), to get an overview of the situation.

Conserving wild plants of economic value

Plants do not just provide food, they provide material for clothing, shelter, medicines and a host of other products of value to people. IBPGR is active in conserving the world's major crops, but conservationists have given little emphasis to the many thousands of other species which are used by people. Plant conservation, with its basis in plant systematics and taxonomy, has tended to give equal weight to each and every threatened plant rather than emphasise the useful ones.

A change in emphasis is needed, with far more effort on wild plants used by people, whether or not these plants enter the monetary economy. Because the subject is so broad, in fact covering all of economic botany, IUCN and WWF have to be selective, but with the aim of doing strategic and model projects that can be used and copied elsewhere.

A number of activities and projects focus on the conservation of medicinal plants, a group selected for its tremendous importance to daily health in the developing countries and of great public concern. For example, collaboration has been established with WHO's Programme on Traditional Medicine. A joint workshop on conservation and sustainable use of medicinal plants is planned for late 1987, in the east Asian region. The objectives will include formulation of joint priorities and identification of select countries for implementation of a model project.

In Sri Lanka, a project has been initiated for *ex situ* and *in situ* conservation of indigenous medicinal plants. Nearly 20% of the about 3,000 indigenous angiosperms of Sri Lanka are used in the Ayurvedic medicinal system, which has been practised in this country for over 2,000 years. These plants are now threatened by accelerated land clearing and by over-exploitation. To protect and conserve these species, which are of direct importance to the quality of life of the people, *in situ* reserves and *ex situ* herbal gardens are to be established. The project is carried out in collaboration with the Sri Lankan Ministry of Indigenous Medicine.

Botanic gardens — a vital link

There are nearly 1,400 botanic gardens around the world and most are keen to do more for plant conservation. Indeed, the movement to conserve plants really began in botanic gardens, where staff were concerned at how little attention conservation bodies gave to plants. Even now, conservation bodies and botanic gardens rarely work closely together. Few conservationists have understood what botanic gardens have to offer, and gardeners have, perhaps, been slow to appreciate the urgency of the crisis facing the plants they know and love.

Yet in recent years botanic gardens and museums have undergone great changes and are keen to change their Victorian image. More and more, they see plant conservation as a principal reason for their existence. In developing countries, conserving genetic resources and finding ways to make better use of wild plants makes the modern botanic garden a vital part of national development.

Gardens accept that the top priority is to protect plant habitats rather than grow the plants in cultivation. In a garden, plants need constant tending and are vulnerable to mechanical breakdown or human error. More seriously, it is rarely possible to maintain more than a very small proportion of a plant's genetic diversity in cultivation, and plants propagated by seed tend to rapidly adapt for conditions in the garden rather than those in the wild. Nevertheless garden collections play a valuable part in conservation, both as a back-up when conservation in the habitat fails, and also to provide material for research, for horticulture and for education.

In the Plants Conservation Programme, IUCN and WWF defined 3 broad aims related to botanic gardens: to draw botanic gardens closer into the conservation network; to coordinate them as the *ex situ* network for threatened plants; and to promote their use to tell the public about plants and the need for conservation. Having discussed these broad aims at its meeting in December 1984, the IUCN/WWF Plant Advisory Group advised IUCN and WWF to focus on 3 specific activities: to call an international conference on how botanic gardens can contribute to implementing the World Conservation Strategy; to prepare *The Botanic Gardens Conservation Strategy* building on the concepts of the World Conservation Strategy; and to develop further the monitoring programme carried out by the Threatened Plants Unit of the IUCN Conservation Monitoring Centre. The present international conference, the strategy document and the outline for developing the monitoring programme, are consequently forming a most important part of the Plants Conservation Programme.

Promoting plant conservation in selected countries

To balance the conceptual and strategic projects, the Plants Conservation Programme comprises a set of on-the-ground field projects.

Since IUCN and WWF cannot realistically develop and implement projects all over the world, 24 countries were selected on criteria of biological importance and of operational, political and socioeconomic considerations. The countries selected are mostly tropical, reflecting the far greater biological diversity of tropical areas compared to temperate ones and the far greater threats facing the natural vegetation of the tropics.

Many of these field projects consist in supporting the establishment of large protected areas and in securing the future of those areas protected on paper but not yet on the ground. As examples, IUCN/WWF are assisting in the establishment of the 130,700 ha Sapo National Park, Liberia's first park located in an untouched rain forest; supporting the creation of three massive parks in lowland Cameroon; and backing the development of park networks in Kalimantan and Irian Jaya in Indonesia. To secure existing parks, IUCN/WWF continue to support the Manu National Park in Peru, a rain forest bigger than Northern Island with more birds, and perhaps more plants, than any other park on earth. In Sri Lanka, the IUCN/WWF involvement has served to bring together the national organisations and agencies responsible for conserving the Sinharaja Forest, the last remnant of tropical lowland forest in the country. The Sinharaja Forest contains about 50% of the endemics of Sri Lanka and is of outstanding scientific interest. At a recent workshop in Colombo, where the foundations for a new management plan were laid, the Sri Lankan Government pledged its full support to conserving this rich forest. The Sri Lankan Minister of Lands and Mahaweli Development stated "we may even decide to import the country's timber requirements if it is deemed necessary to protect the nation's forest cover".

There are also projects to conserve endemic and threatened species, for example, in Mauritius, Nepal, the Canaries, Tanzania, Madagascar and parts of Brazil. In some places, both *in situ* and *ex situ* measures are necessary. In Mauritius, for example, IUCN and WWF are working to save the seriously threatened remnants of indigenous flora and vegetation. To do this, intensively managed pilot reserves are being established in key areas on the island and its dependency, Rodrigues, in close cooperation with the Mauritian Forestry Service. And to save some of the threatened endemics, which have been reduced to a few surviving individual plants, nursery facilities are being established.

Following the recommendations of the Plant Advisory Group and of the working group on *in situ* conservation of plant genetic resources under the Ecosystems Conservation Group, a number of projects deal with floristic and botanical surveys. Typically, such projects are intended to provide the basic knowledge on which to formulate more comprehensive conservation plans for certain areas or plant groups. For example, there are surveys of multipurpose palm species in Latin America and south-east Asia, a survey of the plant genetic resources of Socotra Island (Yemen), a survey of medicinal plants of Siberut Island (Indonesia), and an inventory and assessment of the Campo Rupestre vegetation in Brazil.

Conclusion

Conservation organisations have tended to neglect plants. Yet, as we all know, plants are essential to human and animal life, as well as providing the fabric of vegetation and lanscape. Entire plant communities are in danger of extinction, and the losses of biological diversity are a very real threat to humanity.

This is the background on which the IUCN/WWF Plants Conservation Programme was designed. Plants are not only neglected and in need of better conservation, but represent also a significant opportunity to develop new partnerships for conservation. And the tremendous popular interest in plants and gardens presents a unique opportunity to interest people in wider conservation issues as well.

From 1984 to the end of 1986, during the first 3 years of the programme, WWF and other donors have committed 3 million Swiss francs to 49 activities and projects specially initiated as part of the Plants Programme.

At the same time, WWF has committed some additional 5·6 million Swiss francs to 83 projects on tropical forest conservation, such as the national park projects mentioned earlier on, thereby helping to conserve plant-rich sites. Altogether, this is a substantial investment to help plant conservation gain its proper place in natural resources conservation. However, compared to what needs to be done, the investment in plant conservation is relatively small; so, what can a modest programme like the IUCN/WWF Plants Conservation Programme hope to achieve, and how do we know whether our efforts are actually successful or not? To answer these questions, it is important to reiterate that the main tactic of the programme is to be a catalyst to persuade other organisations with larger resources to work for the same goals and to provide them with the tools to do the job.

And indeed the catalytic impact is starting to show. One prime example is the present conference, generously supported by the Canarian Government, the Island Council of Gran Canaria, WWF and by the UN organisations of Unesco, UNEP and IBPGR. Without such national and international support, it would not have been possible to convene such a full representation from the botanic gardens of the world. Another example is the positive interest displayed by UN organisations for doing further work on *in situ* conservation of plant genetic resources — an interest backed by the allocation of funds for joint activities in this field.

Concerning field projects, some activities can easily be seen to be effective: for example, practical management of a national park can ensure the survival of plant-rich forests, which serve as important water-catchment areas, while neighbouring forests are deteriorating and thereby losing their value both biologically and endangering the livelihood of the inhabitants. Or combined *in situ* and *ex situ* conservation measures actually save plants threatened by extinction, as is the case on Mauritius.

The Plants Programme is having an impact, but in order to be really successful, much more needs to be done, and on a larger scale. IUCN and WWF invite you to join in the efforts to ensure that the essential diversity of the plant kingdom survives into the next century and beyond. As individuals, as professionals and as institutions, you can all help to make the difference that counts.

References

IUCN/UNEP/WWF (1980). "World Conservation Strategy". IUCN, Gland, Switzerland.

IUCN and WWF (1984). "The IUCN/WWF Plants Conservation Programme 1984–85". IUCN, Gland, Switzerland.

2

Botanic Gardens and the Community

A botanic garden for our city and university and a conservatory: the example of Nancy

F. MANGENOT and P. VALCK

Conservatory and Botanical Gardens of Nancy

Summary

This paper describes the experiences which made it possible to create a new botanic garden with an international rôle. The objectives were to create a substantial garden for the public that could also open new doors in science, education and conservation of the plant kingdom.

Botanic gardens are old institutions in Nancy, the Jardin Botanique Sainte Catherine being founded in 1758. The present management established in 1966 the Jardin d'Altitude du Haut Chitelet, in the heart of the Vosges Mountains. It is now the most plant-rich alpine garden in France, with about 2,800 plant species from different mountains. This was followed in 1970 by a new botanic garden, the "Jardin Botanique du Montet".

To achieve this, we first had to find a site and set up a syndicate of all the interested partners. This structure was set up by the town of Nancy and the University of Nancy, with the financial assistance of the Départment of Meurthe and Moselle and the Région Lorraine. For its part, the Ministry of Environment became involved and in 1978 designated the garden as a "Conservatoire national de Botanique".

Work began in 1975. Today, the "Jardin Botanique du Montet" displays about 3,000 plant species in the open air, spread over 25 ha. There are a further 5,600 species in automatic greenhouses, including some of the many threatened tropical plants of French overseas territories. There is still much to be done, notably to develop the collection Horticulteurs Lorrains (Lemoine and Crousse), and the

Botanic Gardens and the World Conservation Strategy

ISBN: 0-12-125462-3

collections of useful plants. Another objective of the Nancy garden is to collaborate with the administration of Réunion in building a Conservatoire Botanique on the island, where the most interesting plants are rapidly disappearing. The project is being undertaken in association with IUCN.

Sommaire

Un jardin botanique pour la ville et l'Université: un conservatoire, l'exemple de Nancy

On décrit les expériences qui nous ont amenées à concevoir puis à créer un jardin botanique de niveau international. Dans un premier temps nous définirons les objectifs que nous nous sommes fixés: réaliser un grand jardin pour tout public, avec des buts scientifiques, pédagogiques et de conservation du monde végétal.

Les jardins botaniques de Nancy sont une vieille institution puisque le Jardin Botanique Sainte Catherine a été crée en 1758. L'équipe de direction, déjà en place en 1966, créa cette année-là le Jardin d'Altitude du Haut Chitelet au centre des montagnes vosgiennes. C'est maintenant le plus riche Jardin Alpin de France: il posséde environ 2.800 espèces de plantes de montagnes différentes. Puis en 1970 germa l'idée de créer un nouveau jardin botanique: le Jardin Botanique du Montet.

Pour cette création, il fallut des terrains et surtout mettre en place une unité de gestion regroupant les partenaires intéressés par cette nouvelle réalisation. Cette unité fut créée par la ville de Nancy et l'Université de Nancy, avec l'aide financiére du Départment de Meurthe et Moselle et de la Région Lorraine. A son tour, le Ministère de l'Environnement s'y est associé et a nommé, en 1978, les Jardins Botaniques de Nancy, Conservatoire national de Botanique.

Les premiers travaux débutèrent en 1975. Actuellement le Jardin Botanique du Montet présente, sur 25 ha, environ 3.000 espèces de plantes de pleine terre et 2.000 m^2 de serres automatisées avec 5.600 espèces de plantes tropicales dont de nombreuses plantes menacées des Départements d'outre mer. Beaucoup de travaux restent encore à réaliser, notamment la collection des horticulteurs Lorrains (Lemoine et Crousse), la collection de plantes utilitaires, etc. Un autre objectif des Conservatoire et Jardins Botaniques de Nancy est de réaliser en collaboration avec les administrations réunionnaises, un grand Conservatoire de Botanique à l'Ile de la Réunion ou les plantes les plus intéressantes disparaissent rapidement. Cet objectif est en cours de réalisation, grâce au concours de l'IUCN.

Resumen

Un jardín botánico para la ciudad y la universidad: una reserva (conservatorio), el ejemplo de Nancy

Se describe las experiencias que nos han llevado a concebir y después crear un jardín botánico a nivel internacional. En una primera etapa definiremos los objectivos fijados: hacer un gran jardín para el público con fines científicos, pedagógicos y de conservación dirigidos al mundo vegetal.

Los jardines botánicos de Nancy son una vieja institución puesto que el de Sainte Catherine fue creado desde 1758. El equipo directivo de 1966 creó allí el Jardín de Altitud del Alto Chitelet en el centro de las montañas vosquienses. Es ahora el jardín alpino más rico de Francia, posee más de 3.000 especies de montaña. Después, en 1970, se originó la idea de crear un nuevo jardin botánico: el Jardín Botánico de Montet.

Para esta creación, fueron necesarios terrenos y sobre todo poner en funcionamiento una unidad de gestión que agrupara a los interesados en esta nueva realización, esta unidad fue creada por la ciudad y universidad de Nancy a las cuales se unieron las provincias de Meurthe y Moselle, de la región de Lorraine. A su vez el Ministerio de Medio Ambiente asociado también, ha declarado, desde 1978, a los jardines botánicos de Nancy, Reserva Nacional de Botánica.

Los primeros trabajos empezaron en 1975. Actualmente el Jardín Botánico de Montet tiene unas 25 Has., alrededor de unas 3.000 especies de tierras altas y 2.000 m^2 de invernaderos automatizados con 5.600 especies de plantas tropicales muchas de ellas amenazadas en las Provincias de Ultramar. Muchos trabajos quedan aún por realizar en particular la colección de los horticultores lorrenos (Lemoine y Crousse), la colección de plantas útiles, etc. Otro objetivo de la Reserva y Jardines Botánicos de Nancy es realizar en colaboración con las Administraciones de la isla de Reunión una gran Reserva Botánica en dicha isla en la que las especies más interesantes desaparecen rápidamente. Este objetivo está en marcha gracias a la cooperación de la IUCN.

Historical Survey

The first garden in Nancy, the "Jardin Botanique Sainte Catherine", was founded in 1758 by Stanislas, former King of Poland and Duke of Lorraine. This garden is preserved in the town centre near the famous Place Stanislas and is an example of the School of Botany set up by Dr Gordon in the middle of the 19th century. Recently, however, a greater emphasis has been given to ornamental plants.

A second garden was founded in 1908 by Pr. Brunotte in the Vosges Mountains, as an experimental garden for the acclimatisation of ornamental and pasture plants. Visitors were admitted for the first time on

30 July 1914. Some weeks later the garden was destroyed by soldiers and their mules during the First World War. Symbolically, in the 1960s, young French and German gardeners worked together to build a new garden, at present known as the "Jardin d'Altitude du Haut Chitelet".

This comprises 10 ha of preserved moorland and 1 ha of rock garden at an elevation of 1228 m. Examples of the natural flora may be found together with collections of interesting or endangered plants, principally from the Vosges but also from other mountain ranges in France (Jura, Alps, Pyrenees, Massif Central) and abroad (North America, China, Himalaya, Japan, Caucasus, etc.). There are about 2,800 species in this garden, with particular emphasis on endangered plants listed by IUCN. About 50,000 people visit the garden during the summer.

In 1970, a new and larger botanic garden was planned for Nancy according to our conception of the goals of a modern botanic garden. Some years later, 15 ha from a former experimental farm were put at our disposal by the University of Nancy, in a scenic place with a beautiful view of the whole town. At present, with a total area of 25 ha, the "Jardin Botanique du Montet" is the largest botanic garden in France.

The article which follows mainly covers the "Jardin du Montet" which, from the beginning, was intended as a garden for a town and a university. As will be shown later, it is also one of the three conservatories of the French flora recognised by the French Ministry of Environment.

The garden's status

From an historical viewpoint, France was the second country in Europe to create botanic gardens. Most of them still exist but are starved of funds and resources. The most famous belong to the universities; they do not have much money and even fewer staff. Others belong to towns; they have sufficient funds and technical staff, but are mainly leisure parks more interested in the picturesque than in scientific goals. In many foreign countries, botanic gardens are engaged in educational programmes, but this is very rare in France.

Thus it was necessary for the new garden to belong both to the town and the university, and what is called under French regulation a "syndicat mixte" was created. A "syndicat mixte" is the result of an association between 2 or more public agencies, in this case the University of Nancy I and the Council of Greater Nancy (District de l'Agglomération Nancéienne). According to French regulations a "syndicat mixte" is almost eternal, in that only a decision of the highest chamber, the "Conseil d'Etat'', can dissolve it. Moreover, the partners are legally required to furnish the necessary funds to the ''syndicat mixte'' for its activities.

Therefore, it is the ''syndicat mixte'' of the Botanical Gardens which administers the 39 ha of the three gardens. The personnel belong to the University and to the town and are permanently assigned to the Botanical Gardens.

Why create a botanic garden on the eve of the 21st century?

It is our belief that a well-conceived botanic garden is even more important in the era of technology than at the time when society was largely agricultural and closely linked to nature. In the past, botanic gardens had principally a scientific objective. Today, in addition, they must offer an outline of the plant kingdom, its laws and its uses, to the widest public possible. The botanic garden has to be a museum where visitors, no matter what their age or the level of their knowledge, can admire and learn to respect nature.

The scientific objective

This requires specific tools: a herbarium, a library, a seed collection, etc., and of course staff. The Botanical Gardens of Nancy do not have their own scientific staff apart from the Director who is required to be a professor from the University. But the gardens are at the disposition of neighbouring researchers: taxonomists, cytologists and physiologists, for example. Some collaboration has been established with more distant laboratories such as the Phanerogamic Laboratory of the National Museum of Natural History in Paris (on the flora of the Amazon region in Peru and the flora of the Seychelles); with the Laboratory of Tropical Botany of the University of Paris VI (on the flora from the understorey of the forests of Malaysia and Thailand); and with the European Institute of Ecology at Metz (on the medicinal plants from North Yemen and certain succulent plants). Personally, we are most especially interested in epiphytic ferns and succulent Asclepiadaceae.

The second goal: education and training

Of course, we supply materials to universities and secondary schools, but it is also necessary for students to benefit from the garden itself. This has been made possible thanks to a team of instructors belonging to the gardens and the University, as well as some volunteer specialists. Thus, in 1985, 186 classes comprising 4,650 students benefited from the "Jardin du Montet" either for general visits or for studying themes as varied as species, medicinal plants, or for the horticultural profession.

Those who are no longer of school age have not been forgotten; guided visits have been organised for groups. For individual visitors there are explanatory labels and documentary brochures, for example on local trees, medicinal plants or cultivated plants of the tropics.

The gardens are participating in the life of the town by taking the name of Nancy to floral exhibitions, by producing radio and television programmes on regional facilities, and even by taking part in the production of a series called "Plant Adventure" to be shown on several television channels throughout the world. Another example is the large international orchid exhibition which will be organised by the garden and the Orchid Society of France, and will be housed in Nancy's buildings in 1987.

What can you see in the "Jardin du Montet"?

Before making the plans for the new garden, we studied the recently created European botanical gardens (Tübingen, Marburg, Hamburg, Haren, Utrecht, Amsterdam) and then we made our own plans.

The garden has the features of a public park with vast lawns surrounding thematic collections. It displays about 3,000 plant species in the open air, and there are also 2,000 m^2 of precisely controlled glasshouses as well as flower beds and nurseries.

The principal collections are as follows.

The historical collection

This is dedicated to plants which have been grown by man since the Neolithic Era, the Bronze Age, the Iron Age, Roman times, etc. Wherever possible, the guide mentions the archeological data gathered in the region.

This collection includes a section of weeds, formerly or recently introduced plants from the list recommended by Charlemagne for cultivation in the cloisters and the towns of the Empire, and finally, ornamental plants presented chronologically according to their dates of introduction into the gardens.

The ornamental collection

This is largely dedicated to a local theme: the Lorraine Horticulturists. From the end of the 19th century up to the First World War, Nancy was an important centre of horticultural creativity.

One of the best known plant growers of Nancy at that time, F.F. Crousse, was celebrated for his Begonias and Pelargoniums. A street in Nancy was named "rue des bégonias" in his honour. Victor Lemoine is well known for his *Syringa, Fuchsia, Philadelphus* and *Deutzia*, and was the first foreigner to be awarded the Veitch Medal. He also received the George Robert White Medal of Honour and won gold medals at the World Fairs of Paris in 1878, of Chicago in 1893, of Amsterdam in 1895, etc.

After the disappearance of their horticultural establishments, horticulture fell into oblivion in Nancy. By creating this collection, we intended to revive this horticultural past, but we have met with some difficulty because many of the plants they grew are cultivated no longer. Therefore, we are taking this opportunity to ask all of you to help us if you have any of the Crousse and Lemoine varieties. We would be very grateful for any assistance.

The alpine collection

When it is finished, this collection will cover about 1 ha surrounding a little chapel. About half of this area has been built up with ponds, waterfalls and streams surrounding granite and gneiss rocks brought from the Vosges. The first plantings, covering 280 species, are finished and have been placed in geographical order. This area benefits from a special microclimate which, at an elevation of about 300 m, permits the growth of plants unable to endure the more severe climate of the "Jardin d'Altitude du Haut Chitelet".

The arboretum

In creating this arboretum-fructicetum, we voluntarily limited ourselves to small trees and shrubs in order to complement the Arboretum of the Forest Research Services of Champenoux (near Nancy) and not to compete with it.

The arboretum is presented to visitors in the form of a geographical map. For each continent, a few regions have been chosen in relation to their climates which approximate the Lorraine climate. Thus, for North America, we have chosen the Cascade Range in the west and the Laurentides in the east.

The systematic collection

This is being planned in the shape of a genealogical tree following the system of Cronquist. The sub-classes branch out from the top of the slope where the Magnoliideae form the starting point of the collection. Along the hypothetical evolutionary branches, the Orders diverge at different levels going from the most primitive to the most highly developed, the latter being at the bottom of the slope.

Other collections will be developed in the future: useful plants, plant morphology, hedges and, at last, a collection of Gallé plants, which are very close to our hearts in Nancy because of the link to our local history.

During the lifetimes of Crousse and Lemoine, Nancy was one of the major centres of modern art. One of the principal artists was Emile Gallé, whose vases, glass lamps and dishes are known all over the world. His decorative themes came principally from plants and he was a great plant connoisseur and president of the local horticultural society. After a long and painstaking investigation, we have established the list of plants he collected and a list of those he used in his art, and we intend to plant them in a faithful reproduction of his garden. This will be a page from the history of art in Lorraine.

The glasshouses

The present area (1,700 m^2) will soon be extended to 2,000 m^2 with the completion of the palmarium now under construction. A collection of 5,600 tropical species is being cultivated in 8 glasshouses. Four of these are open to the public and are dedicated to aquatic vegetation (*Victoria amazonica*, Syn. *V. regia*, mangroves, etc.), to useful plants, to tropical and to arid zone vegetation.

Four glasshouses are reserved for scientific collections and are only accessible to the public for guided visits or by special permission. Little by little, the traditional botanic garden flora have given way to plants collected from their natural habitats, as for example plants collected by Serge Barrier in the Amazonian forests or the most noteworthy plants of the Seychelles collected by Francis Friedmann, author of the "Flora of Seychelles", now being published. Also we have a very important collection of Araceae from tropical America and south-east Asia.

These collections include many plants still in the process of identification plus some new species and many endangered plants.

The conservation of endangered plants in the conservatory and botanical gardens of Nancy, as linked to the IUCN programmes

The Conservatory of Nancy participates in the conservation of traditional French cultivars, such as fruit trees formerly cultivated in Lorraine, and vegetables and cereals from the Hautes Alpes in connection with the Ethnobotanical Laboratory of the Museum of Paris.

In addition, we can report on an undertaking in close accord with IUCN policy for links between temperate and tropical botanic gardens.

The Botanical Conservatory of the Mascarenes

The Botanical Gardens of Nancy have chosen to promote the creation of a Botanical Conservatory of the Mascarenes on Réunion in the Indian Ocean and to help in its implementation.

An agreement between the ''syndicat mixte'' of the Botanical Gardens and the ''Conseil Général'' of Réunion has just been signed. There are 4 main points to this agreement:

Article 1

In order to safeguard the Mascarenes plants in danger of extinction, it has been decided to unite the efforts of the Conservatory and Botanical Gardens of Nancy and the Botanical Conservatory of the Mascarenes.

Article 2

The Conservatory and Botanical Gardens of Nancy will assure: the establishment of studies and plans; their follow-up and advice at time of implementation; the training of specialised personnel (the first staff member has already begun his training in Nancy); supply to the Mascarenes Conservatory of endangered plants grown in its glasshouses; and, in accord with the French University of the Indian Ocean, the elaboration and realisation of a program of *in vitro* cultivation with the help of the Laboratory of Woody Plants Biology of the University of Nancy I.

Article 3

The Botanical Conservatory of the Mascarenes will assure: the upkeep and maintenance of the collections in one or more sectors, and more satisfactory protection of the endangered plants, material of which will be sent to the Botanical Gardens of Nancy for cultivation.

Article 4

Constant liaison will be maintained between the two conservatories by means of meetings, reciprocal visits, field work, etc., and by compiling and editing booklets, leaflets, and other documents to make the public more aware of the problems of the endangered flora of the Mascarenes.

The conservatory and botanical gardens of Nancy and the National Programme of Protection of the Plant Genetic Resources of France

There has been a Bureau of Genetic Resources in France since 1983. It is in charge of promoting, coordinating and initiating work on genetic resources from the scientific standpoint. It also has the duty of advising the government at all levels and of representing it internationally. On the other hand, the Flora and Fauna Service collects data dealing with French wild flora.

There are the brains of conservation but they need arms. Since 1975, the Ministry of Environment has recognised three specialised "Conservatoires nationaux de Botanique".

(1) Nancy, in charge of the flora of Northeastern France, the high mountains (except for those of the Mediterranean area) and overseas France.
(2) Porquerolles, in charge of the Mediterranean area.
(3) Brest, in charge of the Atlantic area and oceanic islands.

These three conservatories have conducted extensive bibliographical studies on endangered plants, have surveyed natural localities and have carried out *ex situ* cultivation when necessary.

The Natural Parks have recently taken part in genetic conservation in France, often in liaison with these conservatories, but until now their interest has been directed mainly toward cultivated plants, especially fruit trees. Also in France, there are a large number of private organisations dedicated to the conservation of traditionally cultivated varieties but rarely to the protection of wild flora.

In 1984, an Association of French Conservatories of Plant Species was created, including the three national conservatories, the Museum of Paris, the Natural Park Federation, the Bureau of Genetic Resources and private associations. Its objective was to compile a list of existing conservatories and their plant holdings. This would enable them to favour the most interesting projects, to promote the creation of conservatories in the most neglected areas such as vegetables, forage plants, cereals, etc., and to organise symposia and seminars. The seat of the Association is located at the "Jardin du Montet" and for the time being its President is the Director of the Botanical Gardens of Nancy.

In conclusion, we wanted to create a garden for a town and a university. We consider that we have succeeded and that we are appreciated by the City and the University. However, our success has led us to go outside the local level to take on some national responsibilities and also to establish a programme in the Indian Ocean.

Appendix 1

Members of the Association of French "Conservatoires nationaux de Botanique"

Association des Conservatoires Français d'Espèces Vététales, 100, rue du Jardin Botanique, F — 54600 Villers-Lès-Nancy. Tél: 83.41.47.47

Bureau des Ressources Génétiques, Museum National d'Histoire Naturelle, 57, rue Cuvier, F — 75231 PARIS Cedex 05. Tél: 47.07.15.75

Conservatoire Botanique de BREST, 52 Allée du Bot, F — 29200 BREST. Tél: 98.02.63.14

Conservatoire Botanique des Mascareignes, Conseil Général, 4, rue de la Victoire, F — 97400 SAINT DENIS, Ile de la Réunion. Tél: 262.21.08.98

Conservatoire et Jardins Botaniques de Nancy, 100, rue du Jardin Botanique. F — 54600 Villers-Lès-Nancy. Tél: 83.41.47.47

Conservatoire Botanique de Porquerolles, Hameau Agricole, Ile de Porquerolles, F — 83400 HYERES. Tél: 94.58.30.80

Fédération des Parcs Naturels de France, 4, rue de Stockholm, F — 75008 PARIS. Tél: 42.94.90.95

Ministère de L'Environnement, 14, Boulevard du général Leclerc, F — 92524 Neuilly s/ Seine Cedex. Tél: 47.58.12.12

Museum National d'Histoire Naturelle, Laboratoire d'Ethnobotanique, 57, rue Cuvier, F — 75005 PARIS. Tél: 47.07.36.25

Museum National d'Histoire Naturelle, Service des Cultures, 43, rue de Buffon, F — 75005 PARIS. Tél: 43.36.12.33

The botanic garden as a vehicle for environmental education

BERNADO NAVARRO VALDIVIELSO

Jardín Botánico Canario "Viera y Clavijo", Las Palmas de Gran Canaria, Canary Islands

Summary

The Jardin Botánico "Viera y Clavijo" has as its principal objective the conservation of the Canarian flora. Conscious that to achieve this it is vital for the Canarian population to understand the importance of the vegetation and flora and the advantages that its conservation can bring for the development and future of the islands, the botanic garden has, over the last 10 years, developed a broad programme of environmental education, which has produced an important response on the island of Gran Canaria.

During the decade of the 1970s, and as an expression of the centralist concept of the Spanish education system, the subjects taught in schools were very general and homogeneous throughout the nation. They did not take into account the local or regional aspects of the environment and culture which were otherwise familiar to the children of the distinct regions. It is possible to state categorically that from an educational point of view and especially from that of natural sciences, the Canarian environment was totally unknown and ignored by both teachers and students alike.

This educational situation is actually undergoing a profound change with the new political development of Spain and the setting up of the Autonomous Regions. The Jardín Botánico "Viera y Clavijo" has, at a local island level, contributed to this change in the field of environmental education.

In this paper the various activities and programmes carried out and those planned for the future are described. These include courses for

Botanic Gardens and the World Conservation Strategy

ISBN: 0-12-125462-3

teachers on the Canarian environment, which have had over 2,000 participants so far; the "Flora y los Niños" project in which 10,000 Dragon Tree seedlings (*Dracaena draco*) were distributed to school children who then looked after them for a year at home before planting them in public areas; the creation of an educational resources centre, school laboratory and school nursery in the garden with the elaboration of audio-visual aids; incorporation of art and handicraft in environmental education, and the creation of the Sventenius Prize for environmental activities amongst school children.

Resumen

El jardín botánico como vehiculo para la educación ambiental.

El Jardín Botánico Canario "Viera y Clavijo", tiene como objetivo primordial la conservación de la Flora Canaria. Conscientes de que para posible es imprescindible el que la población canaria conozca su vegetación y flora y las ventajas que éstas aportan a nuestro desarrollo y futuro es por lo que en los últimos diez años el jardin botánico ha desarrollado un extenso programa de educación ambiental, que ha tenido una amplia acogida en la isla de Gran Canaria donde se encuentra situado.

A mediados de la década de los 70, como expresión de un concepto centralizado del sistema educativo español, las enseñanzas que se impartían eran de carácter muy homogéneo y general, sin incidir en absoluto en aquellos aspectos regionales o locales cuya realidad vivían más de cerca los niños y los jóvenes de las distintas regiones. Podemos afirmar categóricamente que por entonces, el conocimiento de la realidad canaria y en especial el relacionado con las ciencias naturales era con carácter general totalmente ignorado por profesores y alumnos. Esta situación educativa está experimentando actualmente un profundo cambio con el desarrollo politico de España y la configuración del Estado de las Autonomías. El Jardin Botánico Canario a nivel insular, contribuye a ello.

En la ponencia se describen las distintas actividades programadas y llevadas a cabo por el jardín botánico tales como: cursos para docentes sobre Naturaleza Canaria que han recibido más de 2.000 maestros, Proyecto "La Flora Canaria y los niños", que supuso la entrega de 10.000 dragos a niños que los tuvieron durante un año en su casa para luego plantarlos en espacios públicos; creación de un centro de recursos didácticos y laboratorio escolar; vivero escolar, elaboración de audiovisuales, el arte y la artesanía en la educación ambiental, Premio Sventenius que recompensa los trabajos sobro naturaleza canaria y otras actividades.

The World Conservation Strategy, prepared by the International Union for the Conservation of Nature and Natural Resources in 1980, includes a recommendation that botanic gardens, as institutions that bring people and plants together, be used to explain what are the objectives of conservation and how it can contribute to the survival and well-being of the human race. The Jardín Botánico "Viera y Clavijo", which is run by the Cabildo Insular de Gran Canaria (the Island Government), has as its primary objective the conservation of the Canarian flora. We realise that to achieve conservation it is absolutely vital that the Canarian population are aware of the local vegetation and flora, and perceive the advantages that native plants can bring for future development. For the past 10 years and under the slogan "To know is to conserve", the Canary Garden has developed a programme of environmental education which has been widely accepted and praised on the island of Gran Canaria.

In 1766, over 200 years ago, Don José de Viera y Clavijo, the first native Canarian naturalist and to whom our botanic garden is dedicated, wrote in his *Diccionario de Historia Natural de Las Islas Canarias*, "Nobody can ignore the fact that the density of our forests is one of the factors which most attracts the beneficial rains and which as a result contributes to the springs of life-giving water. Therefore, never cut down a tree without previously having planted ten in its place". If these words had been heeded, the present situation of the natural environment of the islands, both in the availability of water and in the conservation of the landscape, would be considerably different. His advice was not followed probably because the population, then uneducated and ignorant, did not receive the message and continued to consider the forests only as a source of firewood and charcoal for their kitchens and as timber for construction.

Scientific advances made during the last 100 years have shown, without a doubt, that an adequate vegetation cover is indispensible for the future of humanity. Scientists can today, with irrefutable evidence, repeat Viera y Clavijo's message, but now we have to give the conservation movement the means and support to reach out and pass the message to the whole population. The support needed is, as indicated in the World Conservation Strategy, in the areas of public participation and environmental education campaigns and programmes.

In the mid 1970s, as an expression of the centralist concept of the Spanish education system, subject curricula were very general and homogeneous throughout the nation. They did not take into account the regional aspects of life and the peculiarities which affect the day-to-day lives of children living in different parts of the country.

This centralisation was very dominant. Indeed it can categorically be said that knowledge and experience of the Canary Islands, especially in the natural sciences, were totally unknown to both teachers and pupils in the islands. However, about this time, people became increasingly interested in the local environment, especially in its history, archaeology, art, literature and, of course, natural science. The fruit of this interest is shown by the appearance of a new generation of young scientists specialising in local island themes. Through them, serious programmes have begun which can transmit to the children and youth the knowledge necessary for the islander to develop in harmony with local history and the environment. At the same time, the political development of Spain and the configuration of the state into autonomous regions has permitted great changes in the education system. The responsibility for education is now directly in the hands of the autonomous Canarian Government, which has officially permitted the development of studies on the problems and realities of life of the Canarian region. In this context the education programme of the Jardín Botánico "Viera y Clavijo" has been developed. The Ministry of Education of the Canarian Government recognises the value of the teaching resources of the garden and has designated 2 full-time teachers to the garden. Their task is to take advantage of and to improve these resources, and to enable the garden to offer a better environmental education service to the schools.

The education programme of the garden covers the subjects discussed below.

Courses for teachers

As noted in the previous section, education in schools on environmental science in general and on the Canarian flora in particular is rather unsatisfactory. Due to deficient teacher training, in many cases not done on the islands, most of the teachers in charge of these subjects have not been properly prepared to teach them. Because of this we started courses to improve the preparation of teachers in this field, providing them with the basic indispensible information from which to update their knowledge and giving them a simple bibliography from which to work.

We have varied the content of these courses according to the level of preparation of the participants. We started with elementary courses that give an overall view of the most important aspects of flora and vegetation and their inter-relationship with the environment; we proceeded to specialised courses in specific fields of botany. The content of these courses has in many cases required the development of suitable teaching methods which allow the teacher to take better advantage of the knowledge obtained during the continued development of his or her experience. In the present courses, such as our first course for environmental animators, one of the 3 arranged for the present school year, we have defined the following objectives:

(1) To make known the facilities of the botanic garden as an education resource.
(2) To improve the participant's knowledge of Canarian nature and natural history.
(3) To make known and to use traditional techniques and methods which help to interpret man's rôle in the environment of the islands through history. These techniques include weaving with natural fibres from native plants, dyeing using natural substances, pottery, stone carving and wood-working.
(4) The discovery and exploration of artistic expression such as in Canarian literature, painting, sculpture, dance and music. These courses include use of natural materials from the Canarian landscape to which children can relate emotionally.

"The Canarian flora and children" project

The urgent necessity to initiate education programmes to guarantee the conservation of the flora obliged us to carry out, at the same time as the teachers' training courses, other projects which would have an immediate effect on local children. In collaboration with a group of ecologists, the Canarian Association for the Defence of Nature (ASCAN), and an enthusiastic group of teachers who participated in our first course, we prepared a project known as "La Flora y Los Niños" (Flora and Children). The objectives of this project were:

(1) To ensure that the maximum number of pupils possible are aware of the benefits that the plant kingdom provides for man.
(2) To ensure that children are aware of the Canarian flora and its importance.
(3) To ensure that children respect nature, in particular plants, and avoid damaging them.

Three audiovisual programmes were prepared and 14 sets of each were distributed for use in the main educational centres of the island. A travelling exhibition was assembled and put on public display in more than 30 schools, where it was visited by over 30,000 people, most of them children. This exhibition, in the form of wall-panels with an educational design, demonstrated the uniqueness of Canarian nature and also the rôle of the botanic garden in its conservation and study.

Ten thousand young dragon trees (*Dracaena draco*) were given to children at the 6th level (12 years old). For 2 years the children maintained and looked after these plants and were able to observe their very slow growth and the difficulties involved in maintaining them. Most of them survived and for these we organised symbolic planting sessions to reintroduce them into natural areas such as the Caldera de Bandama and Montana Almagro and to plant them in the gardens of colleagues and of the schools themselves.

Sventenius prize

In 1980 an annual award was initiated known as the Sventenius Prize, in recognition of the work of the founder of the garden, Dr Eric Sventenius, who made an enormous contribution to the conservation and study of the Canarian flora. This prize is directed towards groups of up to 5 children with the aid of a teacher as assessor and advisor. The children can carry out projects on any theme related to the Canarian environment and the prize is a visit to each of the Canarian national parks. In the last 5 years over 300 projects have been presented for the prize, some of which have had a very important scientific or educational value.

Exhibitions

On the island there are a number of annual exhibitions of plants and flowers in various places. There are also neighbourhood associations and other cultural activities through which fiestas and flower shows can be organised on saints' days and other holidays. Participation in these is, for the botanic garden, a public window which permits us to show our work and to demonstrate various aspects of conservation and protection of the local flora. We can also display collections of living plants that include many endangered species. To some extent this takes the work of the institution outside the physical limits of the garden itself and to the population in general.

Visits to the botanic garden

Because of its location in a natural habitat, the Canary Garden is an ideal place for experimental education in the field. There are natural routes through the garden that show a wide coverage of the endemic flora of the island. Here people can directly observe and study the plants while at the same time seeing other items of interest such as the geology and fauna. For school visits a choice of programmes is available so that in successive annual visits it is not necessary to repeat the same programme.

As a support unit for school visits, the garden has an educational centre, which can be used by visitors, students and schoolchildren. It has the necessary equipment for more structural environmental education. The centre has an exhibition hall with a permanent display of education panels, a school laboratory where experimental work can be carried out and where some of the garden's research programmes can be shown to the public, and a classroom which can be used for various purposes such as audio-visual displays. There is also a library of basic works on the natural history of the Canaries, which is available to visitors.

A school visit can also be completed with practical work in the plant nursery at the school. Here children can experience plants with their own hands and at the same time contribute individually to the stock of plants available for rebuilding the natural vegetation of the island.

These are basically the education activities in the Jardín Botánico "Viera y Clavijo". They could probably be used as a basis for similar programmes in other gardens.

Botanic gardens and community education in Australia

R.W. BODEN and E.A. BODEN

Australian National Botanic Gardens, Canberra

Summary

Although the first botanic gardens in Australia were established almost 150 years ago and there are now 8 major gardens and more than 18 provincial ones in the country, education has been slow to develop as an accepted function.

In the early years of European settlement, Australian botanic gardens concentrated on the introduction and assessment of exotic food and forage plants for local cultivation. More recently some botanic gardens have placed greater emphasis on indigenous plants. Australia's population is becoming increasingly urbanised, resulting in demands on botanic gardens for recreation and leisure use. Education is now being more widely recognised as a legitimate aspect of the public rôle of botanic gardens. This emerging rôle is discussed in relation to the following points.

(1) The diversity of learners in terms of educational background, relevant interests and levels of self-direction of their learning.
(2) The employment of professional educators to make plant sciences relevant to the visitor and to extend the educational use of plants beyond the confines of conventional plant sciences.
(3) The increasing interest in the cultivation of Australian native plants at a time when Australians appear to be consolidating their national ethos.
(4) The expressed need for contact with living things by urbanised citizens.

Botanic Gardens and the World Conservation Strategy

ISBN: 0-12-125462-3

(5) Community involvement through friends groups and volunteer programs.
(6) Recognition by administrators of botanic gardens that meeting the needs of visitors, including their educational needs, is essential for continuing public support.

Resumen

Aunque el primer jardín botánico en Australia fue establecido hace casi ciento cincuenta años, actualmente existen ocho grandes jardines botánicos nacionales, y más de dieciocho jardines provinciales en este país; la educación ha tenido un lento desarrollo hasta ser aceptada como función de estos Centros.

En los primeros años de la colonización europea, los jardines botánicos australianos se concentraban en la introducción y establecimiento de plantas exóticas comestibles y de interés forrajero para el cultivo local.

Más recientemente algunos jardines botánicos han puesto gran interés, en el estudio de las plantas autóctonas. En los últimos años la población urbana de Australia se ha incrementado grandemente, como resultado de esto existe una demanda de jardines botánicos para ser utilizados como lugares de recreo y ocio. Ahora es cuando empieza a ser reconocido el papel público de los jardines botáncos.

Esta nueva función que está emergiendo se discute en relación a diversos factores:

(1) La diversidad de estudiantes en el panorama educativo, su relevante interés y los niveles de autodirección de su aprendizaje.
(2) La utilización de educadores profesionales que hagan aplicable la botánica al visitante y extiendan el uso educativo de las plantas más allá de los confines de la botánica.
(3) A medida que los australianos parecen estar consolidando su identidad nacional, aparece un incremento del interés en el cultivo de plantas autóctonas de Australia.
(4) La mencionada necesidad de los habitantes de la ciudad de estar en contacto con los seres vivos.
(5) La sociedad se compromete por medio de amigos grupos y programas de voluntarios.
(6) Reconocimiento por parte de la junta de administración de los jardines botánicos de la necesidad de los visitantes y su educación medioambiental, como esencial para la continuación del apoyo público.

Australian botanic gardens

The Commonwealth of Australia consists of 6 sovereign States, 2 mainland Territories and 6 small external island territories. Botanic gardens have been established in each State capital city, in Darwin in the Northern Territory, in Canberra, the National Capital, and recently on Norfolk Island, one of the external territories. In addition to the capital city gardens there are 18 provincial ones varying in size, level of professional management, and security of land tenure.

A botanic garden was established in Australia at Sydney soon after European settlement commenced in 1788. By the middle of the 19th century the development of botanic gardens had begun in other State capital cities including Melbourne, Adelaide, Brisbane and Hobart. Unlike many European botanic gardens, none in Australia was established in conjunction with a university, although the site of the gardens in Canberra was selected to be near the University to assist in cooperative research and to provide an educational resource.

The main reasons for the establishment of the early Australian botanic gardens were to duplicate the European practice of developing pleasant open spaces for leisure and to test plants with economic and food potential. Australia's geographic isolation, absence of indigenous agriculture and lack of native plants palatable to Europeans necessitated the introduction and establishment of exotic crop and forage species.

Joseph Banks and Daniel Solander, the first botanists to visit eastern Australia, were impressed by the strange beauty of the flora but not by its food potential. They advised Captain Arthur Phillip, the first Governor of New South Wales, to carry food plants with him in the First Fleet. Phillip took seeds of cereals and other plants from England and collected a range of edible species from Rio de Janiero and the Cape of Good Hope. These were planted at Farm Cove, now the site of the Royal Botanic Gardens, Sydney.

By contrast with the horticultural emphasis on exotics, the botanical efforts in those gardens with associated herbaria were concentrated on the native flora which was largely unknown at that time.

A major exception in horticultural emphasis was the Royal Botanic Gardens, Melbourne. Under the direction of Baron Ferdinand von Mueller, the well-established practice of collecting and sending both herbarium specimens and seeds to European institutions was paralleled by the cultivation of Australian plants in the Melbourne Gardens and elsewhere throughout the State. Mueller introduced over 2,000 Australian species to his gardens.

Overwhelming interest in exotics and a perception that Australian plants were difficult to grow meant, however, that native plants continued to be only minor elements in most Australian botanic gardens for decades.

It was not until about 1950 that the neglect of Australian plants in botanic gardens was identified and a decision made to concentrate on these in the new botanic gardens at Canberra, the national capital. These gardens, now known as the Australian National Botanic Gardens, set about collecting and cultivating a comprehensive collection of native plants representative of the flora of the whole continent. About 6,000 species are now in cultivation there.

Interest in the cultivation of the wealth of native plants occurring in Western Australia was stimulated in 1962 by the allocation of part of Kings Park, Perth, to grow Western Australian plants.

Recently, several of the State capital city botanic gardens have established annexes to enable them to increase their collections of both native and exotic Australian plants in environments different from the parent gardens.

A new impetus to the cultivation of Australian plants in botanic gardens derives from a national survey of the existing plantings in public and private gardens. The report of this survey recommends *inter alia* the establishment of a national core network of 39 botanic gardens and arboreta involving an expenditure of 1,092 million Australian dollars in capital works over a period of 10 years (Royal Australian Institute of Parks and Recreation, 1985).

The community

Although Australia, with an area of 7·7 million km^2 is a relatively large country, about the size of the U.S.A., it has a small population of only 15·8 million people. These people are highly urbanised with almost two-thirds living in the large, sprawling cities of Sydney and Melbourne.

Most Australian families formerly favoured detached houses set in private grounds of about 0·1 ha; however since World War II there has been a substantial shift towards apartment-dwelling. Australian cities therefore now tend to occupy large built-up areas in the heart of which are concentrations of residents deprived of daily contact with nature.

A survey of recreation participation by Australians aged 14 years and over carried out in autumn 1985 showed that 36·8% of people gardened for pleasure (Dept of Sport, Recreation and Tourism, 1986). This was the seventh most common activity with watching television at home (93·6%), listening to music (65·4%) and visiting friends and relatives (66·8%) being the 3 most common activities for all age groups and both sexes. If participation in those activities carried out only outdoors is considered, gardening for pleasure had the highest participation rate followed by walking for pleasure. Participation in gardening for pleasure and walking for pleasure increased with age for both males and females with the highest participation rate being by those aged 55 and over.

The Australian population is an ageing one and if present birth rates and migration patterns continue, the proportion of older people will increase markedly. In 1985, 19·5% of the population was 55 years of age and over. By the year 2000 this is projected to increase to 20·9%, by 2010 to 24·6% and by 2020 to 28·2% (Dept of Immigration and Ethnic Affairs, 1986).

Thus the combination of an ageing population and the interest in gardening provides a guide to the likely increase in community interest in botanic gardens.

For residents and tourists alike, botanic gardens in mid-city locations have important recreational and educational potential. However, the large area occupied by Australia's major cities justifies decentralising botanic gardens through the development of outer urban and rural annexes which are more readily accessible to suburban and provincial residents. At present the annexes which are open to the public are recreationally attractive but educationally under-developed and under-utilised. It is hoped that in time the educational potential of all annexes will be achieved through the development of facilities and services of the type existing at their parent institutions.

Education

Until relatively recently Australian botanic gardens tended to concentrate on 3 of the major accepted functions of botanic gardens — plant introduction, botanical studies and recreation. A fourth accepted function, namely education, was only poorly developed. This appears to confirm the views of Simmons (1981): "Compared however to museums and zoos it is noticeable that botanic gardens have been generally slow to develop the potential educational opportunities of their work ....". The appointment of education officers was pioneered as late as 1971 by Adelaide Botanic Gardens. Within the following decade the 6 State botanic gardens and the Australian National Botanic Gardens adopted the practice of appointing an education officer, usually an experienced classroom teacher with qualifications in botany and education. By contrast, the appointment of professional educators in non-botanical museums including zoos was initiated as early as 1949.

Predictably the establishment of education services in these museums during periods of greater economic stability has created a situation where their staffing levels are considerably higher — even up to 15 times higher — than those currently prevailing in their botanical counterparts.

However, while the education services of Australian botanic gardens have modest staffing levels, their teaching strategies are generally innovative and emphasise both experiental and discovery learning. In this way Australian botanic gardens parallel national moves away from didactic teaching styles in primary and secondary education systems.

Before the appointment of professional teachers, education in Australian botanic gardens was largely a passive process relying on the plants themselves and simple labelling. These labels carried little more than botanical, and sometimes common names and broad geographic occurrence. Information on economic uses, habitat and associated species was usually lacking. Simple leaflets describing the gardens' history and facilities were sometimes provided. These processes are probably better defined as information rather than education. They presume levels of self-direction probably lacking in all but the most highly motivated adult visitor.

More recently the broad educational rôle has expanded to include interpretive signs, frequently supplemented by leaflets. These often relate to thematic walks such as the Aboriginal Trail at the Australian National Botanic Gardens. Educationally the Aboriginal Trail is a popular ethnobotanical resource for social studies at primary level (to 12 years of age) and for courses in Australian history, heritage studies and human biology at secondary level (12 to 18 years of age). Another trail, the Nature Trail, enables student visitors to these inner-urban botanic gardens to observe a relatively unaltered sample of the local open eucalypt forest and to gain an understanding of inter-relationships between plants, animals and their environment.

Thematic indoor displays play an important educational role. Their value for student visitors may be enhanced by the preparation of worksheets for completion during the visit, with suggestions for follow-up classroom activities as reinforcement. These displays may also form the basis of experimental activities during the visit. For example, associated with a display at the Australian National Botanic Gardens entitled "Bushfire!", students discover the relationship between fire and the Australian flora by igniting *Banksia* fruits to observe their heat-induced dehiscence. They also treat *Acacia* seeds to simulate the fire-related effects on germination.

Education programmes at the Australian National Botanic Gardens may be classified as responsive or innovative. The responsive programmes meet specific curriculum needs readily perceived by teachers and often written into curriculum documents. For example, Australian senior secondary biology courses usually include a topic on plant classification at the family level, with Myrtaceae, a significant Australian family, as the specified example. Teachers look to botanic gardens to enrich school-based studies of this topic.

Educational innovation includes botanical programmes for high achievers, and horticultural and plant-craft programmes for a wide range of students including those with social or learning disabilities. Innovative plant awareness programmes for adults and in-service programmes for teachers are presented during school holidays and after school-hours.

Educational outreach is achieved by means of poster kits, videotape, slide-tape and annotated slide kits, the provision of teaching specimens, promotional displays at public venues and publishing programmes in house journals and the public press.

The Royal Botanic Gardens, Sydney, introduced an isolated-schools programme to assist students living in remote areas who have little opportunity to visit the Gardens. The programme involves the supply of live plants, dried specimens and printed material. The Education Officer also visits schools in rural areas.

Modern learning theory encourages the professional teacher in the rôle of facilitating learning rather than exposing facts. It also emphasises the teacher's rôle in identifying students' levels of knowledge and so leading them from the known to the unknown. For these reasons, most Australian botanic gardens maximise the teaching skills of their education officers by providing direct contact between the education officer and visiting students, rather than restricting their rôle to the production of resource materials. The consequence of this is that many students are denied the experience because of the small number of teachers in Australian botanic gardens.

Education officers in botanic gardens naturally have a botanical focus in their programmes. However, as teachers, they recognise the validity of using botanic gardens as a means of achieving learning beyond the confines of conventional plant sciences and across the maximum span of the curriculum. For example, the universal dependence of people on plants has been used as the basis for visits to the Australian National Botanic Gardens by adult migrant students learning English as a second language. Such visits achieve vocabulary enrichment, practice in English language conversation and social skills in the Australian way of life. At the same time, newcomers to Australia learn about the plant heritage of their adopted country.

Funding and support

All major botanic gardens in Australia are funded by Governments. Unlike zoos and some art galleries, Australian botanic gardens do not charge entrance fees. Cash revenue from non-Government sources is minimal at present. However, corporate funding and support by way of voluntary labour as guides and herbarium assistants are emerging as important contributions to gardens' operations. Dependence on Government funding means that political, and therefore public, support is essential for both survival and growth. Currently in Australia there is only limited public funding for scientific work in botany and horticulture, and it is difficult to foster political support for botanic gardens on their scientific roles alone. Botanic gardens' administrators are becoming increasingly aware that education both within the gardens themselves and extending into the community, is a key element in popular support and therefore financial survival.

Taylor (1971) identified museum education as a means of producing a new generation of taxpayers responsive to museums' financial needs: "The children of today are the taxpayers of tomorrow. Inculcating a love of museums in the young will not only help to broaden their education but might also ensure more generous support in the future!" Botanic gardens administrators may be wise to support both Taylor's philosophy and pragmatism.

References

Department of Immigration and Ethnic Affairs (1986). "Australia's population trends and prospects 1985". Australian Government Publishing Service, Canberra.

Department of Sport, Recreation and Tourism (1986). "Recreation participation survey — April/May 1985". Department of Sport, Recreation and Tourism, Canberra.

Royal Australian Institute of Parks and Recreation (1985). "A report on the collection of native plants in Australian botanic gardens and arboreta". Royal Australian Institute of Parks and Recreation, Canberra.

Simmons, J.B.E. (1981) Developing acceptable information and interpretation systems for botanic gardens. *Bot. Jahrb. Syst.* **102**, 81–95.

Taylor, W.W. (1971). Museums and education. *Museum* **23**(2), 125–133.

The Córdoba Botanic Garden in the local community

J.E. HERNÁNDEZ BERMEJO

Jardín Botánico de Córdoba, Spain

Summary

The Córoba Botanic Garden was instituted with 2 basic objectives in mind: (1) to keep a collection of plants with economic and particularly agricultural interest (fruit trees, cereals, garden and ornamental plants, weeds, etc.); and (2) to collaborate in the conservation and protection of the Spanish flora, especially from Andalucia. The first of our 2 objectives (conservation of endangered flora) has led to a series of activities, as follows.

(1) The specialisation of our *Index Seminum*, which since 1982 has included an assorted offer of endemic Iberian-Balearic species for cultivation in other botanic gardens.
(2) A plant germplasm bank specialising in endangered Iberian-Balearic species.
(3) The development of a research programme to propagate these endangered species, to acquire new populations and to reintroduce them into potential habitats.
(4) The promotion of high altitude gardens and ''satellites' in the area. The first one is already being set up in the Sierra Nevada with the collaboration of the Regional Government.
(5) Education and public awareness through radio programmes, publications, lectures, school visits and the establishment of living collections of the native flora of the region.

These duties have been undertaken as a result of a regional responsibility — that of being the only botanic garden in Andalucia. The Córdoba Botanic Garden also advises local authorities in the

Botanic Gardens and the
World Conservation Strategy

ISBN: 0-12-125462-3

botanical and landscape reconstruction of historical and archaeological sites. Our help has already been asked for the reconstruction of the Ducal Palace Gardens in Fernán Nuñez, and more recently for the landscape reconstruction of the archaeological ruins of Medinat-al-Zahara.

To summarise the Córdoba Botanic Garden's experience within the community, we feel that it should play an important social and political rôle in the protection of endangered species. Unduly centralised solutions hinder active participation from local and regional governments. The rôle of an international agency should be that of promoting, coordinating and managing resources, in this way letting the local communities feel that they are not only participants but also direct beneficiaries of conservation.

Resumen

El Jardín Botánico de Córdoba en su comunidad local.

El Jardín Botánico de Córdoba ha sido diseñado con dos objetivos principales: (1) colaborar en la conservación y protección de la flora española y, más especialmente de Andalucia; (2) conservar colecciones de plantas de interés económico, especialmente agrícolas (frutales, hortícolas, cereales, ornamentales, malas hierbas....). El primero de los objetivos (conservación de flora amenazada) ha inspirado acciones como:

(1) especialización del *Index Seminum* que incluye desde 1982 siempre una variada oferta de especies endémicas ibero-baleares para le extensión de su cultivo a los restantes jardines botánicos;
(2) puesta en marcha de un banco de germoplasma vegetal especializado en la flora ibérico-balear amenazada;
(3) desarrollo de un programa de investigación dedicado a la multiplicación de especies amenazadas, consecución de nuevas poblaciones e introducción de las mismas en sus hábitats potenciales;
(4) promoción de Jardines de altura y "satélites" en la región (el primero de los cuales está siendo ya instalado en Sierra Nevada en colaboración con la Junta de Andalucia).
(5) difusión mediante programas de radio, publicaciones, conferencias, documentación de las visitas de los centros de enseñanza, elaboración de colecciones vivas, etc., de la flora autóctona regional.

Estas funciones han sido asumidas como resultado de la responsabilidad regional que implica ser el único jardín botánico existente en Andalucia. Además, el Jardín Botánico de Córdoba asesora a las autoridades locales en la restauración botánica y paisajística de conjuntos históricos y arqueológicos. Así, ya fue demandada su ayuda con ocasión de la restauración de los Jardines del

Palacio Ducal de Fernán Núñez y recientemente en la restauración paisajística de la ciudad califal de Medina Zahara.

Como resumen de la experiencia adquirida en las relaciones del Jardín Botánico de Córdoba con su propia comunidad regional, mantenemos el criterio de que en la protección de la flora amenazada debe de jugar un importante protagonismoen el entorno social y politico regional. Soluciones excesivamente centralizadoras obstaculizan la colaboración activa de los pueblos, regiones y gobiernos autónomos. El papel de los organisimos internacionales deberá ser el de promover, coordinar y gestionar recursos pero siempre haciendo sentir a las comunidades locales no sólo participes sino también beneficiarias inmediatas de la conservación.

When, in 1979, we took the first steps towards the creation of a botanic garden for the City of Córdoba, which would be the only one in Andalucia, we did not suspect the enthusiasm with which the various parts of the local community would welcome and support the idea. This support led immediately to the setting up of a public foundation formed by the Municipal Government for this purpose, and to the signing of an agreement between the University and the City Council of Córdoba for the joint direction and management of the Foundation. Various agreements and financial assistance from the central, regional and local administrations followed. But above all, support from the local citizens has given us considerable economic resources, considering the modest financial capacity of our city, towards the creation of its botanic garden.

We have sensed this warm, popular response through the interest shown by the man in the street, in the modest but significant donation of plants for the garden, in the prolonged voluntary collaboration of a retired nature-lover, in the persistent interest shown by educational centres in the progress of the work, and in the interest of local journalists who have informed the public regularly and objectively of each significant step in the construction of the garden. What is the reason for all this interest and community sympathy towards the garden? Perhaps it is partly because in our community there is a profound sense of culture dating back to Caliphal Córdoba where one of the most enlightened scientific and cultural communities of the Middle Ages was focussed; perhaps it is also due to the deep interest in and appreciation of the plant world, of gardens and of ornamentals, which has been practised by the Córdobés over several centuries. It may also be because of those failed attempts in the past to revive a botanic garden which the historical records indicate once existed in Córdoba, constructed by Abd-al-Rahman I in the 8th century; this was 700 years before the creation of the botanic gardens of Padua and Pisa, normally accepted as being the first in Europe. Centuries after, in the reign of Carlos III, there were also two frustrated attempts to create acclimatisation gardens in Córdoba. So the people of Córdoba appreciate the idea of having a place for the study and contemplation of the plant world in all its diversity. It was certainly because of this that our original proposal of 1979 became a reality in only 5 years.

With this background and using the knowledge and experience of the staff of the University and of the Architect's Department of Córdoba City Corporation, a profile of the characteristics of the project was drawn up. The garden was conceived as a cultural setting in which scientific, educational and environmental activities could take place in an aesthetic and landscaped framework. It would use a garden style in keeping with the Hispano-Arabic traditions of the region, where such notable examples as the Alhambra of Granada and the gardens of the Alcázar of Córdoba are to be found. However, a formula was sought which had to integrate these traditional styles with the multiplicity of elements that characterise a modern botanic garden and with the fundamental intention of orientating the functions and design of the garden to its regional context.

To achieve this, the 2 main objectives of the garden were defined.

(1) As a consequence of its location in a predominantly agricultural region, an objective is the maintenance of collections of representative species and varieties of economic interest for Andalucian agriculture. These consist of the open-air collections of cereals, vegetables, aromatic plants, textile plants, weeds, citrus, *Prunus* and *Malus*, olive varieties and Mediterranean fruit species. They also include subtropical and tropical species of economic interest grown under glass and in shade houses.

(2) As a result of being located in an extremely rich floristic region and a country with a high concentration of endemic species, a second sphere of activity is to take on a serious commitment to the protection of the threatened or endangered local flora. Of about 1,600 species endemic to the Iberian peninsula, about 1,000 occur in the Andalucian region and of these about a quarter are exclusively endemic to Andalucia. More than 150 of them are in some danger of extinction.

These two principal objectives guided the design of the garden. The main constituent elements were the Arboretum, the rock-garden, the collections of economically interesting plants, the systematic collection of mainly Spanish plants, the glasshouses for subtropical and tropical plants of agricultural value, the glasshouse for endemics of Spanish territories, the palaeobotanical museum (with a collection of over 150,000 specimens of plant fossils from the Carboniferous basins of Spain donated by Professor

Wagner), the rose garden, the garden for the blind and the ecological plantings.

Independently of the design of the garden as such, it is useful to look in more detail at the specific functions taken on by the Córdoba Botanic Garden at the same time as its physical structure has developed.

Index Seminum

Since 1982 4 seed-lists have been published. From them about 9500 seed samples have been distributed to the greater part of the world's botanic gardens. Priority is given to seed-collection in Andalucia and, on a wider scale, to the more botanically interesting southern and Mediterranean regions of the Iberian peninsula. We hope, by specialising in this way, to cooperate in the conservation and diffusion of the native species of southern Spain, so assuming a previously unattended responsibility. Numerous taxa from Sierra Nevada, Sierra Morena, Cazorla, Grazalema and the Cadiz–Huelva–Algarve littoral region have been offered and distributed over the past 3 years along with others from the Balearic Islands, the Central Cordillera and other Iberian localities.

Setting up an Iberian-Balearic endangered plant germplasm bank

The collection of seeds of biologically valuable species either directly from the wild or as a result of multiplication in the botanic garden cannot be directed solely towards exchange of material via a seed-list or simply serve for obtaining living plants for the garden's collections. Not to complete the conservation programme by developing and maintaining a germplasm bank under such circumstances would mean the waste of a good deal of the initial effort involved.

To this effect we have now started to conserve several hundred seed samples of narrowly distributed Iberian species, either field-collected or from the garden, at low temperature and reduced humidity. Our intention is to direct this bank primarily towards Andalucian endemics. The new cold-rooms which will eventually be used for the permanent home of the germplasm bank at the garden are under construction and in the meantime the most important seed samples have been encapsulated and stored in refrigerator cabinets.

In addition to the function as an internationally orientated germplasm bank with wide-ranging objectives, our responsibility within the Andalucian community inevitably obliges us to maintain a germplasm bank specialising in the protection of the genetic resources of the region. The functions of the bank are threefold:

(1) long-term conservation;
(2) to allow a responsible exchange programme to be undertaken without the need for repeated and often deleterious resort to wild resources;
(3) to supply the requirements of our own living plant collections.

Multiplication of endangered species and natural habitat reintroduction programme

Using our germplasm bank and our plant propagation installations, we have been working to obtain garden populations of various endangered species. We have been successful with several Balearic Islands species such as *Lysimachia minoricensis, Silene hifacensis, Naufraga balearica* and also some Andalucian endemics, such as *Aquilegia cazorlensis* and *Antirrhinum charidemi*. A cooperation agreement has recently been signed with the Directorate-General of Research and Agricultural Extension of the Agriculture Department of the Andalucian Regional Government to develop a research project on the propagation of Bétic Region high-mountain species of special biological interest. Populations are being obtained not only for naturalising in the garden but also for reintroduction into natural habitats where they are either very scarce or have already become extinct. This is a continuation of a programme which we started several years ago when based in another institution.

Promotion of associated montane gardens

We recognise the impossibility of conserving, in optimum conditions in the natural state, many of the plant communities and local species of the Bétic mountains. Therefore we have become recently involved in promoting the creation of associated gardens in the areas of highest natural diversity within the region.

At the moment, among various projects initiated, the outstanding one is the construction of a conservatory and mountain garden in the high valley of Lanjarón at the foot of the Cerro del Caballo, the third highest peak after Mulhácen and Veleta in the Sierra Nevada. An agreement has been signed with an Autonomous Regional Government and a technical design prepared. The area chosen is the home of many of the most significant endemics of the Sierra Nevada. It is situated at 2,700 m and will be 2·5 ha in size. Construction began simultaneously with the biological research project, in May 1986.

We consider this to be one of the most important models for the projection of our garden within the community. It is a very effective way of encouraging local councils and other organisations to participate in the functions of the garden, which itself requires valuable complementary aid for its nature conservation and cultural extension missions.

Education and public awareness

Probably one of the most gratifying experiences in our short history as a garden has been that of having to respond to the demand for education and teaching on a scale which we did not even suspect existed in our local community. We have been asked to help, and have responded, in public activities such as lectures, round-table discussions, radio programmes and so on. For a year we have organised a weekly radio programme on botany for school-children. We have prepared an audiovisual programme on the major features and objectives of the garden for use by neighbourhood associations in the suburbs of Córdoba and are now making a video film on the same subject.

In fact it was necessary to begin school and college visits to the garden long before the official opening. These visits take place by means of small groups, guided by our University intern students who are based at the garden. The initial visits allowed us to estimate the future demand on the garden and to prepare a suitable cultural content for the visits.

Several departments of the University of Córdoba use the installations of the garden for the practical and experimental parts of their courses. In addition to these collective activities, there are 25 intern students of the University in the garden, carrying out annual projects. The content of these projects and practical exercises is not only botanical but takes into account the agricultural vocation of our University and of the region in general. The School of Agricultural Engineers uses the installations of the garden (irrigation systems, water collection systems, glasshouses heated by solar energy, techniques of construction of rural buildings, etc.) in its teaching programmes. Some of the infrastructure such as the heating system for the exhibition glasshouse is unique in the region and constitutes a focal point of interest for the agricultural community of the Guadalquivir valley.

Landscaping assessment and restoration

The advice of the Córdoba Botanic Garden has been sought during the construction of various new public gardens within the municipality of

Córdoba. Even more significant, however, has been its participation in the restoration of other historic gardens.

In the case of the gardens of the Ducal Palace at Fernán Nuñez, it was the architect designated by the Directorate General of Fine Arts along with the town council of Fernán Nuēz who proposed that we should direct the restoration of the palace gardens. This was carried out during 1984. We have also been asked to participate in the conservation of the gardens of the Viana Palace in the City of Córdoba, but perhaps the most ambitious and creative of the projects in which we are involved is at the request of the Cultural Delegation of the Autonomous Regional Government. This is the landscaping and restoration of the historical site at Medinat-al-Zahara, a city built by Abd-al-Rahman III in the 10th century and adorned with outstandingly beautiful gardens. Situated 10 km from Córdoba at the foot of the Sierra Morena, this is one of the most important archeological sites in Europe comparable only with the Alhambra of Granada or the Mesquite Córdoba. The restoration project requires a comprehensive study of the bibliography and botany of the period and we have been working on this for some months.

The Director of the Botanical Garden of Córdoba has participated in and introduced a book edited by the City Council on the ornamental trees of Córdoba and has supervised a doctoral thesis on the history and styles of the parks, gardens and patios of Córdoba. Amongst the future projects is that of centralising in the garden the education programmes on these subjects at a provincial level.

Through our short but intense experience of local community participation which has helped, conditioned and finally seen the birth of our garden, have we learned anything of value towards the search for a world strategy for plant resource conservation? We think so: our experience indicates that endangered species conservation can and should be carried out by and for local communities. In other words, the protection of the plant kingdom, of its endangered species, should not be conceived only as a universal heritage and responsibility, and much less as the exclusive reserve of a reduced elite. Communities should:

(1) be made aware of the need to conserve their local plant heritage;
(2) cooperate actively in its conservation;
(3) and be made to feel, at all times, the beneficiaries of conservation.

These three components can be converted precisely into the principal functions of a botanic garden in each community. Gardens should develop this public awareness role and channel local civic participation. Simply by their presence and by their educational programmes, gardens can continually demonstrate to the public authorities and to the man in the street the benefits of an adequate level of genetic resource conservation.

The international organisations (IUCN, IABG, UNESCO, etc.) cannot be the only active leaders in this field and even less the capitalisers on conservation. Their function should be that of coordinating strategies, seeking solutions and resources, and driving forward and coordinating activities. It should be local autonomous geographical communities personified culturally or politically who should assume the active leadership and take the necessary measures as integrated elements in a "world mosaic". Excessively centralised measures (world collections, international germplasm banks, etc.) may well be viable, but have the considerable inconvenience of not relating to local populations, regions or governments, without whose cooperation a globally effective strategy for the conservation of endangered flora can never be secured.

Botanic gardens and education in South Africa

J.N. ELOFF

National Botanic Gardens, Kirstenbosch, South Africa

Summary

In South Africa there are several types of botanic gardens. The National Botanic Gardens is an autonomous state-supported organisation with a head office at Kirstenbosch and with 7 regional gardens situated throughout the country. There are also botanic gardens which are totally supported by the state such as the botanic garden of the Botanical Research Institute, several university botanic gardens and a number of botanic gardens associated with cities such as Durban and Johannesburg. In addition, many small towns have wildflower gardens which play an important nature conservation rôle.

The educational function of all these gardens in the context of nature conservation, and also the relationship between these gardens, conservation agencies and conservation societies, is highlighted in this paper. The impact of a coordinated effort by the National Botanic Gardens, the South African Nature Foundation and the Botanical Society during the Flora '83 festival on conservation attitudes in South Africa is also discussed. Attention is also focused on formal and non-formal educational activities in the National Botanic Gardens.

Resumen

Jardines botánicos y educación en Sudáfrica

En Sudáfrica hay varios tipos de jardines botánicos. El Jardín Botánico Nacional, es una organización autónoma subvencionada por el Gobierno con una oficina central en Kirstenbosch y siete jardines

Botanic Gardens and the World Conservation Strategy

ISBN: 0-12-125462-3

regionales situados a lo largo de todo el país. Existen también jardines botánicos con un 100% de subvención estatal, tal como el jardín botánico de algunas Universidades y jardines botánicos asociados con ciudades como Durban y Johannesburgo. También existen jardines de flores silvestres en pequeñas ciudades que juegan un importante papel en la conservación de la Naturaleza.

Es destacada la función educativa de todos estos jardines, dentro del contexto de conservación de la naturaleza, asi como las relaciones entre estos jardines y los organismos y sociedades para la conservación. Aquí se exponen y discuten las actitudes tomadas para la conservación en Sudáfrica durante el Festival de Flora 83, en el que se coordinan los esfuerzos del Jardín Botánico Nacional, la Fundación Sudafricana de la Naturaleza y la Sociedad Botánica. También se enfocan las actividades educativas del Jardín Botánico Nacional desde un punto de vista tanto formal como informal.

Introduction

If one asks why education is important in botanic gardens, several reasons could be given. In the first place, most accepted definitions of botanic gardens stress the importance of an educational function. As early as 1914, Gager said that a botanic garden is "a collection of growing plants, the primary purpose of which is the advancement and diffusion of botanical knowledge" (Lawrence 1969). This was expanded somewhat by Warner in 1975 when he stated that "a botanical garden is a scientific and educational institution with an area of plants arranged according to some system of botanical classification. The purpose is the advancement and diffusion of knowledge and love of plants. Its primary functions are research, display, education, public service and recreation, but not necessarily in that order". Bailey and Bailey in *Hortus III* (1977) stated "a botanical garden is a controlled and staffed institution for the maintenance of a living collection of plants under scientific management for the purposes of education and research ... The intention of the enterprise is the acquisition and dissemination of botanical knowledge."

From the above it is clear that education should form a major component of the activities of every botanic garden. It is, therefore, not surprising that when Warner (1975) determined the relative importance of several functions in 50 botanic gardens and arboreta in the United States, education was found to be the most important according to their directors. Recalculation of the data of Warner gives the following relative degrees of importance

(Eloff 1985a), education 87%, display 85%, public services 81%, research 64%, education through recreation 46% and recreation 32%.

Secondly, education is seen as an important component of botanic gardens because in many countries the tax exemption status of a garden is determined by whether it is an educational institution or not. The very survival of many gardens is largely determined by the degree to which they are successful as educational organisations. This is especially the case in the United States, where generally gardens are not supported by the state or by local authorities to the same extent as in many gardens in Europe. Consequently, gardens in the United States have tended to be more educationally-orientated than many gardens in Europe.

Finally, education is extremely important in helping botanic gardens to fulfil their nature conservation function. More and more botanic gardens see their plant conservation function as one of their main reasons for existence. Botanic gardens have a special function in the whole nature conservation scene in protecting threatened plants by means of *ex situ* conservation. The furthering of conservation through education should consequently be a major function of most botanic gardens (Eloff 1984, 1985b).

Botanic gardens in South Africa

If one considers the various defining features of a botanic garden, many of the institutions considered to be botanic gardens in South Africa do not qualify. Most of them, however, fulfil at least two of the functions that can be ascribed to a botanic garden. The botanic gardens in South Africa can be divided into four groups, as follows.

Municipal botanic gardens

Most municipalities have planted areas set aside for recreation. However, only 2 municipal gardens in South Africa can really be considered to be botanic gardens, the Johannesburg Botanic Garden and the Durban Botanic Garden. The main purpose of these gardens, however, is recreation and not education or research. Consequently there is hardly any educational activity at any of these gardens. The very small municipally-controlled garden in Johannesburg, The Wilds, which only grows indigenous plants, does however receive up to 1,000 schoolchildren on visits annually. There is no programme available for them but the curator of the garden or the teacher accompanies the children on trails through the garden. This is usually on an unplanned basis.

University botanic gardens

Many of the universities in South Africa have planted areas where plants are grown for demonstration purposes and where material is obtained for use in practical classes by students. At the Universities of Stellenbosch, Potchefstroom and Pretoria especially, the planted areas have grown to such an extent that they may be considered as small botanic gardens. Many other universities are in the process of building up botanic gardens. In general these botanic gardens are open to the public and are visited by enthusiasts, while students make use of them. There is no, or very little, planned use of facilities by the public.

Botanic garden affiliated with the Botanical Research Institute

The Pretoria botanic garden, Brumeria, was developed by the Botanical Research Institute, which also houses the National Herbarium. Being governed by the Department of Agriculture, it is totally funded by the state. This garden was established in 1946, on the eastern side of Pretoria, on 60 ha of land. The nursery in the garden includes 2,800 species and in the garden approximately 2,200 other species are represented, including one of the world's largest collections of succulents from the Republic of Malagasy. The purpose of this garden is mainly to provide material for research by the staff of the Botanical Research Institute and other scientists in South Africa. It does not have any organised educational function at present. Nevertheless, many people visit the garden and, because most of the plants are labelled, it is used to a large extent by enthusiasts interested in indigenous plants in South Africa.

National Botanic Gardens

Harold Pearson, born in England in 1870 and who was a student at Cambridge, was appointed as the first Professor of Botany at the University of Cape Town in 1903. The interest in botanic gardens, which he developed during the 4 years that he spent at Kew, ultimately led to the establishment in 1913 of the first National Botanic Garden in South Africa, on the Kirstenbosch estate, which had been bequeathed to the nation by Cecil John Rhodes in 1902.

It was discovered quite soon that not all South African plants could be grown at Kirstenbosch and consequently the policy of evolving regional gardens was established. This has led to the present situation whereby the National Botanic Gardens consists of 8 gardens totalling 1,352 ha and situated in various ecological areas throughout South Africa. The first regional garden was established in 1921 at Whitehill, near Matjesfontein, and subsequently transferred to Worcester in 1946, as the Karoo National Botanic Garden. The Harold Porter Botanic Garden in Betty's Bay was placed under the control of the National Botanic Gardens in 1959. The Orange Free State Botanic Garden, situated in Bloemfontein, was established in 1967. The Drakensberg Botanic Garden was established at Queen's Hill in Harrismith in 1967 and was transferred to the present site, which was donated by the Municipality of Harrismith, in 1969. The Lowveld Botanic Garden in Nelspruit was established in 1969. The Natal Botanic Garden in Pietermaritzburg, which was established by the Botanical Society in Natal in 1870, was placed under the control of the National Botanic Gardens in 1969. The Witwatersrand Botanic Garden was established in 1982 on land donated by the Roodepoort Municipality.

With the exception of the Natal Botanic Garden, all the national botanic gardens concentrate on indigenous flora of Southern Africa. At present more than half of the 22,000 species of flowering plants native to Southern Africa are grown in the different botanic gardens.

Shortly after its inception, the National Botanic Gardens was declared a state-aided institution and eventually became subject to the Cultural Institutions Act of 1969. Up to 1967 the National Botanic Gardens fell under the authority of the Ministry of Education, Arts and Sciences, was then transferred to the Department of Agriculture and Fisheries and from 1981 to the Department of Environment Affairs. The State at present provides approximately 79% of the total budget needed to manage the 8 gardens with a staff complement of 402 people.

In the Forestry Act of 1984, the National Botanic Gardens may (1) collect and cultivate plants indigenous to the subcontinent in national botanic gardens; (2) undertake and promote research in connection with plants and related matters and make indigenous plant material available for research; (3) study and cultivate endangered plant species; (4) investigate and utilise and promote the utilisation of the economic potential of indigenous plants; (5) promote an appreciation of indigenous plants among the public; (6) establish non-indigenous plants for comparative studies and educational purposes.

There are 2 ways in which the National Botanic Gardens can promote the conservation of indigenous flora. Firstly, the botanic gardens have a vitally important rôle to play in trying to save critically endangered plant species. In a publication by the World Wildlife Fund and the International Union for the Conservation of Nature and Natural Resources (IUCN and WWF 1984), the importance of botanic gardens has been stressed by its author, H. Synge, as follows:

> There are 600 or so botanic gardens around the world and most are keen to do more for plant conservation. Indeed the movement to conserve plants really began in botanic gardens, whose staff were concerned at how little attention conservation bodies gave to plants. Even now conservation bodies and botanic gardens rarely work closely together. Few conservationists have understood what botanic gardens have to offer and gardeners have perhaps been slow to appreciate the urgency of the crisis facing the plants they know and love.
>
> Yet in recent years botanic gardens and museums have undergone great changes and are keen to change their Victorian image. More and more they see plant conservation as a principal reason for their existence. In developing countries conserving genetic resources and finding ways to make better use of wild plants make the modern botanic garden a vital part of national development. Gardens accept that top priority is to protect plant habitats rather than grow the plants in cultivation. In a garden plants need constant tending and are vulnerable to mechanical breakdown or human error. More serious, it is rarely possible to maintain more than a very small proportion of a plant's genetic diversity in cultivation and plants propagated by seed tend to rapidly adapt for conditions in the garden rather than those in the wild. Nevertheless, garden collections play a valuable part in conservation, both as a back-up when conservation in the habitat fails, and also to provide material for research, for horticulture and for education.

Secondly, botanic gardens can fulfil their conservation function through education. The first task of botanic gardens formulated by IUCN and WWF is "to educate the public on why plants are important and why they need conserving". Because botanic gardens are situated in urban areas, they can reach many people. They bring nature to people, whereas in many of the national parks the people have to be brought to nature. The challenge to the horticulturist is to make the artificial seem natural. If a botanic garden has a reserve or natural area attached to it, it is very fortunate, but it is certainly not an essential component of a botanic garden. Consequently, education is one of the prime ways in which a botanic garden can fulfil its conservation function.

Present education situation in South Africa

Introduction

In the de Lange Report on Education in South Africa, three education types were identified (Anon. 1984), i.e. Formal, Non-formal and Informal. Because hardly any educational activities take place at other botanic gardens, only the National Botanic Gardens will be discussed here.

Formal education

This is education that takes place in a planned way at recognised institutions such as schools, colleges, technical colleges, universities, etc. Lectures given at a botanic garden as part of a school or university curriculum, or as a part of a series of lectures which lead to a certificate or diploma, may be considered as formal education.

The important rôle that Kirstenbosch could play in the education of schoolchildren was recognised shortly after it was established. As early as 1923 the Department of Education of the Cape Provincial Administration seconded a teacher to Kirstenbosch and classes were given to primary schoolchildren in the lecture hall, which still stands today. In 1969 the Provincial Adminsitration built the Nature Study School and appointed two teachers. Pupils from Standards III to V, mainly from southern suburbs of Cape Town, visit the Nature Study School for a period of 4 hours, during school hours, if the weather is favourable. In 1984 approximately 8,000 schoolchildren visited the school.

In the Natal Botanic Garden at Pietermaritzburg, the Natal Provincial Administration built a school and appointed a teacher soon after the garden was transferred to the National Botanic Gardens in 1969. In 1984, 12,000 schoolchildren visited this school.

With the exception of a few courses run on subjects such as nature photography and dyeing with indigenous plants, hardly any formal adult education at all takes place at the National Botanic Gardens.

Non-formal education

This is education that proceeds in a planned but highly adaptable way in institutions, organisations and situations outside the sphere of formal education.

More than 33,000 schoolchildren visited the botanic gardens on planned excursions of a duration of 1 to 2 hours, during 1984. These children were either supplied with work sheets or were guided by the Curator or educational staff through the different exhibits in the garden or along trails running through the natural area of the garden. This activity was quite pronounced at Kirstenbosch, which had more than 18,000 visitors from schools.

Competitions that are held for schoolchildren on indigenous plant or conservation-related topics is another component of non-formal education. During 1984 more than 2,000 entries from schools throughout South Africa were received for this type of competition. A competition which entailed making a Christmas card using indigenous plant material was especially popular among primary schoolchildren. Children from secondary schools have competitions such as designing a jacket for a book on plant conservation or an essay on a conservation or botanical subject.

The Botanical Society of South Africa, which was established at the same time as Kirstenbosch in 1913 with the sole aim of lending financial and moral support to the National Botanic Gardens, has its head office in Kirstenbosch. This can be considered as a ''Friends of the Garden Society'' and, with a membership of more than 14,000 at present, this society also plays a very important rôle in education. Apart from the quarterly journal, *Veld & Flora*, they have also produced a whole series of wild flower guides, aimed at making people aware of our indigenous plants and promoting nature conservation in general.

There are about 40 voluntary guides who make a significant contribution towards non-formal education by guiding tours through the garden for the casual visitor or for groups who ask for such tours. The guides are selected from a much longer list of applicants and then are given a course of lectures and examined before they are accepted as official guides for the National Botanic Gardens (NBG).

Informal education

This is education that is given in situations in life that come about spontaneously, for example within the family circle or within the neighbourhood. When people visit a botanic garden and they are educated passively as far as the botanic garden's staff is concerned, one could consider this a form of informal education. This is not a planned exercise and people are able to pick up information on their own by looking at exhibits or at named plant collections.

The informal education components of the National Botanic Gardens take place both inside and outside the gardens. More than 700,000 people visited the gardens during 1984 and all have been influenced in several ways to appreciate the beauty and the value of the indigenous plants of southern Africa. Indeed, Kirstenbosch is now one of the prime tourist attractions in South Africa with 583,000 visitors in 1984. Even the famed Kruger National Park was less popular, with 400,500 visitors during 1984. All the exhibitions in the garden influence these visitors to a greater or lesser degree.

Outside the garden, there are several aspects with which the NBG is involved. The most important of these is probably the wild flower shows that are held from time to time. Flora '83 is a prime example of a wild flower show which was cooperatively organised by the National Botanic

Gardens, the Botanical Society and the South African Nature Foundation. The theme of the show was Conservation through Education. The aim of this 3-day South African wild flower display was to make South Africans aware of the tremendous and unparalleled wealth of wild flowers in this country and to recognise the threats facing this fragile floral heritage. It has been said that this was the largest wild flower show ever held in the southern hemisphere and few of the more than 70,000 people who visited the show could have been left unmoved by the sheer beauty of the spectacle they witnessed. The show was held in the Good Hope Centre, the biggest exhibition hall in Africa, and yet it proved too small for this great number of people who visited it over 3½ days.

Plants of more than 4,000 species were exhibited. During the naming of the exhibits 1 new species was found and 2 which were considered to be extinct were rediscovered. Twenty-three exhibitors from throughout South Africa were involved in Flora '83. This included all four provincial Nature Conservation Departments, the cities of Cape Town, East London and Port Elizabeth, as well as Escom and the National Parks Board. One of the major attractions was a combined presentation by 9 rural towns of the western Cape, including Caledon, Clanwilliam, Darling and Hermanus, all famous for their spring wild flower shows. That year several of these towns did not hold their individual shows but brought their flowers fresh from the veld to Flora '83. It was the first time in 20 years that these regions pooled their displays, giving the public a rare opportunity to see the enormous variety of majestic proteas, delicate ericas and graceful reed-like restios in a single venue, without having to travel hundreds of kilometres to each show.

Apart from the flower exhibitions, there were also flower arrangements by the best flower arrangers. There was also a series of morning and evening lectures on southern Africa's flora and conservation, a thematic display of stamps featuring flowers, as well as wine-tasting and many commercial exhibits, all helping to demonstrate the importance of plants in the economy. Many of the endangered plants found only in the Cape Fynbos and some of the plant invaders threatening the local flora were exhibited. Flora '83 received tremendous media coverage and was also attended by large numbers of special tour groups from overseas who came to see this unique event.

If one measures the impact of Flora '83 on nature conservation, its most important contribution is that so many people were made aware of the value and the beauty of plants and of the threat of extinction. It has, for example, been reported that the enquiries about wild flower areas to visit for viewing has risen by 37% since Flora '83, the seed of indigenous plants supplied by NBG has increased by 37·2% between 1982 and 1985 (46,350 to 63,600 packets), and Botanical Society membership has increased by 27% (from

8,922 in 1982 to 11,321 in 1985). Part of the funds of Flora '83 was used to establish new floral nature reserves in South Africa. Although there have been one or two dissenting voices claiming that the veld was raped to provide plants for such a display, this was scornfully dismissed by farmers who stated that comparable areas of flowers were devoured by sheep within a week.

The attendance at the wild flower shows held in different regions has also grown remarkably. These wild flower shows are supported by the National Botanic Gardens and the nature conservation authorities of the province. In the Cape there are more than 50 wild flower reserves under the management of very small rural communities as well. All of these have also become much more generally accepted due to the impact of Flora '83.

Another result of Flora '83 is that the interest in the growing of indigenous plants in private gardens has increased dramatically. For example, at the annual indigenous plant sale held in Kirstenbosch in 1985, the plants were sold for a net value of R65,000, which is a tremendous increase on the sales of previous years.

Possible future developments

Formal education

Primary school

The present system will have to be continued and extended to incorporate more children of all races. In order to reach a significant percentage of schoolchildren, (1) the number of teachers will have to be increased, (2) the system will have to be extended to all botanic gardens, and (3) the education programme will have to be extended to after school hours and during school holidays.

Secondary school

The botanic gardens should become involved in school biology projects by making material and facilities available. Special exhibitions that fit into school curricula should also be set up. Lectures on specialised topics could also be arranged during holidays or after school hours.

Tertiary education

We should ensure that technical colleges, teachers' training colleges and universities know that our facilities are available to them. We should also liaise with them to ensure optimal use of our facilities. The teachers' training colleges have a very high priority because if we can have a good programme that is well supported, we can continue to influence hundreds of thousands of schoolchildren by giving future teachers an even more positive attitude towards nature conservation. There is also a wide scope for cooperation with technical colleges and universities. At Fairchild Tropical Garden in Miami, for example, a course is given on tropical plants, which is taken by students from elsewhere in the United States and the course is accredited by those universities as part of their curricula.

NBG diplomas or certificates

Many of the overseas botanic gardens in the U.K. and the U.S.A. have programmes which lead to a diploma in horticulture. There is a great need for horticulturists with experience and knowledge of indigenous plants. Possibly Kirstenbosch could run a programme for people who already have a diploma in horticulture in which accent is placed on the use of indigenous plant material. With the increasing realisation by different local authorities of the value of indigenous plants, this could fulfil an important need.

Certificates could also be given to people who have completed a number of courses in the adult education programme (see below).

Non-formal education

Field and Wager (1973) suggested that the following principles should be kept in mind for the effective interpretation of the resource to the visitors: (1) visitors are diverse; (2) visitors anticipate a relaxed and enjoyable atmosphere; (3) interpretative information must be rewarding; (4) interpretative information must be understood; (5) the effectiveness of information must continually be evaluated. This clearly shows the complexity of the situation. Evaluating the effectiveness (i.e. to what degree the goal is attained) of the non-formal education effort is especially difficult. It requires clear objectives and the design of a feedback procedure. Different target groups should also be addressed.

It is practically impossible to offer something in one exhibition that will satisfy all visitors. In visits to overseas botanic gardens we were perturbed to find that visitors more or less ignored what we considered to be excellent exhibitions. It is easy to fall into the trap of presenting an exhibition that widely misses the audience. The most difficult exhibition is therefore the one aimed at all visitors. Fortunately the work done on interpretation of natural parks and museums is applicable to botanic gardens as well. It is important to get the attention of the audience, e.g. by giving information that may be valuable to the audience. People also respond better to questions than they do to statements, e.g. "What do you do when you are trapped in the mist on Table Mountain?". People generally are more interested in things that move, make a noise or that require reaction from the visitor than in reading a long legend explaining an exhibition. Because we do not have an interpretative centre yet, apart from plant labels in the garden, we cannot even start with the interpretative services that are urgently needed.

More specialised visitors may be divided into pre-school children, schoolchildren and adults. Regarding pre-school children, a garden could be developed in which children could also be left under supervision while their parents visit the garden. This would fulfil a need and the children could form positive attitudes about plants.

For schoolchildren, worksheets should be used, corresponding to different projects at different levels of sophistication for different age groups. Summer school programmes, which could entail adventure trails on a full-day basis, would probably be popular. Different projects undertaken by schoolchildren on a competitive basis in association with nature conservation organisations should also be encouraged. We should also consider the possibility that we may move out of the garden as well, either by using video films on aspects important to children, or through practical demonstrations and/or lectures at different schools.

In the United States adult education has become an important factor in enriching the lives of people. In South Africa many people would also be interested in courses related to environmental aspects, horticulture or general biological matters. This is an important area in which the NBG could make a contribution, especially by making use of guest lecturers.

Education programmes should also be organised for special interests, e.g. handicapped people, youth clubs, senior citizens, special interest clubs. In the New York Botanic Garden, for example, a "Speakers Bureau" has been set up in which garden staff make their services available to organisations needing a guest lecturer.

Another approach is the telephone answering service, as is for example being used by Wisley at present. They handle approximately 20,000 telephone enquiries per annum on matters ranging from plant identification, pathology and entomology to general information.

Informal education

With botanic gardens becoming more and more important as tourist attractions, the opportunities for informal education increase. It is important to try to think ahead and to plan exhibits that would interest as diverse a visitor population as possible. There is such a tremendous number of aspects relating to plants or horticulture, that it should be possible to have something of interest for every visitor.

One difficulty that is going to face managers of botanic gardens is determining how effective a display is in teaching people, in catching their attention or, most difficult of all, in changing their attitudes. One has to be careful not to direct one's attention to the wrong audience, e.g. one's peers or financial backers.

It may be much more effective to have a series of smaller thematic gardens instead of having enormous gardens with plants based on a taxonomic grouping. The following thematic gardens are to be considered in the future planning of Kirstenbosch: arboretum, demonstration of use of indigenous plants in horticulture, threatened plants, economically important plants, home garden demonstration area, plant improvement demonstration, children's play area/kindergarten, do-it-yourself garden, waterbird garden, glasshouse for plant-insect interaction, parasitic plant demonstration, historical plant garden, Bible plant garden, orchid collection.

If the aim is to change the attitudes of a wide section of the public, education through recreation is going to become more important. The following facilities are planned in Kirstenbosch: electric transportation system for visitors, trails through forest areas, part of garden accessible in evening, indoor and outdoor art exhibition areas and an area for the performing arts in the garden.

Informal educational activities will also be promoted outside the garden by supporting regional wild flower shows, tours and by supporting conservation-orientated organisations. The success achieved by Flora '83 led to demands for more of these types of shows. Consequently the next national wild flower show will be held in 1988 and it will be called Kirstenbosch '75 to commemorate the 75th year of the founding of the National Botanic Gardens.

What do we need to make these dreams come true?

Facilities

In every garden we should have an interpretive centre and a lecture hall, otherwise we cannot really hope to realise our educational aims. The new facilities being planned for Kirstenbosch include a number of lecture halls and other requirements. In the meantime, discussions have been held with the Education Department of the Cape Provincial Administration, and the Nature Study School will probably soon be available for use in our educational programmes.

Staff

The Department of Environment Affairs have agreed that the staffing situation with regard to the education function should be re-examined. There is also the possibility that additional teachers could be seconded to botanic gardens. The management of the NBG has also been altered in response to these needs.

Finance

If we could get confirmation from the Treasury that donations towards our educational and scientific projects could be income tax deductible, we should be able to generate enough financial support from the private sector. If we can get the public to accept that they will have to pay for the educational benefits they obtain it may also cover a part of the costs.

What will we gain from an improved education programme?

Attaining the mission of the NBG

To effectively manage an organisation as complex as a botanic garden and to use its resources optimally, it is extremely important to determine the goals and objectives clearly.

The mission of the NBG is: "To promote knowledge and appreciation of southern African flora and to undertake the *ex situ* conservation of threatened plants."

The goals of the four functional areas are:

Horticulture

To establish collections of largely indigenous and threatened plants according to aesthetic, scientific and educational considerations in gardens accessible to the public.

To give technical cooperation and make information available on all aspects of horticulture.

Research

To identify, study and evaluate threatened species, to conserve them in a genetically stable state for possible re-establishment in the wild.

To investigate problems experienced with the cultivation and utilisation of indigenous plants.

Plant utilisation

To identify and develop indigenous plants with economic, horticultural or medicinal potential.

To make available and promote the use of indigenous plant material, in order to foster an appreciation of and reduce pressure on plants in the wild.

Education and information

To promote a greater appreciation of our indigenous flora by conveying knowledge of plants and the environment to as wide a public as possible.

Because all the functional areas are integrated in the organisation an improved education programme will not only promote the goal of education but also that of horticulture, research and plant utilisation and promote the attainment of the mission of the NBG.

Rôle in promoting nature conservation

If people knew of the importance of plants as sources and potential sources of food, medicine, energy and industrial chemicals, they might value them much more. If the importance of preserving genetic diversity and the serious threat of extinction to a large proportion of plants were common knowledge, the man in the street might become a strong supporter of conservation in general.

Botanic gardens can play a very important rôle in this regard. With the large numbers of visitors they attract, they can provide a major contribution to environmental education. Teaching environmental education to visitors to natural parks or conservation areas in the wild is somewhat like preaching to the converted. Botanic gardens in urban environments could reach the people without a conservation orientation. As has been shown with Flora '83 the action could also be outside botanic gardens. If conservation societies supported the educational aspects of botanic gardens to a larger extent, nature conservation in general could benefit enormously.

Social role in environment

In Brooklyn Botanic Garden education programmes have been used very effectively to interest culturally deprived children in plants. In the process, this has had an influence on the whole community. Botanic gardens in South Africa have an important rôle to play in this regard because plants may be used to build bridges between people of different cultural backgrounds. All the activities of the NBG, including the educational programmes, will be run on a non-racial basis. Because plants and nature conservation are an apolitical cause, the activities of the NBG may lead to closer cooperation between people of different political viewpoints or cultural backgrounds, but with similar interests in plants.

References

Anon. (1984). Working documents for a meeting on A National Policy on Environmental Education. Midmar Dam, March 1984: Education Committee, Council for the Environment.

Bailey, L.H. and Bailey, E.Z. (1977). "Hortus III: A Concise Dictionary of Gardening and General Horticulture". MacMillan, New York.

Eloff, J.N. (1984). Kirstenbosch, Quo Vadis? *Veld & Flora* **70**(3), i–iv.

Eloff, J.N. (1985a). "Botanic Gardens: Victorian Relic or 21st Century Challenge?" Inaugural lecture, March 13, 1985. University of Cape Town.

Eloff, J.N. (1985b). Conservation through Education — a prime function of the National Botanic Gardens. *Veld & Flora* **71**(4), i–vii.

Field, D.R. and Wagar, J.A. (1973). Visitor Groups and Interpretation in Parks and other Outdoor Leisure Settings. *J. Environmental Ed.* **5**(1), 6 pp.

IUCN and WWF (1984). "The IUCN/WWF Plants Conservation Programme 1984–95". WWF-U.K.

Lawrence, G.H.M. (1969). "Historical roles of the Botanic Garden. The Longwood Seminars", Vol 1. University of Delaware, Newark, U.S.A.

Warner, W.S. (1975). "A National Study of the Attitudes of Arboreta and Botanical Gardens' Management towards Recreation within their institutions". M.Sc. Thesis. University of Wyoming, U.S.A.

3

Research and Rescue

What the conservationist requires of *ex situ* collections

DAVID R. GIVEN

Botany Division, DSIR, Christchurch, New Zealand

Summary

Ex situ and *in situ* conservation are interdependent, but there has been a tendency for "pure" botany to separate from its more applied branches such as horticulture and forestry. This sometimes results in field and research botanists, and horticulturists, being unaware of, or inadequately satisfying, each other's needs. An essential part of *ex situ* conservation must be adequate documentation of accessions with respect to provenance and field data, and subsequent history of maintenance, propagation and distribution to other collections. Another requirement is a long-term commitment to maintenance of threatened species in cultivation with availability of material both for research and establishment in the wild. Problems can occur with disease and predators and the potential for their accidental introduction into the wild or into other *ex situ* collections. It is necessary to maintain an adequate representation of genetic diversity in *ex situ* collections and to minimise genetic erosion. There are limitations to the conservation rôle of gardens because of the need to satisfy a range of conflicting aims.

Resumen

¿ Que necesita el conservacionista de las colecciones ex situ?

Conservación *ex situ* e *in situ* son interdependientes, pero en botánica pura ha existido la tendencia para separar esto de sus ramas aplicadas

Botanic Gardens and the World Conservation Strategy

ISBN: 0-12-125462-3

como horticultura y silvicultura. Como resultado de ello, investigadores botánicos, botánicos de campo, y horticultores, no se han dado cuenta, o de forma inadecuada, de las necesidades de unos para con otros. Una parte esencial de la conservación *ex situ* deberia ser la adecuada documentación de acceso con respecto a datos de campo y "provenance", y subsiguientemente datos históricos sobre la mantención, propagación y distribución de otras colecciones. Otro requisito seriá la necesidad de un compromiso a largo plazo para la mantención de especies amenezadas en cultivo haciendo disponible el material para la investigación o el establecimiento en la naturaleza. Pueden surgir problemas con enfermedades y predadores y con el potencial para su introducción accidental en la naturaleza o en otras collecciones *ex situ.* Es necesario mantener una representación adecuada de la diversidad genética en colecciónes *ex situ* y minimizar la erosión genética. Exister limitaciones en la función conservacionista de jardines botánicos debido a la necesidad de satisfacer a una serie de objectivos en conflicto.

Introduction

"I want what I want when I want it" — Henry Blossom

These few words encapsulate the basic requirements of conservationists with respect to *ex situ* collections of threatened plants. However, to want is not necessarily to get, and in order to determine whether the conservationist can reasonably expect to obtain what he or she wants one must be more explicit about the requirements of conservation and the expectations of conservationists. Gardens concern themselves primarily with living plants. There are 5 main reasons for botanic gardens to grow plants for conservation purposes.

(1) To have as many threatened species in cultivation as practicable as an insurance against their loss in the wild.
(2) To cultivate critically threatened species in sufficient numbers so as to prevent significant genetic erosion.
(3) To have material available for research, and for assessment for economic use.
(4) To have collections of plants available for education programmes and for public displays.
(5) To propagate and maintain plants suitable for use in programmes to reintroduce species into the wild, or to reinforce wild populations.

The various rôles of botanic gardens have been spelt out on many occasions (e.g. Soepadmo 1979, Mohan Ram 1983, Simmons 1976,

Bruinsma 1976). It is unfortunate that inappropriate management, conflicts arising from competing aims of gardens, and the inability to maintain long-term commitments to conservation, have the potential to negate the conservation value of *ex situ* collections.

Impediments to achieving conservation aims

Inadequate documentation

An efficient documentation system is fundamental to a botanic garden conservation policy (Simmons 1976). Minimal documentation must include place of origin of plant material, collector and date of collection, and type of material collected (e.g. seeds, cuttings or whole plants). Too often locality information is imprecise, although Womersley (1981) has suggested that it should include "sufficient detail to enable another person to return, as accurately as possible, to the place from which the plant was obtained". A corollary is that such data should not be simplified in a recording system unless full, original details are retained in a supplementary file. It is of course essential that specimens be accurately identified and, where possible, vouchers deposited with a permanent herbarium. This is particularly important with groups such as carices and grasses, for which flowering or fruiting specimens are often required for accurate identification. Other data which greatly enhance the conservation value of *ex situ* specimens include size of the parent population (area covered and number of plants), whether material came from atypical individuals and whether the samples are all from one plant or from several. This information is particularly important for critically threatened species where total genetic variation may be almost confined to *ex situ* collections.

A record of cultural history greatly enhances the conservation value of captive plants. Even commonly cultivated plants may represent clones derived initially from single introductions to horticulture. Until recently the kiwi fruit (*Actinidia deliciosa*) industry in New Zealand was based on a single seed collection probably from one plant (R.L. Bieleski *in* Hughes *et al.* 1981). Records of propagation techniques are important in estimating the genetic variation of a captive species; on the one hand continued vegetative propagation may perpetuate a narrow genetic base, but on the other hand uncontrolled outcrossing may lead to introgression by related species growing nearby.

A further deficiency is that full records are not always maintained to show what has been lost from cultivation and why, or to where progeny has been distributed. A related problem arises where *ex situ* material is reidentified; new identifications should be passed on to those to whom the species concerned has been distributed but one suspects that, as for herbaria, this rarely occurs. Snogerup (1979) notes the allied problem of correct labelling, observing that "we have repeatedly delivered plants to an associated garden, only to see how sooner or later they disappeared or became mis-labelled during weeding or by exchange of labels". One instance in *Celmisia* where this may have occurred and the error perpetuated in successive gardens concerns *C. ramulosa*. In the British Isles two forms are generally grown: *C. ramulosa* "type" which has white cobwebby tomentum, and *C. ramulosa* "green form". In fact the green form is *C. ramulosa sens. strict.* and the other is probably of hybrid origin.

A New Zealand example of the frustrations of inadequate documentation concerns *Carmichaelia prona* (A.W. Purdie, pers. comm.). Presumed extinct in the wild for many years, plants were recently located in Edinburgh Botanic Garden, from which progeny have been distributed back to New Zealand gardens. The Edinburgh specimens differ slightly from herbarium collections of *C. prona*. However, no provenance data exist for the cultivated specimens. They cannot be correlated with known collections of the species from the wild; nor can garden-propagated material be planted back into the wild at a site from which the plant was originally taken.

Lack of genetic variability

Where species are repeatedly collected in the wild, there is a tendency to take propagating material from the same sites. In horticulture, the tendency is to collect and select the biggest, brightest and most vigorous plants, or those with peculiarities particularly suitable for garden culture. There is a distinct danger that such procedures will lead to survival *ex situ* of highly selective portions of the total genetic variation of a species. This can become a matter for concern when species become critically endangered or extinct in the wild. In view of some current predictions of large species losses in the future (e.g. Raven 1976, Lucas and Synge 1978, Ehrlich and Ehrlich 1981), it may become a serious problem in the next few years for a large number of plant species.

It is important that botanic gardens take steps to minimise genetic erosion. This should start at the collecting stage. Rarely do collectors indicate whether propagating material is taken from a single plant, a few adjacent individuals or from a selection of individuals throughout the population. Despite the recommendation for seed collection that, "50 capsules be taken from separate specimens over one square km" (J.G. Hawkes in Simmons 1979), most seed is probably taken from whatever is available as rapidly as possible. At the distribution stage, it is sometimes customary to distribute small numbers of seeds (or other propagating material) to a large number of institutions, which may also lead to reduced genetic variability within a particular botanic garden collection. This is analogous to critical reduction in a natural population to only a few individuals. Continued inbreeding at this low population size is likely to lead to progressive genetic erosion. Temporary genetic bottlenecks are unavoidable where only single individuals are known in the wild, for example, the New Zealand species *Tecomanthe speciosa, Pennantia baylisiana* and *Hebe breviracemosa*.

There is no reliable method which can predict with certainty the level of inbreeding that will be tolerated by a species (Frankel 1983). However, Frankel and Soulé (1981) suggest that an effective population size of 50 individuals is probably of minimal size whether *in situ* or *ex situ*, and they regard this as *the basic rule of conservation genetics*, because it serves as the basis for calculating the irreducible minimum population size consistent with short term preservation of fitness. Levin and Kerster (1974) in their review of gene flow in seed plants have concluded that the level of gene flow between individuals in most species is very low and that most breeding unit size calculations are gross over-estimates of those actually existing in natural populations. On the one hand small populations in the wild and in gardens should not be abandoned or ignored on the assumption that they cannot be maintained (Given 1983), but on the other hand every effort should be made to build up populations in each instance to a genetically acceptable size, perhaps (in the absence of a clear alternative) adopting Frankel and Soulé's rule.

High mutation rates can occur in meristem and single-cell cultures, and there is evidence that some long-lived perennial plants may be genetic mosaics with one part of the plant genetically distinct from another (Cherfas 1985). However, continued cloning will not markedly alleviate the problem of genetic bottlenecks in *ex situ* collections. Rapid increase using seed from controlled crosses probably provides the most efficient means of overcoming the immediate dangers of genetic erosion.

The maintenance of genetic variability in *ex situ* collections is important, but botanic gardens may find it hard to justify collections containing fewer species but greater numbers of individuals. A possible solution may lie in greater use of plantings for landscaping such as roadside plantations. It is important that "we are, when talking about conservation in gardens ... talking of conservation of representative gene pools for practical research and rehabilitation in natural ecosystems and not just as museum collections" (Bramwell in Taylor and Synge 1979). Seed banks offer means of maintaining relatively large representative collections of some species. Snogerup (1979) suggests that genetic erosion may be minimised by keeping the number of generations as low as possible by seed banking for long intervals, although this must take into account the need to store seed under conditions which will not reduce viability markedly and increase genetic damage (Frankel 1974).

Finally, skewed genetic representation may inadvertently result from differential survival of plants under cultivated conditions. Cranston and Valentine (1983) describe variations in the survival of individuals of species transplanted from Upper Teesdale to two *ex situ* sites. In the case of *Saxifraga aizoides* vigorous plants rapidly swamped those which were less strong, thus reducing the genetic variation of the sample.

Introduction of disease

Myosotidium hortensia is endemic to the Chatham Islands east of New Zealand. It is a giant forget-me-not well established in horticulture but is included in the New Zealand threatened plant list as "vulnerable". In 1976 plants growing at Botany Division, DSIR, developed leaf mottling and lost vigour. Subsequent examination and tests indicated that the plants were affected by cucumber mosaic virus (Thomson 1981).

This example serves to warn that plants taken from the wild can acquire diseases and pathogens under cultivation. These can be transferred from one garden to another, or back into the wild when plants from the garden sources are reintroduced into the wild. This question has been posed recently in New Zealand in relation to proposals to take garden-propagated young plants of *Tecomanthe speciosa* back to the Three Kings Islands to plant in proximity to the one known wild plant. The Three Kings Islands are not known to harbour *Phytophthora cinnamomi*, although this fungal pathogen is common on the New Zealand mainland. Its introduction to these islands could occur unless stringent precautions were taken to ensure fungal-free stock is used (R. Beever, pers. comm.).

Attempts to re-establish plant populations in the wild through replanting can turn out to be a good intention gone wrong if diseased material is used (Broembsen 1979). Stringent measures should be taken by botanic gardens to ensure that threatened plant stock in cultivation is disease-free, not only if intended for export to another country but even for widespread distribution within the country, and especially if any re-establishment in the wild from *ex situ* stock is contemplated.

Conflicting aims of botanic gardens

In New Zealand, Paterson (1983) has suggested that the potential rôle of botanic gardens as a complementary effort to other conservation strategies is immense. However, he also summarises the results of a survey of botanic garden and local authority "parks" departments aimed at identifying any conservation rôle being undertaken or planned. This indicated "a spectrum of awareness and effort ranging from a complete lack of knowledge of the subject, to those willing to play a rôle in the future, and to those with written policies and active programmes". New Zealand gardens have developed along different lines from those in many other countries. They generally lack a research emphasis and clear conservation rôle, and maintain a passive rôle as attractive places for the public to visit for picnics, recreation and relaxation. An education function is emerging in some gardens; for example Dunedin has established a visitor centre and employs a botanist, and Timaru has set up a small but effective conservation display area. Collections of threatened species are established at several gardens, but in most instances few plants of any one species are held. The gardens generally do not have any formal association with research or conservation agencies and are almost funded by municipal authorities. The country has no national botanic garden or formal commitment to foster development of a national network of coordinated regional gardens, although this has been the subject of several meetings and has been recommended by the Nature Conservation Council. It is not surprising that in New Zealand conservation aims frequently conflict with other perceived rôles of botanic gardens.

Dissatisfaction with gardens far distant from New Zealand is expressed by Shaw (1976) who observes that "botanic gardens in Europe come under the control of a number of authorities ... and botanic gardens perform many rôles — some conflicting with others. The rôles are so varied that the single term 'botanic garden' is confusing ... I am not so sure that all the activities I am referring to are rightly placed under the single heading 'botanic garden' ". Bruinsma (1976) considers that modern botanic garden organisation represents simply adjustments to traditional ideas.

For the purist, whether horticulturist, scientist or conservationist, the primary purpose of a botanic garden may *appear* obvious. To the casual user and to the politicians, rôles may not be so clear cut. In whimsical fashion Wilfred Blunt (1978) comments on the problem as it concerns the Royal Botanic Gardens at Kew, for "to the botanist, Kew is a scientific institution, but one whose work is sometimes hindered and whose funds are excessively squandered, to provide ignorant Londoners with a day in the fresh air among flowers whose names they neither known or greatly wish to know".

The problem of accommodating the aspirations of funding bodies and politicians, as well as the public, is well demonstrated by the example of the recently established botanic gardens at Xalapa in Mexico (Vovides 1979). Justification of botanic gardens in newly developing countries can be difficult because of socioeconomic and political factors, especially when there are urgent problems with housing, food production and provision of public services. Consequently these gardens have inverted the traditional priorities into education as the main priority, conservation through education as an immediate second priority, then scientific research. This includes tree propagation to assist reafforestation programmes, land use demonstrations and education programmes from kindergarten upwards. These aims elicit a sympathetic response from outsiders, particularly as the gardens are linked to an institute with the principal objectives of promoting rational land use and investigating food production.

Curiously, one area of potential conflict lies with horticultural interests, perhaps traditionally seen as allies in the use of *ex situ* collections for conservation. There is concern that the increasing emphasis on conservation of threatened wild plants will result in decreasing attention to cultivated plants. It has been suggested of botanic gardens that, "traditionally sedate and slow moving organisations, they are dumping cultivars at such a rate that it is impossible for the limited number of gardens capable for maintaining their collections to sort the wheat from the chaff before valuable cultivars disappear for ever" (Swindells 1981). A reasoned case for conservation of cultivated plants is set out by Brickell (1977) who asks the challenging question, "But what of our cultivated plants, in particular those used in horticulture — are they equally in danger?", in response to the emphasis being put on conservation of wild species. As he points out there are good reasons, aesthetic, economic and historical, for conserving cultivated plants alongside wild ones as part of an overall strategy of genetic conservation. Certainly, the conservation of cultivars and botanical species should not be seen as competing, mutually exclusive exercises.

It is relatively easy to plan and even start a comprehensive programme to save the plant species of the world. However, enthusiasm can become dampened as the magnitude of the task becomes apparent and as other agencies indicate their unwillingness to become involved. Funding may be reduced as initial expectations are not achieved. Priorities can change according to public demand. Individuals whose foresight and energy maintained momentum can leave. This means deliberately setting out to develop conservation programmes sufficiently modest to be sustained despite unexpected setbacks, yet large enough to be significant in their contribution. It also means assessing and satisfying the real needs of garden clients, while convincing the public and funding bodies that useful functions are being performed.

No one likes to see precious plants which have been obtained at great cost and effort, languish through lack of care, lack of commitment, or inability to sustain an *ex situ* conservation programme. This is a distinct possibility if long-term commitments cannot be met or momentum maintained.

Emphasis on non-conservation rôles of gardens may be necessary in the short term, at least, as there are increasing demands made for portions of a smaller financial cake in many countries. However, this is no reason to abandon conservation functions entirely, or to relegate them to a back corner of the garden where they can be conveniently ignored.

Relationships and interconnections

In the plant sciences there is a tendency, sometimes deliberate, to separate "pure" botany from its applied branches such as horticulture and forestry (Wagner 1972). The key issue of "interconnections" is identified by Wagner who suggests that "botany as such is no longer so solid an umbrella for the applied sciences related to it as it might be', but that 'nevertheless, botany and its applied aspects have many obvious interconnections and these should be fostered".

Attempts to make a garden carry out a wide range of functions may result in pointless duplication. For example if a well-organised independent herbarium and associated taxonomic unit is located within reasonable distance of the garden every effort should be made to ensure the two organisations complement each other so that there is eventual cooperation, not competition. In this way each can draw on the strengths of the other while countering its own inherent deficiencies.

In the narrowest sense, botanic gardens do those things best which can be done nowhere else, that is, growing and studying live plants (Wagner 1972). However, in the area of conservation, they must relate to a number of widely different institutions and interest groups. For the gardens to simply expand and engulf all these is to risk setting up a large bureaucracy with perpetuation of inefficiencies and communication blocks within the system. It also does not take account of the fact that these institutions and groups will, in turn, have their own particular sets of relationships with other agencies.

Although emphasis in this discussion has been on botanic gardens, these are not the only places where threatened species are cultivated *ex situ*; indeed, they may not always be the most appropriate sites. Sanctuaries and parks set up for conservation and public display of animals may also serve a valuable secondary role in plant conservation, for example, Palmitos Park in the Canary Islands (Bramwell and Bramwell 1984). The role of private gardens should not be overlooked, although this can involve problems in documentation and collection maintenance beyond a few decades. In the United Kingdom, the National Council for Conservation of Plants and Gardens (NCCPG) national scheme demonstrates that it is possible to take account of a wide range of garden types and a diversity of plant groups in building up a national network of *ex situ* collections (Stungo 1982). A central coordinating staff based at the Royal Horticultural Society Garden at Wisley is supported by a nationwide membership of volunteers, both amateur and professional, who belong to local groups affiliated to the NCCPG. By April 1985 approximately 250 national collections had been designated. In New Zealand, two Royal Society of New Zealand *ad hoc* committee reports recognise the desirability of a conservation network incorporating regional botanic and private gardens (Forde *et. al.* 1985), and designating assemblages of genetic material, including endangered taxa, as collections of national importance (Knox *et al.* 1985).

The significant role of *ex situ* collections outside botanic gardens is well illustrated by the cultivation of *Hydatella inconspicua*. This little-known threatened species is of interest as New Zealand's only member of the genus, especially as 2 of the remaining 3 species are 'presumed extinct' (Leigh *et al.* 1984). *Hydatella* and the small genus *Trithuria* are the sole members of the order Hydatellales. Plants of *H. inconspicua* have been maintained *ex situ* at the Dargaville High School laboratory allowing observation of growth rate and longevity (Pledge 1974). A flexible attitude attuned to the complete range of options for *ex situ* conservation is necessary if resources are to be used efficiently.

When species are disappearing weekly from the face of the earth is it crucial that botanic gardens and related groups concentrate on developing effective channels of communication, cooperative programmes making best use of limited resources, and mutual understanding and exploitation of each others weaknesses and strengths. G. Paterson (pers. comm.), Director of the Timaru Botanic Gardens in New Zealand, has suggested that a ''sister-city'' relationship with a research-scientific institute and garden of similar size and climate, could be helpful to gardens with limited resources. For the conservationist an important requirement is that like well-fitting pieces of a jigsaw puzzle, the work of botanic gardens and associated institutions and interest groups should lock together to form an integrated picture of conservation strategies, both *in situ* and *ex situ*.

Reciprocal requirements

Finally, it is important to note that obligations are not all in one direction. The botanic garden makes certain demands on the conservationist and botanist. One such obligation is the need to build up adequate collections of living plants. Research botanists sometimes seem to have a notorious reputation for being unable to provide *ex situ* collections with adequate living plant material. In writing of plant conservation in relation to horticulture, Melville (1970) relates that Joseph Hooker was greatly impressed by the profusion of orchids, and particularly *Vanda caerulea* as he toured the Khasia Hills of India in 1847. He quotes Hooker as recording that 'we collected seven men's loads of this superb plant for the Royal Botanic Gardens at Kew, but owing to unavoidable accidents and difficulties, few specimens reached England alive'. The comments of some botanic gardens staff and botanists working with living material suggests that this might be an accurate commentary on some field trips today. In part this is due to botanists in the field having to put priority on collections and data for their own research, to the limited amount of time available for processing collections, and to the lack of clear guidelines on what is required for *ex situ* collections. Sometimes quantity of collections as judged by collecting numbers becomes more important than quality of individual accessions. It has been suggested that codes of conduct for conservation of wild plants (e.g. Given 1981) should be extended to cover collection, documentation, and maximising of genetic variance when specimens are gathered for horticulture (G. Paterson, pers. comm.).

There may also be an unwillingness for botanic gardens and conservation organisations to adequately recompense botanists for their efforts, or to provide finance and personnel to accompany botanical field parties when the opportunity arises. In New Zealand there is urgent need to adequately represent the threatened and endemic flora of the Chatham Islands in *ex situ* collections. Although there has been opportunity for horticulturists and propagators to accompany scientists in the field, this offer has not been taken up and finance to allow this has not been forthcoming from botanic gardens and conservation organisations.

Nevertheless, botanists themselves could well study to some profit, guidelines such as those of Simmons (1979) regarding collection of living plant material from the wild. Having stated what they require from *ex situ* collections it will be ironic if conservationists and botanists are in part responsible for their requirements not being met, because of ignorance, carelessness and slipshod habits.

Conclusion

The World Conservation Strategy identifies 3 main objectives of living resource conservation.

(1) To maintain essential ecological processes and life support systems.

(2) To preserve genetic diversity.

(3) To ensure the sustainable utilisation of species and ecosystems.

Although it is generally agreed that, where feasible, *in situ* conservation is the most desirable way to achieve these objectives, the 1,400 or more botanic gardens of the world provide a substantial resource to reinforce and supplement *in situ* methods. In reality *ex situ* and *in situ* conservation are complementary and not clear cut alternatives.

However, efficient use of *ex situ* collections will require greater recognition of conservation as a proper objective by those administering and funding such collections. It will require wider use of accurate and comprehensive documentation systems, an appreciation of the biological limitations of *ex situ* collections and the need to adopt techniques to maintain and where possible enhance genetic variation, and the fostering of effective interfacing with allied institutions such as herbaria, research institutions and education establishments.

Although planning for *ex situ* conservation must be long-term, implementation is an immediate concern, if botanic gardens and related groups are to adequately play their part in preserving genetic diversity in the face of accelerating habitat depletion and species loss. Commitment to the principles and action plans of a botanic gardens conservation strategy will be a significant step towards the effective use of *ex situ* resources for conservation of the botanical diversity of our planet. It will cost money, time and resources but there can be little doubt that the effort will be of immense value to — and even necessary for — the survival of the world as we know it, and perhaps mankind.

References

Blunt, W. (1978), "In For a Penny". Hamish Hamilton, London.

Bramwell, D. and Bramwell, Z.I. (1984). "Palmitos Park". Palmitos Park, Canary Islands.

Brickell, C. (1977). Conserving cultivated plants. *The Garden* **102**, 197–201.

Broembsen, S.L. von (1979). Phytophthora cinnamomi — a threat to the Western Cape flora. *Veld & Flora,* June 1979, 53–55.

Bruinsma, B.F. (1976). Past: aims and objectives. *In* "Conservation of Threatened Plants" (J.B. Simmons *et al.*, eds), New York pp. 19–26.

Cherfas, J. (1985). When is a tree more than a tree. *New Scientist*, 20 June 1985; 42–45.

Cranston, D.M. and Valentine, D.H. (1983). Transplant experiments on rare plant species from Upper Teesdale. *Biological Conservation* **26**, 175–191.

Ehrlich, P. and Ehrlich, A. (1981). "Extinction. The Causes and Consequences of the Disappearance of Species". Random House, New York.

Forde, M.B., Burdon, R.D., Dawes, S.N., Dunbier, M.W., Given, D.R. and Park, G.N. (1985). Report of the *ad hoc* committee on conservation of plant genetic resources in New Zealand. *Proceedings Royal Society N.Z.* **113**, 117–133.

Frankel, O.H. (1974). Genetic conservation: our evolutionary responsibility. *Genetics* **78**, 53–65.

Frankel, O.H. (1983). Genetical principles of *in situ* preservation of plant resources. *In* "Conservation of Tropical Plant Resources" (S.K. Jain and K.L. Mehra, eds) pp. 56–65. Botanical Survey of India.

Frankel, O.H. and Soulé, M.E. (1981). "Conservation and Evolution". Cambridge University Press, Cambridge.

Given, D.R. (1981). "Rare and Endangered Plants of New Zealand". Reed.

Given, D.R. (1983). Conservation of island floras. *In* "Conservation of Tropical Plant Resources" (S.K. Jain and K.L. Mehra, eds) pp. 82–100. Botanical Survey of India.

Hughes, H., Harris, G., Collins, C., Garrett, K., Graham, A., Johnston, K., Simpson, P. and Trotman, I. (1981). "Integrating Conservation and Development. A proposal for a New Zealand conservation strategy". Nature Conservation Council.

Knox, G.A., Beu, A.G., Crosby, T.K., Forster, R.R., Given, D.R. and Yaldwyn, J.C. (1985). Report of the *ad hoc* committee on national collections. *Proceedings Royal Society N.Z.* **113**, 107–117.

Leigh, J., Boden, R. and Briggs, J. (1984). "Extinct and Endangered Plants of Australia". Macmillan, Sydney.

Levin, D.A. and Kerster, H.W. (1974). Gene flow in seed plants. *Evolutionary Biology* **7**, 139–220.

Lucas, G.L. and Synge, H. (1978). "The IUCN Plant Red Data Book", IUCN.

Melville, R. (1970). Plant conservation in relation to horticulture. *Journal Royal Horticultural Society* **95**, 473–480.

Mohan Ram, H.Y. (1983). Role of botanic gardens in the conservation of tropical plants. *In* "Conservation of Tropical Plant Resources" (S.K. Jain and K.L. Mehra, eds) pp. 154–158. Botanical Survey of India.

Paterson, G. (1983). The role of horticulture and botanic gardens. *In* "Conservation of Plant Species and Habitats" (D.R. Given, ed.). Nature Conservation Council.

Pledge, D.H. (1974). Some observations on *Hydatella inconspicua* (Cheesem.) Cheesem. (Centrolepidaceae). *N.Z. Journal Botany* **12**, 559–562.

Raven, P.H. (1976). Ethics and attitudes. *In* "Conservation of Threatened Plants" (J.B. Simmons, *et al.*, eds) pp. 39–47. Plenum Press, New York.

Shaw, R.L. (1976). Future integrated international policies. *In* "Conservation of Threatened Plants" (J.B. Simmons *et al.*, eds) pp. 39–47. Plenum Press, New York.

Simmons, J.B. (1976). Present: The resource potential of existing living plant collections. *In* "Conservation of Threatened Plants" (J.B. Simmons *et al.*, eds) pp. 27–38. Plenum Press, New York.

Simmons, J.B. (1979). The collection, establishment and distribution of natural source plant material. *In* "Survival or Extinction" (H. Synge and H. Townsend, eds) pp. 75–84. Royal Botanic Gardens, Kew.

Snogerup, S. (1979). Cultivation and continued holding of Aegean endemics in an artificial environment. *In* "Survival or Extinction" (H. Synge and H. Townsend, eds) pp. 85–90. Royal Botanic Gardens, Kew.

Soepadmo, E. (1979). The role of tropical botanic gardens in the conservation of threatened valuable plant genetic resources in South East Asia. *In* "Survival or Extinction" (H. Synge and H. Townsend, eds) pp. 63–74. Royal Botanic Gardens, Kew.

Stungo, R. (1982), National reference collections. *The Garden* **107**, 472–474.
Swindells, P. (1981). Who will conserve. *Gardeners Chronicle and Horticultural Trade Journal,* September 4, 1981, 34.
Taylor, N. and Synge, H. (1979). Report from the working parties. *In* "Survival or Extinction" (H. Synge and H. Townsend, eds). pp. 8–14. Royal Botanic Gardens, Kew.
Thomson, A.D. (1981). New plant disease record in New Zealand: cucumber mosaic virus in *Myosotidium hortensia* (Decne) Baill. *N.Z. Journal Agricultural Research* **24**, 401–402.
Vovides, A.P. (1979). Practical conservation problems of a new botanic garden. *In* "Survival or Extinction" (H. Synge and H. Townsend, eds) pp. 117–124. Royal Botanic Gardens, Kew.
Wagner, W.H., Jr. (1972). Botanical research at botanical gardens. *In* "A National Botanical Garden System for Canada" (P.F. Rice, ed.). Royal Botanical Gardens, Hamilton, Technical Bulletin 6.
Womersley, J.S. (1981), "Plant Collecting and Herbarium Development: a Manual". FAO.

Biological considerations in *in situ* vs *ex situ* plant conservation

PETER S. ASHTON

Arnold Arboretum, Harvard University, Massachusetts, U.S.A.

Summary

After a prolonged period of neglect by the scientific community, botanic gardens, particularly in the species-rich Mediterranean and tropical regions, face a new and perhaps their greatest challenge. Owing to the relative ease and low cost of introduction, cultivation, and propagation of plants, botanic gardens provide irreplaceable resources for furthering research into those aspects of endangered plant taxa relevant to the future conservation of these plants, and for educating decision-makers and the public on the often critical state of a nation's plant genetic resources. But *ex situ* cultivation is in no measure a substitute for *in situ* conservation. Even where taxa are on the point of extinction, or where they putatively represent genetically uniform biotypes, history indicates that living botanical collections by themselves serve as unreliable long-term alternatives to *in situ* conservation. In the majority of taxa, which are genetically variable, even when reduced to small numbers, the expense of maintaining adequate genetic representation, combined with the impossibility of simulating natural selection, must inevitably limit the value of gardens in species conservation for periods exceeding one generation. In short, though the possibility of *ex situ* conservation of species is remote, this realisation should enforce a clearer definition of purpose for *ex situ* conservation of plant material. With the methods now available, the *ex situ* conservation of particular heritable attributes, particular genotypes and even "provenances", is realisable. Meanwhile, *ex situ* culture in botanic gardens can provide the stimulus to increased research vital in the future management of *in situ* populations, and can simultaneously

Botanic Gardens and the
World Conservation Strategy

ISBN: 0-12-125462-3

heighten public concern over continuing extinctions, through publicity and education.

Resumen

Consideraciones biológicas in situ *contra la conservación* ex situ

Tras un prolongado periodo de descuido por la comunidad científica, los jardines botánicos, particularmente en las abundantes especies mediterráneas y regiones tropicales, afrontan un nuevo y quizás su más grande desafío. Debido a la relativa facilidad y poco costo de la introducción, cultivo y propagación de las plantas, los jardines botánicos proporcionan recursos irremplazables para ayudar a la investigación en aquellos aspectos de táxones en peligro relativos a su conservación. También proporcionan recursos para educar a políticos y al público concernientes al frecuente estado crítico de los recursos genéticos vegetales de las naciones.

Pero el cultivo *ex situ* no tiene capacidad para sustituir la conservación *in situ*. Aún donde los táxones están a punto de la extinción o donde se piensa que representan biotipos genéticamente uniformes, la historia indica que las colecciones botánicas vivas por sí mismas son alternativas inciertas para la conservación *in situ*. Para la mayoría de los táxones, que son genéticamente variables, aún cuando estén reducidos a un número pequeño, el costo del mantenimiento adecuado de una representación genética combinada con la imposibilidad de simular la selección natural ha de limitar inevitablemente de forma grave el valor de los jardines en la conservación de las plantas para períodos que exceden a una generación. Aunque la posibilidad para la conservación de las especies *ex situ* a corto plazo es remota, este proyecto debería presentar claros própositos para la conservación *ex situ* de material vegetal. Se podría realizar la conservación *ex situ* de algunos atributos heredables, como genotipos particulares o incluso "provenances" con los métodos actualmente disponibles. Al mismo tiempo, cultivos *ex situ* en jardines botánicos pueden estimular un aumento de la investigación en el control futuro de poblaciones *in situ*, y simultaneamente puede aumentar el interés público sobre continuas extinciones de especies, a través de educación y publicidad.

Introduction

Plants frequently exist in small populations in nature. Many rare endemics, such as those of mountain peaks and many in the Mediterranean dry sclerophyll scrublands, particularly of Cape Province and south-west Australia, and those of certain rain forests in the northern Andean foothills and Madagascar for instance, may always have existed thus. Over the relatively short time scale with which we realistically have to work as conservation managers, extremely small stands have been found to persist in nature. Thus Worthington recollected *Sonneratia apetala* from apparently the same and only group of 5–6 trees in Sri Lanka from which it had first been collected about a century earlier (MacNae and Fosberg 1981), and Kostermans did likewise from the only known stand of the endemic shrub *Stemonoporous moonii* (Kostermans 1981–2). *In situ* conservation must always be the highest priority, for biological diversity world-wide is infinitely too vast to be addressed by any other means. Higher plants, being sessile, are often highly site-specific. This promotes development of logical plans for demarcating minimal areas for *in situ* conservation, based on ecological knowledge and island biogeographic principles. On the whole, a few environmentally heterogeneous reserves, of moderate to large size to reduce edge effect and chance extinction, and ideally loosely connected by small stepping stones or corridors to allow gene-exchange to be maintained, are favoured (Diamond 1975). This must be our highest priority in the absence of even the most general information upon which to plan, including basic inventory let alone distributional and ecological data from many of the richest biota. This underlines the vital necessity of increased inventory as a prerequisite to any logical plan for conservation. In practice, of course, the luxury of regional planning seldom exists. The conservationist only succeeds in raising awareness when the plant is reduced to an endangered situation in one or a few isolated localities or, at best, is offered a patchwork of lands that the farmer and the planner have failed to find other use for. As development proceeds and natural habitats become increasingly fragmented extinction accelerates (Wilcox and Murphy 1985). The most endangered floras are those of the arable lands; the current distribution of preserves takes little account of this.

Even when the luxury of planning does exist and centres of species richness and endemism can be identified and conserved, many locally endemic plant species in floristically rich regions refuse to follow the rules and occur in isolated areas where, overall, conservation priorities are low. Even under ideal circumstances, though, decrease and fragmentation of natural areas is certain to lead, through island biogeographic effects, to substantial increases in extinction rates, though whether on the scale calculated for large animals (e.g. Schonewald-Cox 1983) is uncertain (Soule *et al.* 1979, Simberloff and Abele 1984). A case thus exists for some form of selective program of *ex situ* conservation.

Too little is known of the ecology of any, let alone rare and endangered, plant species to consider transfer of species from one natural community to another; but plants have many practical advantages over animals for *ex situ* conservation, in addition to their immobility, which are discussed below.

Plants are, generally, easy to propagate asexually and a variety of methods are available which include division of the rootstock, cuttings, and tissue culture. For conservation of heritable character traits, gene cloning is proving to be a realistic possibility, based on experience on a variety of crop plants.

Collection of propagants, be they vegetative or seed, can be accomplished with minimal disturbance to the wild population, and they are cheap and easy to transplant.

Management of *ex situ* populations, when compared with animals, is relatively simple and inexpensive. Plants do not require caging, and genotypes can often in practice be maintained for prolonged periods, though probably not permanently, through propagation or forced reinteration (e.g. see Rackham 1976 on the effect of pollarding on trees recorded over half a millenium). Plants need less constant care than animals: the majority of higher plants are bisexual which implies, broadly, that minimum population sizes for maintenance of heterosis can be half those of populations comprised of two sexes. Their modular construction allows of considerable phenotypic plasticity. Their habitat requirements can generally be reasonably accommodated *ex situ* provided competition is excluded. They do not manifest demanding behavioural traits and are rarely dangerous to humans.

Added to these attributes, plants are both attractive and unobtrusive; they are more often scented than smelly; and they are generally perceived by humanity as benign. In short, they are welcomed adornments to the human environment.

It is certainly true, then, that the rapidly increasing demand for *ex situ* conservation, occasioned by the inexorable destruction of natural habitats, faces botanic gardens with both a challenge and an opportunity, the likes of which has not arisen for over a century.

Methods of *ex situ* conservation

Flowering plants have the disadvantage over animals, from a conservationist's perspective, in the extraordinary diversity of their reproductive systems and notably their sexual differentiation. These are considered to be major determinants of genetic pattern within populations. The kind of reproductive system thereby influences minimal viable populations sizes for conservation (Wilcox and Murphy 1985) *in* and *ex situ*, and controlled pollination strategies for stock regeneration. In brief, self-compatible and apomictic species will vary more genetically between than within breeding populations (Allard 1960); but among outbreeders and especially dioecious species the reverse will hold, though differences in gene frequency between reproductively isolated populations will increase over time. In self-compatable species, representative samples of a wide range of breeding populations need to be sampled, but individual samples need be represented by comparatively few individuals. Facultative apomicts must be treated with particular care, for they are generally outbreeders and, if cross-pollination occurs at all frequently, will need large population subsamples as well as many sampled populations if within-population genetic variety is to be adequately represented. For outbreeders, individual populations should be well sampled, but fewer representatives of different populations will generally be necessary.

Methods of *ex situ* conservation now available can conveniently be classified according to what part of the plant is conserved — the whole organism, seed, or tissues or genetic material in culture.

Whole plants when kept *ex situ* in living collections have advantages in education and display and, in species taking time to reach reproductive maturity, mature specimens on hand are advantageous for research. They have the relative disadvantage of high maintenance costs, including high spatial requirements especially in the case of trees. Conversely, annuals would require to be subjected to frequent controlled pollination and re-establishment, unless methods of vegetative propagation are available. Whole plants conserved in gardens and plantations often readily hybridise with related taxa. Controlled pollination is therefore obligatory for regeneration from seed among outbreeders.

Whole plants, when grown in single species plantations, are more susceptible to communicable diseases than when scattered, as in nature, in a matrix of other species. Also, plants will as a rule prosper outside their natural range in the absence of co-evolved pathogens. This explains why crop plants, particularly long-lived tropical species such as *Hevea* rubber, have only flourished in plantations outside their region of origin. Clearly, this biological reality confronts political problems, which must be faced and overcome if *ex situ* conservation is to succeed and its subjects are to be exploited to human benefit. It also identifies a conflict between the needs of conservation *per se*, and display or education when demonstration of indigenous flora has high priority. Specimen plants though, in contrast to masses in plantations, will not incur this danger.

The preferred method of *ex situ* conservation, with present technology, is through storage as seed. The principal advantage of seed banks is their economy of space and the larger sample sizes that then become possible and, in countries with high labour costs, their low labour demands. The principal practical disadvantage is their reliance on dependable power supply, the need for meticulous monitoring of germinability over time, and the need for periodic regeneration under conditions which minimise selection among the residual seed stock. Added to this is the fact that research in seed storage has overwhelmingly been done with crop plants. Many of these are plants of early succession, a habitat in which many species in nature possess seed dormancy. It can be expected that species of closed mature phase plant communities, especially of moist equable climates such as the humid tropics where the majority of endangered species now occur, will more often and even predominantly possess seeds which lack dormancy and which will prove recalcitrant to induction of dormancy using current methods. It will not pass unnoticed that it is in these same habitats that long-lived species, often large at maturity, also predominate.

Tissue cultures, especially meristems, can maintain genotypes unaltered over long periods (Henshaw 1975, Wilkins and Dodds 1983). They also provide an economic means of suspending, at least temporarily, changes in gene frequency. Somatic mutations will continue to occur, particularly in cell and callus cultures, but are less frequent than in whole plants. They can often be culled if identifiable. Management of genetic variation is therefore a practical possibility. Here, then, it is the *in situ* populations alone that are likely to diverge genetically over evolutionary time, away from those *ex situ*.

Each taxon has its own requirements for successful establishment in tissue culture and regeneration. At present the basic mechanisms are poorly understood, so that successful techniques must be arrived at through tedious trial and error. However, experience is accumulating. It now seems possible that established techniques are only applicable to at most half the taxa of higher plants, though response can be predicted from phylogeny (Einset 1985).

Tissure culture, nonetheless, provides an invaluable alternative for conservation of multiple lines of taxa not easily conserved *ex situ* on this scale by other measures, such as rain forest trees with recalcitrant seeds.

Currently, then, all 3 methods require periodic regeneration, and sexual reproduction, of the stock. The latter presupposes knowledge of the breeding system and pattern of genetic variability of the species concerned.

Though cryogenic storage of seed may in future provide a solution for long-term preservation of natural patterns of genetic variation within population samples *in vitro*, DNA libraries are probably the most stable form in which genetic information can be stored (Weissinger, in press). Methods for cloning DNA for construction of gene libraries are described in detail by Maniatis *et al.* (1982).

Gene libraries provide a means to conserve DNA sequences, even whole genomes, for research. The genetic material thus conserved can be introduced to other extant genotypes, but whole individuals cannot be regenerated independently. Looking further ahead, Paabo's (1985) recovery of relatively large clonable DNA fragments from a 2,600-year-old desiccated mummy, suggests the prospect of the eventual reintroduction of valuable extinct traits into extant populations. In this context, herbaria may one day prove of enormous long-term value in conservation.

Though genomic libraries are subject to the same population genetic problems as other methods of *ex situ* conservation, they can at present more easily be stored indefinitely. This can be done by a variety of techniques available for reducing metabolic activity or, if desired, rendering them fully inert.

The genetic consequences of *ex situ* conservation

It is generally agreed that a majority of gene loci do not vary at population, or even species, level. Most variable alleles are sufficiently abundant to be adequately sampled, and conserved without danger of chance extinction through random drift, in artificial populations as small as 50 (Marshall and Brown 1975) to 100 (Frankel and Soulé 1981) randomly selected individuals. Specifically Marshall and Brown find that, if samples are selected at random, 50–100 individuals provide a 95% probability of including all alleles with a frequency of at least 5% in the population sampled.

However, the importance of rare variable alleles in the long-term survival of species is unknown. It can be expected that a few alleles rare in one population, or ecotype, will be common in another. Selection varies quantitatively and qualitatively within the life of a population. Some of these changes will be unidirectional, and are the stuff of evolution. Alleles will therefore vary in fitness over time scales exceeding the life cycle of the plant in conservation. Rare alleles at one moment will become abundant at another. At the least then, the sampling procedure should be repeated over a range of surviving populations if they exist.

Even more important, the character combinations by which we distinguish races and even some individual populations (Clausen and Hiesey 1958) may often represent differences in gene complexes rather than at individual loci. Such variation between populations in the representation of gene complexes can be very local (Antonovics and Bradshaw 1970, Hamrick 1983), though they can be remarkably constant within populations (Clay and Antonovics 1985, Morishima *et al.* 1984). We know little about the heritability of such complexes and nothing concerning their reconstruction. New information concerning the heritability, and particularly the transposition of polygenic elements, will probably alter estimates of optimal sample sizes (Weissinger, in press).

I know of no case where anecdotal evidence, suggesting that little random evolutionary change or inbreeding depression has occurred in captive populations, has been tested through reintroduction of siblings to the vicissitudes of natural selection in closed communities in the wild (see Senner 1980, Weissinger in press). However, we certainly need many more experimental reintroductions into nature before concluding that it is unfeasible.

Though for practical reasons, including current nomenclatural custom, we fall into the habit of giving priority to species conservation, it is hardly necessary to remind readers of the elusive nature of the biological species. The genetic mechanisms underlying the maintenance, increase and loss of biological diversity proceed at the scale of the breeding population. For conservation planning, the unit of the breeding population is therefore clearly more desirable on biological grounds.

For very rare and local endemic species this should not prove as serious a problem as among widespread and variable species for, *ipso facto*, they are generally confined to one or a few populations and restricted habitat ranges, in which outbreeding opportunities are already severely limited. Where widespread, variable species are to be conserved a strong case can be made for preserving infra-specific variation (Antonovics in press), particularly when eventual reintroduction into the wild is visualised. The restriction of useful heritable attributes to specific populations and races is well known among progenitors and land-races of crop-plants, and has been demonstrated in unique populations of wild taxa (Bradshaw 1984). Fortunately, at this scale of variation phenetic and genetic variation is at least partially correlated, even in plants (Langlet 1971), permitting relatively simple field sampling procedures. It is still important, wherever possible, to distinguish genetic from environmental effects by observation of variation among plants grown in standard garden conditions (Briggs and Walters 1984).

A more serious problem is that natural selection cannot be simulated *ex situ*. If the artificial population is established from seed, the progeny have been released from natural selection from the start. Whether *ex situ* conservation takes the form of plants, seed, or vegetative propagules including tissue culture, genetic rejuvenation through controlled cross-pollination of grown-on plants is periodically obligatory with present technology. New unselected gene combinations are inevitably introduced every time. However much care is taken and seed is collected from controlled pollinations, natural selection cannot be simulated, particularly if it is mediated by inter-specific competition. Some artificial selection is therefore unavoidable, being imposed through the methods adopted for pollination, germination and continuing storage *ex situ*. Random genetic drift, in particular, can be minimised, though, by growing replicate samples at several sites and periodically interbreeding.

On the other hand pooling of seed or cross pollination of samples among distinct populations is undesirable except where there is a need to increase heterosis, as in the case of samples from some populations on the verge of extinction. Such crossing leads to loss of unique desirable gene combinations, and the accumulation of new undesirable combinations. Both lead to reduction of the proportion of fit progeny. If reintroduction into nature is the aim, pooling of differentiated population samples is therefore less economical than maintenance of several separate samples as it will require disproportionately large samples to ensure an adequate supply of fit genotypes.

Ex situ conservation of small samples therefore can be expected to lead inevitably to unpredictable changes in gene frequencies and loss of gene combinations. Hybridisation between different population samples grown for more than one generation in isolation, and more particularly between samples from different populations, will increase the rate of these changes. The proportion of fit genotypes in progeny intended for reintroduction into the wild can thereby be greatly reduced, particularly if the original wild donor populations are already small and in decline, and hence already suffering from increasing drift, gene fixation and accumulation of deleterious gene combinations.

Methods of reconstructing gene complexes are yet to be developed; the use and identification of induced mutations to rebuild genetic variation requires large populations and extensive resources, and can therefore be considered impractical (Antonovics in press).

I conclude, then, that for all practical intents *ex situ* species conservation leads irreversably to domestication. But *ex situ* samples of breeding populations can effectively serve as means to conserve unique or important character combinations rather than species as such, as has been recognised by forest geneticists whose *ex situ* conservation programmes for "provenances" represent just this (Zobel 1978). Among crop plants, unique alleles and gene combinations, which may have as great importance in the survival of endangered species as they do to the protection of domestic crops, can presently only be recovered, if lost from a culture, by resorting to wild populations. *Ex situ* conservation should also serve the reverse function: reintroduction of desirable alleles and gene combinations which become lost in wild populations. In cases where extinction in the wild becomes reality, the means to control artificial selection is greatly reduced, and the means to reintroduce alleles lost in culture is eliminated. *Ex situ* conservation must, nevertheless, often be deemed preferable to loss (Raven 1976). But it is imperative for long-term conservation that managers of *ex situ* population samples define as carefully as information allows what characteristics they are intended to conserve: all alleles in the subsample, or only some; and, if the latter, which? Examples might be the diagnostic characters of a taxon, or utilitarian attributes. Once a clear set of priorities have been defined the role of what, in effect, will become a form of gene libraries will be considerable.

At this point it is worth recalling that the role of inter-specific competition in speciation has never been rigorously demonstrated in animal, let alone plant populations. In plant populations it remains unclear whether pair-wise natural selection is of common occurrence. Shugart and West (1981) indicated through mathematical modelling that the species composition of southern U.S. hardwood forests could arise purely as a consequence of the hazards of island biogeography — the balance of immigration and extinction over time. Hubbell and Foster (1983) have argued that the floristic composition of guilds within species-rich rain forest could arise purely as a consequence of the hazards of island biogeography. Unanswered by these authors, though, is why the high expense of sex and genetic variability is maintained if the environment remains constant over time, as it would according to their interpretation of community structure. Certainly their view, if correct, would make the task of *ex situ* conservation considerably easier. But the presence of some plants which act as "keystone mutualists" is indisputable. Oaks cannot be conserved without their rich ectotrophic mycorrhizal flora which differs quantitatively, and sometimes qualitatively, between species. Neither can the durian tree of the Far Eastern tropics be conserved without *Eonycteris spelaea*, the far-flying bat which pollinates it and acts as the fragile "mobile link" which unites breeding individuals over tens of kilometers. These plants and their symbionts will, in some cases, prove exceptionally difficult to conserve in perpetuity *ex situ*. Further, it is these keystone mutualists and mobile links whose minimum viable population areas are being used as a principal criterion for assessing the minimum area of *in situ* preserves (Wilcox and Murphy 1985).

Some practical issues

It takes an outbreeding species at least 200 generations to diversify to the point of speciation (Stebbins 1950); some outbreeding forest trees do not start flowering, even when in cultivation, for a half-century (Ng 1966). We need to think in evolutionary time scales if it is really our intention to conserve species as opposed to alleles or gene complexes *ex situ* (Frankel and Soulé 1981). It is questionable whether seed dormancy can be induced for comparable periods in many species. What is the likelihood of a major power cut (a war perhaps?) during such a period? The average annual cost of maintenance for each individual woody plant grown in the Arnold Arboretum is $64; population samples of 50 individuals would cost $3,200 annually. Could not most be conserved *in situ* more cheaply? (Obviously, these figures for herbaceous taxa may be more favourable for *ex situ* conservation.) How is the cost of *in situ* conservation to be estimated?

The record to date for *ex situ* culture does not give grounds for optimism. As Peter Raven (1981) stated so eloquently:

> Unfortunately, such collections are often dismantled or simply deteriorate after the specialists who built them up are no longer active at the respective institutions. Although they are often of very great value internationally, they may if they are not actively utilized come to be viewed as a drain upon the limited resources of the institution where they are housed. Even when financial considerations are not limiting, it is difficult to provide for such collections the meticulous and sustained care that is essential for their survival without the attention of a specialist who is deeply concerned for them.

Similarly, Peeters and Williams (1984) estimate that, of the 2,000,000 accessions of plant germplasm in seed gene banks world-wide, 65% lack even basic data on provenance; 80% lack data on useful characteristics, including methods of propagation, 95% lack any evaluation data such as responses to germinability tests; leaving only 1% with extensive data. It goes without saying that a substantial proportion of these accessions which have yet to be tested for germinability may be dead.

What rôle, then, for botanic gardens?

I am arguing here that permanent *ex situ* conservation is a refuge of last resort: a high-risk refuge of no escape. But I am not arguing against the need for *ex situ* conservation in general: the immediate rôle of *ex situ* culture of rare and endangered species in botanic gardens is rather in research and education and not conservation *per se*. Peeters and Williams (1984) have emphasised that the greatest use is made of gene banks by workers at the same site. In 1981, for example, 4,376 samples were distributed to outside users world-wide from the 57,027 seed accessions at the International Rice Research Institute gene bank, whereas 29,056 samples were requested by Institute staff (IRRI, 1983). I deduce from this that there will be a rapid increase in research on rare and endangered species once population samples are conserved *ex situ* in botanic gardens with active research programmes.

The basic data on methods of seed storage can be best accumulated where there is ready access to captive populations. Many other aspects of reproductive biology will be advanced most rapidly where there are botanic gardens with both captive populations, good laboratory and library facilities, and research staff on site. All aspects of research which can be advanced through comparative study of closely related species, generally of different provenances, are aided when plants are grown side by side and at hand. Even when, as in the case of demographic and population genetic research, the work must be carried out with populations *in situ*, much time and expense is saved if captive populations are available close by for use in the refinement of field and laboratory techniques prior to implementation (Ashton 1984). The standard procedures for diagnosing the genetic component in phenotypic variation within and between populations, through culture of subsample populations under uniform garden conditions, would be an essential adjunct to *ex situ* cultivation methodology, and one for which the botanic garden is best equipped. For all these purposes the long-term maintenance of genetic integrity of the captive population, and indeed the long-term sustainment of the captive populations, are lesser priorities. The aim of the research is always to improve the techniques for managing wild populations *in situ*. Again, *ex situ* is foremost ancillary to *in situ* conservation, not an alternative. Of course, the fact remains that plant species are going into extinction in the wild, and some can and should be conserved in botanic gardens. As emphasised earlier, purpose must then be strictly defined as a prerequisite to management planning. The result will likely be a programme for conservation of chosen heritable attributes, rather than of species or even demes.

Basic to the role of botanic gardens in conservation research is the development of meticulous, consistent, accessible and hence computer-based, record systems. Not only are details of provenance, and information on the ecology of the wild population, essential if this research is to be successfully applied, but plant material *ex situ* must be carefully inventoried together with comprehensive records of all treatments and the ensuing responses (Frankel and Soulé 1981). Results of previous research must be retained with the records of *ex situ* living accessions, together with methods adopted and, preferably, copies of unprocessed results for evaluation and reanalysis by later workers intending to build on them.

By the same argument, botanic gardens can be superb tools in conservation education, but this subject is covered by other contributors to this symposium.

Acknowledgements

This paper derived much inspiration from a symposium organised by Gordon Orians and Bill Kunin at the Institute for Environmental Studies, University of Washington, on economic and biological aspects of genetic resource conservation (proceedings in press). A paper there by Janis Antonovics, on genetically based measures of uniqueness, was particularly helpful. I have also had useful comments from Frank Thibodeau and his colleagues in the Center for Plant Conservation, Inc.

References

Allard, R.W. (1960). "Principles of Plant Breeding". John Wiley, New York.

Antonovics, J. (in press). Genetically based measures of uniqueness. *In* "Proceedings of the Lake Wilderness Conference on Genetic Resources" G. Orians, W. Kunin, G. Brown and J. Swierzbienski, eds). University of Washington.

Antonovics, J. and Bradshaw, A.D. (1970). Evolution in closely adjacent plant populations. VIII. Climal patterns at a mine boundary. *Heredity* **25**, 349–362.

Ashton, P.S. (1984). Botanic Gardens and Experimental Grounds. *In* "Current Concepts in Plant Taxonomy" (V.H. Heywood and D.M. Moore, eds) pp. 39–48. Academic Press, London and New York.

Bradshaw, A.D. (1984). The ecological significance of genetic variation between populations. *In* "Perspectives in Plant Population Ecology" (R. Dirzo and J. Sarukhan, eds) pp. 213–228. Sinauer Press, Sunderland, Mass.

Briggs, D. and Walters, S.M. (1984). "Plant Variation and Evolution", 2nd ed. Cambridge University Press, Cambridge.

Clausen, J. and Hiesey, W.M. (1958). "Experimental Studies on the Nature of Species, IV. Genetic Structure of Ecological Races". Carnegie Institution of Washington, Publ. 615. Washington, D.C.

Clay, K. and Antonovics, J. (1985). Quantitative variation of progeny from chasmogamous and cleistogamous flavors in the grass *Danthonia specata. Evolution* **39**, 335–368.

Diamond, J.M. (1975). The island dilemma: lessons of modern biogeographic studies for the design of natural reserves. *Biol. Conserv.* **7**, 129–146.

Einset, J.W. (1985). Chemicals that regulate plants. *Arnoldia* **45**, 28–34.

Frankel, O.H. and Soulé, M.E. (1981). "Conservation and Evolution". Cambridge University Press, Cambridge.

Hamrick, J.L. (1983). The distribution of genetic variation within and among plant populations. *In* "Genetics and Conservation: A Reference for Managing Wild Animal and Plant Populations" (C.M. Schonewald-Cox, S.M. Chambers, B. MacBryde and L. Thomas, eds) pp. 335–348. Benjamin/Cummings, Menlo Park, California.

Henshaw, G.G. (1975). Technical aspects of tissue culture storage for genetic conservation. *In* "Crop Genetic Resources for Today and Tomorrow" (O.H. Frankel and J.G. Hawkes, eds) pp. 349–357. International Biological Programme 2. Cambridge University Press, Cambridge.

Hubbell, S.P. and Foster, R.B. (1983). Diversity of canopy trees in a neotropical forest and implication for conservation. *In* "Tropical Rain Forest: Ecology and Management" (S.L. Sutton, T.C. Whitmore and A.C. Chadwick, eds) pp. 25–42. British Ecological Society Special Publication 2.

International Rice Research Institute (1983). "Annual Report for 1981". Los Baños, Philippines.

Kostermans, A.J.G.H. (1981–2). *Stemonoporus* Thw. (Dipterocarpaceae): a monograph. *Adansonia* **3**, 321–358 (1981); 2: *Bull. Mus. Nat. His. Paris* 4, **3**, 373–405 (1981).

Langlet, O. (1971). Two hundred years of genecology. *Taxon* **20**, 653–722.

MacNae, W. and Fosberg, F.R. (1981). Sonneratiaceae. *In* "A Revised Handbook to the Flora of Ceylon" III (M.D. Dassanayake, ed.) pp. 450–453. Amerind, New Delhi.

Maniatis, T., Fritsch, E.F. and Sambrook, J. (1982) "Molecular Cloning — A Laboratory Manual". Cold Spring Harbor Laboratory, New York.

Marshall, D.R. and Brown, A.H.D. (1975). Optimum sampling strategies in genetic conservation. *In* "Crop Genetic Resources for Today and Tomorrow". (O.H. Frankel and J.G. Hawkes, eds) pp. 53–80. International Biological Programme 2. Cambridge University Press, Cambridge.

Morishima, H., Sano, Y. and Oka, H.J. (1984). Differentiation of perennial and annual types due to habitat conditions in the wild rice *Oryza perennis. Plant Systematics and Evolution* **144**, 119–135.

Ng, F.S.P. (1966). Age at first flowering in dipterocarps. *Malay. Forester* **29**, 290–295.

Paabo, S. (1985) Molecular cloning of ancient Egyptian mummy DNA. *Nature* **314**, 644–645.

Peeters, J.P. and Williams, J.T. (1984). Towards better use of genebanks with special reference to information. *Plant Genetic Resources Newsletter (FAO)* **60**, 22–32.

Rackham, O. (1976). "Trees and Woodland in the British Landscape". Dent, London.

Raven, P.H. (1976). Ethics and attitudes. *In* "Conservation of Threatened Plants" (J.B. Simmons, R.I. Beyer, P.E. Brandham, G.Ll. Lucas and V.T.H. Parry, eds) pp. 155–179. Plenum, New York.

Raven, P.H. (1981). Research in botanical gardens. *Bot. Jahrb.* **102**, 53–72.

Schonewald-Cox, C.M. (1983). Conclusions: guidelines to management: a beginning attempt. *In* "Genetics and Conservation: A Reference for Managing Wild Animals and Plant Populations" (C.M. Schonewald-Cox, S.M. Chambers, B. MacBryde and W.L. Thomas, eds) pp. 414–445. Benjamin/Cummings, Menlo Park, California.

Senner, J.W. (1980). Inbreeding, depression and the survival of 300 populations. *In* "Conservation Biology: Evolutionary — Ecological Perspective" (M.E. Soulé and B.A. Wilcox, eds) pp. 209–224. Sinauer Associates, Sunderland, Mass.

Shugart, H.H., Jr and West, D.C. (1981). Long-term dynamics of forest ecosystems. *Amer. Sci.* **69**, 647–652.

Simberloff, G. and Abele, L.G. (1984) Conservation and obfuscation: subdivision of reserves. *Oikos* **42**, 399–401.

Soulé, M.E., Wilcox, B.A. and Holtby, C. (1979). Benign neglect: a model of faunal collapse in the game reserves of East Africa. *Biol. Conserv.* **15**, 259–272.

Stebbins, G.L. (1950). "Variation and Evolution in Plants". Colombia University Press, New York.

Weissinger, A.K. (in press). Technologies for germplasm preservation *ex situ. In* "Proceedings of the Lake Wilderness Conference on Genetic Resources" (G. Orians, W. Kunin, G. Brown and J. Swierzbiewski, eds). University of Washington.

Wilkins, C.P. and Dodds, J.H. (1983). The application of tissue culture techniques to plant genetic conservation. *Sci. Prog.* **68**, 259–284.

Wilcox, B.A. and Murphy, D.D. (1985) Conservation strategy: the effects of fragmentation on extinction. *Amer. Nat.* **125**, 879-887.

Zobel, B. (1978). Gene conservation — as viewed by a forest tree breeder. *Forest Ecology and Management* **1**, 339–344.

A strategy for seed banking in botanic gardens

J.G. HAWKES

University of Birmingham, U.K.

Summary

In this contribution the need for conserving a broad range of genetic diversity in botanic gardens is emphasised. For effective conservation a large range of infraspecific diversity is needed, and 1 or 2 living specimens, although useful for display and education, cannot be considered to represent a species that needs to be conserved. It is therefore essential to build up banks of living seeds that represent by their numbers and method of collection, as large a proportion as possible of the genetic diversity of the species in the field throughout its distribution range.

Since species diversity is partitioned within populations, between populations and along eco-geographical gradients, collecting strategy must take this into account. Adequate sampling methods for these various types of diversity have been devised by population geneticists on a sound theoretical basis, and these methods are described in detail. The number and size of samples are defined and a strong plea is made for the use of standard data sheets for data recording and computer processing.

Storage methods are described, in which *base* (long term) and *active* (short to medium term) methods of storage are defined. The differences between seeds of orthodox and recalcitrant storage types are also explained, dormancy type and dormancy breaking is considered and the differences between crop and wild species are dealt with. Details of storage techniques and equipment are outlined, as well as the maintenance of genetic integrity in collections and the need for and type of sequential germination tests. Regeneration is dealt with at

Botanic Gardens and the
World Conservation Strategy

ISBN: 0-12-125462-3

length and the differences in methods between inbreeding and outbreeding species are described so that the genetic integrity of the samples may be preserved.

General seed banking strategies include a flow-chart of gene bank activities, problems of large quantities and size of seed and the need for duplicate collections. Special storage problems with species possessing recalcitrant seeds and those which reproduce by vegetative organs are treated. *Ex situ* plantations, seedling banks and meristem or shoot tip banks are discussed, particularly relating to tropical forest and fruit trees. Cryopreservation is also briefly touched upon.

Resumen

Una estrategia para bancos de semillas en jardines botánicos

Las necesidades para conservar un rango amplio de la diversidad genética en los jardines botánicos es importante. Para una conservación eficaz es necesario un amplio rango de la diversidad intraespecífica, y uno o dos especimenes vivos aunque util para su exhibición y su uso en la educación, no deberían considerarse como representantes de especies para ser conservadas. Por esto debe ser necesario la creación de bancos de germoplasma vivo donde esté una amplia representación de la variabilidad de la diversidad genética a través de todo el rango de distribución de la población, ya que la diversidad intraespecifica es amplia dentro de las poblaciones, entre las poblaciones y a lo largo del gradiente ecogeográfico; la estrategia para la recolección debe tener en cuenta el factor anterior. El método de muestreo adecuado para estos tipos de diversidad ha sido ideado por genetistas de poblaciones con una base teórica sólida, y estos métodos se describen en detalle. El número y el tamaño de las muestras son definidas y se recomienda fuertemente el uso de fichas estandard para el registro de datos y para la informatización.

Se han descrito métodos de almacenaje, uno *base* (a largo plazo) y otro *activo* (corto y medio plazo), se explican también estos distintos tipos de almacenamiento. Las diferencias entre semillas ortodoxas y recalcitrantis también se explican así como los tipos y rupturas de la latencia entre especies silvestres y cultivadas. Se perfilan detalles sobre técnicas y equipos para almacenamiento. También el mantenimiento de la integridad genética en las colecciones y la necesidad de un tipo de prueta secuencial de germinación. La regeneración se trata con amplitud y se describen las diferencias de los métodos entre especies de autofecundación y cruz-fecundacíon, de esta manera puede preservarse la diversidad genética de las muestras.

Generalmente la estrategia de los bancos de semillas debe incluir un organigrama de actividades dentro del banco de genes, problemas de tamaño y grandes cantidades de semillas y la necesidad de duplicar las colecciones. Existen problemas especiales de almacenamiento con especies que poseen semillas recalcitrantes y su se propagar a través de órganos vegetativos. También se habla de las plantaciones *ex situ*, bancos de plántulas y conservación de brotes meristemáticos, particularmente relacionados con plantas tropicales y árboles frutales. La crioconservación también se toca brevemente de pasada.

Introduction

Amongst the varied activities of botanic gardens it has now become abundantly clear that conservation should be accorded a very high priority indeed. The problem, however, is to agree on the correct scientific methods for carrying this out.

It is already well recognised that a broad range of genetic diversity of threatened species should be preserved, both in their natural habitats (*in situ*), and in botanic gardens or some other suitable institutions (*ex situ*). In this chapter I shall try to set out the scientific basis for botanic garden conservation strategies.

In the past, botanic gardens have often contented themselves with growing one or only a few specimens of each species. This is useful for display or education but is valueless for conservation, unless these one or two specimens are all that remain of a nearly extinct species. The work of biosystematists and population geneticists during the past half century has made it abundantly clear that a great deal of genetic diversity exists *within* species, as well as between them. This within-species diversity is not only of a morphological nature but comprises genes which condition ecological adaptation, protein, iso-enzyme and other biochemical diversity, and much variation in resistance to pests and diseases.

Such a range of diversity cannot possibly be conserved in the form of living plants, since the cost of growing the plants and the space for accommodating them would be impossibly large. It is therefore essential to build up banks of living seeds that represent, by their number and methods of collection, as large a proportion as possible of the genetic diversity of the species in the field and throughout their distribution range. Such widely representative collections of each species are generally described as possessing a broad genetic base; and all collecting and storage strategies should keep this concept in mind.

Therefore, although this communication is concerned primarily with storage, it is essential to make sure that correct field sampling strategies are used, so that the stored material represents a satisfactorily wide range of genetic diversity.

Sampling strategies

In the past, botanical collectors have often limited their collections of seeds or living plants to a rather restricted range of samples, conforming to their ideas of what the species in question ought to look like. Concepts have now changed very much and our experiences in collecting crops and crop relatives have demonstrated the need for diversity rather than uniformity. We have realised that diversity within a species can be partitioned in 3 ways:

(1) Within-population diversity
(2) Between-population diversity
(3) Eco-geographical diversity

This latter may be of a smoothly changing clinal nature, or conditioned by sudden changes of altitude, soils, rainfall and other climatic variables. Collecting strategies must therefore bear in mind such concepts.

Strategies for sampling crop and crop relatives may equally well apply to wild species in general, and the slogan "maximum diversity for minimum sample size and number" is also equally applicable. Of course sampling should, in theory, differ between inbreeding and outbreeding species, and between amphimictic and apomictic ones. However, attempts to over-complicate the techniques of collecting may well result in no collecting being done at all. Thus, it is better to adopt a generalised sampling strategy that is suitable for nearly all situations and to leave the population geneticists to use more complex ones for their own special purposes.

The following sampling techniques, based on theoretically sound foundations, have been established, by workers in crop plant exploration, and by population geneticists (Jain 1975, Marshall and Brown 1975, Frankel and Hawkes 1975, Hawkes 1980).

A sampling site is first selected for an identifiable population of plants. Sometimes this population may be quite large, at other times restricted. A population is assumed to possess genetic continuity through exchange of genes by cross-pollination generation by generation. Even if the material is predominately inbreeding, gene exchange will almost certainly occur from time to time through occasional outcrossing. In tropical forests where individuals are very scattered it is often very difficult to identify a population as such. Knowledge of the pollination biology and similarity of phenotype will often help to decide this (see Allen 1984, referring to *Theobroma* (cacao) collecting strategies). In temperate and subtropical areas the population size may be as small as 5 × 5 m to as large as 50 × 50 m. Where there is some doubt several collections each from a smaller area may be better than one collection from a very large one.

The general strategy is to walk backwards and forwards through the population (as is done in ploughing a field), taking seed heads *at random* in all the transects until about 50 to 100 heads are collected. This consists generally of taking a seed head every so many paces, according to the size of the area that needs to be sampled. Ideally, about 2,500 to 5,000 seeds should be obtained (see Table 1) but this is not always possible. Even seeds from 25 plants are better than nothing, and where there is doubt it is better to collect smaller amounts from more sites than larger amounts from fewer sites.

Table 1
Suggested seed numbers per population sample for each collection (Hawkes 1980)

Population type	Plants	Seeds from each	Total seeds per sample
Fairly uniform	50	50	2,500
Highly variable	50	100	5,000

Where only 1 or 2 plants are to be found collections should be made from these and an appropriate note made in the field notebook. Whatever choice is made, all the seed heads from a site *should be put together as a single sample*. They should be collected *randomly*, or at least *non-selectively*, so that bias is not exerted for particular morphotypes. The more polymorphic the population the larger the sample should be. According to Marshall and Brown (loc. cit., p. 73), 50 seeds from 50 plants should be sufficient to enable one to collect all alleles at a particular locus with frequencies greater than 0·05 with a probability of 0·95. Thus a 2,500 seed sample is likely to capture most of the diversity present in a population.

Additional *non-random* or *selective* samples may be made if the collector sees any interesting variants present in small numbers, which were not included in the sample made by strictly random sampling. It is better, however, not to mix these with the random sample but to keep them distinct and give them another collection number.

The above method will satisfy our requirements for "within-population diversity". We now come to "between-population diversity", in what is often called the "target area". To do this it is best to take a group of about 5 population samples in a fairly restricted region, making certain if possible that each is taken from a slightly different microhabitat. Thus, the explorer should be aware of local variations within a local area of, perhaps, 1 × 1 km, and make about 5 population samples accordingly.

Finally, we come to the choice of sites over the total distribution area of the species. If it is a restricted endemic, found only in a few areas, then the choice is already made. A few population samples for each small area where it occurs will be sufficient. If the species is very restricted indeed, common sense will indicate that care must be taken not to destroy or impoverish the last remnants and so hasten their extinction. If, on the other hand, there still remains a fairly extensive distribution area, then even sampling throughout that area is probably best. The method should be based on biological good sense. The target areas can be widely spaced if environmental conditions seem to be changing very gradually over a wide area; they should be rather close together if soil, altitude, rainfall, etc.. conditions are changing rather quickly over a more restricted area (see Allard 1970 for a more detailed discussion).

If seeds cannot be found and circumstances permit, as in a largely vegetatively propagated species, one should collect propagules (e.g. bulbs, bulbils, rhizomes, cuttings, etc.) from 10–15 individuals (or more if time allows) as a bulk sample from an area of about 100 × 100 m or less. These are then brought back to the botanic garden, grown together in a screen house or glasshouse, and sib-mated with bulked pollen. The seeds so produced may then be put together as a single sample under one collection number. The rest follows as for seed sampling, as concerns inter-population and eco-geographical diversity. In sampling vegetatively propagated material it should be remembered that sometimes clonal propagation in the field is very extensive. To prevent taking more than one sample from a single clone the individual subsamples should be as widely spaced as possible.

After the seeds are collected it is important to dry them as soon as possible in paper bags at temperatures not higher than 15°C. It should be noted that seeds should not be dried under very hot conditions as these will result in a marked decrease of viability or even in complete destruction of the sample.

An extremely important part of collecting strategy is the taking of an adequate amount of collecting data on a standardised form, such that it will be easy to transfer this to a computerised data bank in due course.

The need has clearly arisen for a standard minimum data sheet that is not too complex, but is adequate enough for all practical purposes (Hawkes 1980, 1983 pp. 120, 121). This form is used for cultivated and wild samples, and might be further simplified for wild plants only (Figs 1 and 2). The sheets are bound up into books of 100 and provided with tear-out labels for adequate field handling. Although some workers advocate the use of a much more complex collecting form, the time taken to complete it accurately will diminish the time available for making collections. Thus complex forms are not recommended.

Storage methods

Storage of genetic resources samples requires firstly, that they should contain as large a proportion as possible of the diversity of the original populations from which they were sampled; secondly, that the genetic integrity of these samples is conserved, and by this we mean that they will as nearly as possible have the same genetic composition after long periods of storage as when they went into the seed bank; and thirdly, that the storage methods are such that the seeds can be conserved for many decades, perhaps up to 100 years or even more. What are the strategies and techniques that render these requirements possible?

In the first place, we must consider 2 types of seed storage collections, *base* and *active*. By a *base collection* we mean one that is to be stored under optimum conditions and not interfered with until the reduced seed viability indicated by periodic germination tests, requires a new seed generation to be grown out. This latter process is spoken of as "regeneration" or "rejuvenation". However, the word "rejuvenation" is to be avoided, as it does not describe this process accurately.

By an *active collection* we mean one that is stored under less than optimum conditions; subsamples can be taken out for experimentation, evaluation and display whenever required. If not enough seed remains the sample can be multiplied up by growing a new generation of plants and collecting the seeds. Multiplication is essentially the same process as regeneration, but less care need be taken to grow the seeds under optimum conditions. Even the regeneration process may cause some genetic changes, due to the very different selection pressures of the field or glasshouse conditions compared with those of the natural habitat of the original sample. Seed bank managers are advised to regenerate under isoclimatic conditions or even in the place whence the sample was originally gathered, but this is a counsel of perfection, rarely attainable in practice. Nevertheless if the seed storage conditions are really good, regeneration cycles may not be necessary more than every 20 or 30 years, or even longer in some cases.

To sum up this section, every sample needs to be divided in 2, one part to be designated as a base and the other as an active collection, with their respective treatments and uses.

We now come to the problem of *orthodox* and *recalcitrant* seeds (Roberts 1975). Orthodox seeds are those that will store well without loss of genetic integrity for longer periods when the temperature and humidity is reduced. All the major seed crops such as wheat, maize, rice, barley, etc. are of this type, as well as related wild species. The generally agreed storage conditions for orthodox seeds are reduction of seed moisture to about 4%, followed by a lowering of the temperature to about $-18°$ C. Many seeds will store

Expedition/Organization: ....

Country: ....

Team/Collector(s): .... *Collector's Number:* ....

Date of Collection: .... *Photo Number(s):* ....

Species Name: ....

Vernacular/Cultivar Name: ....

Locality: ....

....

....

Latitude: ....°..........′ *Longitude:* ....°..........′ *Altitude:* ....m

Material: Seeds Inflorescences Roots/tubers Live Plants Herbarium

Sample: Population Pure Line Individual Random Non-Random

Status: Cultivated Weed Wild

Source: Field Farm Store Market Shop Garden Wild vegetation

Original Source of Sample: ....

....

Frequency: Abundant Frequent Occasional Rare

Habitat: ....

....

Descriptive Notes: ....

....

....

....

Collector *Collector* *Collector* *Collector*

No: *No:* *No:* *No:*

Fig. 1 Standard collectors' form (front).

Uses: ....

....

Cultural Practices: Irrigated Dry

Season: *Approximate sowing dates:* ....

Approximate harvesting dates: ....

Soil Observations: *Texture:* ....

Stoniness: ....

Depth: ....

Drainage: ....

Colour: ....

Soil pH: ....

Land Form: *Aspect:* ....

Slope: ....

Topography: Swamp Flood Plain Level Undulating

Hilly Hilly Dissected Steeply Dissected

Mountainous Other (specify)

Plant Community: ....

....

Other Crops grown near or in rotation: ....

....

Pests/Pathogens: ....

Name and Address of Farmer: ....

....

Taxonomic Identification: ....

by .... *Date:* ....

Name of Institution: .... *Accession No.* ....

Fig. 2 Standard collectors' form (back).

under even lower temperatures, but −18° C, which is the average temperature of food freezers, is reasonable. Temperatures of −10° C or −5° C are also possible.

The other type of seed is termed *recalcitrant* because it will not store under reduced temperature and humidity. To this class belong various tropical timber and fruit crops and some temperate trees (*Quercus, Castanea*) as well as certain economic plants of the tropics such as rubber, cocoa and nutmeg. Studies on certain seeds that were once regarded as recalcitrant have shown that they are not of this type (Ellis, in Holden and Williams 1984). In the article by Ellis the latest research results are given as to which seeds are orthodox and which recalcitrant. Tropical grain legumes, *Citrus* spp., tropical grasses, and in general species with small seeds, seem to be of orthodox type. Hence botanic gardens in the tropics may not have such a problem as was at one time thought possible. Even so, the long-term conservation of tropical timber and fruit trees has not yet been solved. For crop seeds and related wild species 3 further publications by Hanson (1985) and Ellis *et al.* (1985a, 1985b) are relevant.

The establishment of a seed bank in a botanic garden does not involve many insurmountable technical difficulties, since cold storage and drying equipment is fairly easily obtainable and is not too expensive for the amounts of seed that most botanic gardens need to conserve. A short leaflet on deep freeze chests for medium- and long-term storage of small seed collections is available (Ellis and Roberts 1982) and larger, more comprehensive ones have also been published (Cromarty *et al.* 1982, IBPGR 1979, 1984b, Dickie *et al.* 1984). A special report on *Malus* and *Pyrus* seed storage by Ellis (1982) and procedures for long-term storage by Yndgaard (1983) are also of considerable interest. It must be stated here very strongly, however, that all these publications are concerned with the storage of seeds of crop plants and crop plant relatives, for which there is a considerable background knowledge of storage physiology.

For wild species not or only remotely related to crops our knowledge is not so plentiful. To begin with, we must realise that crop plants have been subjected to many thousands of years of selection for non-dormant seeds or those with easily-understandable dormancy mechanisms (lettuce, for instance, which needs a light shock and temperatures below 20° C before it will germinate). Wild species normally possess seeds that possess a mixture of dormancies so that some seeds remain viable in the soil whilst others which germinate more rapidly might be killed by adverse seasonal conditions. Dormancy in the seeds of many wild species is broken only by alternating temperatures, thus encouraging them to germinate when near the soil surface but not when much lower down where diurnal temperature fluctuations cannot penetrate. Again, other seeds may not be able to survive under low humidity (4%) and temperature (−20° C) recommended for most crop species.

Ideally, of course, lengthy research programmes would be needed to determine the best methods of storage for each species. In practice, however, now that botanic gardens need to conserve species that are in danger of immediate extinction, methods of conservation must be devised that are reasonably effective. When research results later become available the methods may be modified as required.

I would therefore like to make the following seed conservation proposals for which I am also much indebted to my colleague Dr Pauline Mumford.

(1) Look for domancy problems in the seeds. Thus, since a germination test is necessary on a small subsample before the rest go into storage, if, say, only 50% germinate, this may be due *either* to the fact that only 50% were viable, *or* to the fact that those which did not germinate were dormant. When judging the viability of seeds after storage (see section 10), this first germination figure should be taken into account. Dormancy breaking chemicals are also available, again see section 10.

(2) To determine which seeds are dead and which are dormant a tetrazolium test may be used. However, this does not work well with all species.

(3) Before placing seeds in storage they should be dried and cooled. Seeds of wild species should not be dried to a humidity level of lower than 5–6%, and better, only to 7–8%. Larger seeds, in particular, suffer physical damage if dried too quickly and to a level of below 8% humidity.

(4) Most, if not all, seeds should be stored to begin with at +4° C until more is known of their capacity to withstand lower temperatures. Oily seeds, in particular, may be more sensitive to subzero temperatures, and preliminary tests should be made. If, however, experimentation in the gene bank or results from the literature indicate that the seeds are not likely to be damaged at subzero temperatures then the seed temperatures can be reduced to about −18° C and the seeds can be stored in an ordinary food freezer, available commercially.

(5) Before germination the seed coat or the pericarp may be removed, but only in certain cases where, as in beet, germination inhibitors are to be found in the pericarp walls. Grains of grasses should be removed from spikelets where possible, but in all cases it is very important not to damage the seeds themselves.

(6) After a certain period of storage the seed viability will begin to diminish. If this is allowed to drop too far this will almost certainly result in

genetic changes and differential seed death, some genotypes dying more quickly than others. This results in loss of genetic integrity and is clearly to be avoided at all costs (Roberts 1975).

(7) Samples with unknown storage qualities should be given a germination test every 3 years, and thenceforward after longer storage periods. These tests should be carried out on base collections.

(8) A problem with seeds of wild species may be that they require special conditions for dormancy breaking and germination (see Papp and Szabo 1984). Various chemicals such as gibberellic acid, hydrogen peroxide and environmental treatments such as cold shocks and fluctuating temperatures should be tried. Germination may be better at the time of year normal for seedling growth (Mumford, pers. comm.), whilst different temperatures may be tried by using a thermogradient bar where small samples of the same accessions are tried out at different temperatures for maximum germination.

(9) Sequential testing should be carried out so as not to waste too much seed in germination testing (Table 2, from Ellis and Roberts 1984, p. 72).

Table 2
Plan for a sequential germination test where seeds are tested in groups of 40

No. seeds tested (cumulative)	Regenerate if no. of seeds germinated is equal to, or less than	Continue sequential test of accession if no. of seeds germinated between	Maintain in store if no. of seeds germinated is equal to, or greater than
40	29	30–40	
80	64	65–75	76
120	100	101–110	111
160	135	136–145	146
200	170	171–180	181
240	205	206–215	216
280	240	241–250	251
320	275	276–285	286
360	310	311–320	321
400	345	346–355	356
440	380	381–390	391
480	415	416–425	426
520	450	451–460	461
560	485	486–495	496
600	520	521–531	532

(10) Viability should not be allowed to fall below about 85–90% of what it was when first tested before being placed in store. Dormancy problems

raise many difficulties with wild species and much research is needed in this field. Nevertheless, the advice given here is of general value and can be refined later after more is understood about individual species (see also, discussions in Dickie *et al.* 1984).

(11) The seed storage receptacles may be: (i) metal cans (preferably aluminium, so that corrosion does not become a problem; (ii) glass tubes, phials or bottles with a metal (aluminium) screw top and a rubber or soft plastic seal incorporated in it; (iii) pyrex glass ampules heat-sealed, for very long term storage, and (iv) plastic-aluminium foil bags such as are now in common use as food containers or commercially produced vegetable and flower seed packets. These should be heat-sealed with a commercial-type sealer and preferably one that evacuates the air before sealing. It is essential that these foil bags be of high duty plastic foil that is not easily damaged (see Yndgaard 1983, Mumford and Freire 1984). From Mumford and Freire's work it is concluded that laminated aluminium foil bags and well-sealed metal cans are the best containers for the preservation of seed viability. It is also emphasised that seeds should be entered into storage with all possible speed after harvest. "Just a few months in poor storage conditions can seriously reduce the potential life of any seed sample", the authors conclude.

When seeds begin to lose their viability regeneration must be effected by growing out a number of seeds and harvesting a new seed lot to be placed back in the seed store. Many workers have emphasised the need for great care during regeneration, since it is at this stage more than any other, that loss of genetic integrity may be in danger of taking place.

We have mentioned earlier that regeneration under the identical conditions where the species sample was collected is ideal but rarely attainable in practice. However, roughly isoclimatic conditions are generally considered to be satisfactory. One important factor is to reduce as much as possible inter-plant competition so that less well-adapted genotypes will still have a chance of producing good seed of sufficient quantity, and will not be killed by competition with more aggressive genotypes.

To prevent inbreeding depression in cross-pollinated species and genetic drift it is necessary to grow out as many as 30 plants and if possible up to 50, which is the number of plants from which the original sample would have been taken. If the material is completely inbreeding it will be necessary only to take roughly similar amounts of seed from each plant and bulk these together into one sample again. Even inbreeders can undergo some very small percentages of outbreeding from time to time, and it is therefore wise to separate sample plots from each other and from those of other related species by plants with which there is no possible danger of their intercrossing.

With outbreeding species there may be many other problems, of which some are outlined by various authors in a recent symposium (Porceddu and Jenkins 1982). By and large, it is essential to isolate outbreeders in insect-proofed glasshouse compartments, in screenhouses or in gauze cages. When practicable, cross-pollination may be effected by the use of insect pollinators (bees, blow-flies, etc.) or by the collection of pollen from all plants, bulking this together and using it to hand pollinate them. Equal numbers of seeds should be collected from each plant and bulked for storage, as mentioned above.

Finally, both for inbreeding and outbreeding species, an adequate number of seeds should be banked for each sample. The approximate ideal number may be seen in the suggestions given in Table 3 (modified from Hawkes 1980).

Table 3
Suggested seed numbers per population sample for collection and storage (modified from Hawkes 1980)

Population type	For collection	For storage			
		Base collection*	Duplicate of base collection*	Active collection*	Total for storage
Fairly uniform (homogeneous)	2,500 (50 plants, 50 seeds from each)	3,000	1,000	3,000	7,000
Highly variable (heterogeneous)	5,000 (100 plants, 50 seeds from each)	7,000	2,000	3,000	12,000

*By *base collections* we mean those that are to be kept under ideal storage conditions for really long periods (up to 50 or 100 years or even perhaps more in some instances).

**Duplicate storage* of base collections in another institute in another country or even continent is needed as an insurance to prevent total loss.

*By *active collections* we mean those that are kept under less than ideal storage for medium to short-term (5–20 years) and are used for regeneration, multiplication and distribution, evaluation and documentation

Seed banking strategies

Clearly, the running of a seed bank cannot be regarded as a spare-time activity for an otherwise hard-pressed member of staff engaged in many other duties. A certain sequence of events must be followed, and this is shown in Fig. 3. This, although designed for crop seed banks, applies well for seed banks in botanic gardens, except that in most instances breeding and evaluation of material may not be so important as it is with crops. However, data gathering and data storage are both necessary, and will be considered separately in another paper in this symposium.

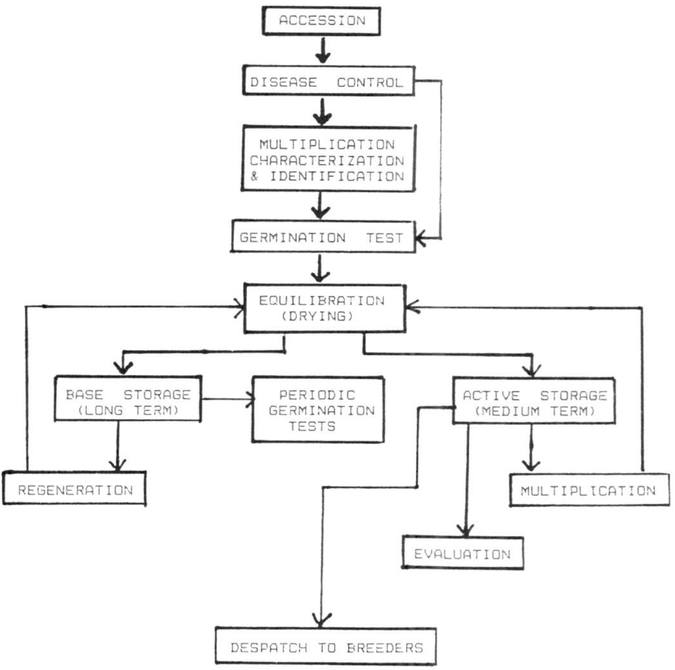

Fig. 3 Gene bank flow chart for seed collections.

It will also be seen from Table 3 that apart from base and active collections, duplicate storage is considered strongly advisable. If possible this should be under base collection conditions. The duplicate storage collection should, wherever possible, be situated in another country and ideally in another continent. Thus, a garden in a temperate region might send its duplicate collection to a tropical garden for storage and vice-versa.

In this way it would be hoped that total loss of a collection through natural or man-made disasters might be avoided.

We have been up to now thinking of the storage of small seeds. However, some species clearly produce large seeds, and if these are of the orthodox type, although storage is no problem they will take up much more space than can be afforded by a chest freezer or normal refrigerator. For large seeds or with large collections "walk-in" stores are necessary. However, although the initial expense is greater, if they are well insulated the running cost should not be too high (see Cromarty *et al.* 1982).

It may be valuable here to discuss the relation of seed banks to *in situ* conservation. For small-seeded species of orthodox type a seed bank is ideal *except* that it "freezes" evolutionary change. In other words, the materials in the bank are kept as nearly as possible in the same genetic form as when they were entered into the bank. They are not subjected to the year-by-year process of natural selection to which natural populations need to be constantly reacting. This may be particularly serious with crops and crop relatives, which should remain if possible to some extent in balance with the races or pathotypes of their pests and parasites.

The conservation strategy for wild species, however, may need to be two-fold, both *in situ* wherever possible, and *ex situ* in seed banks. This latter storage method will serve as an insurance against sudden or man-made disasters in the field, especially when the natural populations are very restricted indeed. An additional advantage of seed bank materials is that they are readily accessible to research workers and evaluators, whilst material in the field may not be immediately available, particularly if it was collected in far-away mountains or tropical forests, where a new collecting expedition would be necessary every time material was needed.

There are still, however, many problems connected with long-lived perennial trees and shrubs. If they provide orthodox seeds the problem is that although the seeds will store well, when seed regeneration is needed it may take anything from 5 to 20 or even up to 80 years for the seeds to grow into sexually mature seed-bearing trees (Hawkes 1982). For such materials *ex situ* plantations are probably the best solutions to the storage problem, though there is no reason why seed banks might not also be considered.

For recalcitrant seeded tropical timber and fruit trees *ex situ* plantations and meristem banks are the only solution to the storage problem, since seed storage is impossible. It must be realised also, that the amount of genetic diversity that can be conserved in meristem banks and *ex situ* plantations is only a fraction of the amount possible for seed banks because of the much greater space needed for the meristem banks and *ex situ* plantations in comparison with seed stores. Thus 1,000 seeds of *Malus* sp. (apple) which would have a seed volume of only 70 cm and a weight of 20–50 g (Ellis

1984) would need a plantation area of about 25 ha (assuming a planting distance of 10 × 10 m. This is only a single sample! Another suggestion for recalcitrant seeds is to store the material as seedling banks (Hawkes 1982), but this has not yet been realised in practice and needs considerable research initiatives.

Problems of vegetatively propagated species

Many species of plants possess alternative reproductive strategies — seeds and vegetative storage organs. It is very clear that seed storage is by far the best method of genetic conservation with both diversity/cost ratio and diversity/sample size ratio being very high. In all other types of storage the ratio is much lower, and this includes the diversity/sample size ratio for vegetative propagules, since each propagule represents only a single genotype. Nevertheless, it may be essential to conserve individual genotypes, particularly with material that possesses some economic or ornamental advantage.

One method of conservation is to maintain a collection of clones, that would normally need to be planted out in the field or glasshouse every year or two. Losses due to disease or pest infection or human error can be great, and costs are high due to the labour-intensive nature of such work. It is thus far better to conserve such materials in meristem banks as is done with many root and tuber crops such as potatoes, sweet potatoes, cassava and others (Henshaw 1975, IBPGR 1984a, Withers 1980). When seeds can be obtained and stored in the orthodox way large meristem collections are really not needed. However, their advantages are important since they can be kept disease-free, subculturing is easy and quick, they do not take up too much space, they lend themselves to genetic engineering research and they preserve exact genotypes that may be needed from time to time for a variety of purposes. For a whole range of materials it seems quite possible to submit the meristems to cryopreservation techniques under very low temperatures (liquid nitrogen), and this may reduce the amount of labour necessary to maintain them as well as the risk of genetic change. Thus the use of meristem or shoot tip banks, although perhaps not always necessary for every species, should always be borne in mind by directors of botanic gardens. It is, however, very easy for administrators and fund raisers to insist in their enthusiasm that meristem banks are the only "scientific" way of genetic conservation. Such a view must be resisted at all costs, and meristem banks should be considered only as possible alternatives to seed banks under the circumstances which we have discussed above.

Conclusions

To those unfamiliar with gene bank activities even this rather brief description given above may seem impossibly over-complex and beyond the scope of most botanic gardens. This is not really so. Provided that adequate funding is available and that a scientist with a good basic training in genetic resources methods and in seed technology is appointed, together with 2 or 3 assistants, much can be done. Data management on a computerised basis would clearly be dealt with by another officer, and the existing botanists would easily adapt to the field sampling techniques now required.

It is hoped that enough has been said to convince the participants at this meeting that the conservation of rare and threatened species is a problem that must be faced immediately. Clearly, once these species disappear they can never be replaced. Their conservation in seed banks is well worth the effort, especially if materials can at a later stage be reintroduced into newly created reserve areas. If that cannot be done, and if the fragile ecosystems which once supported these species can never again be created, botanic gardens will still have the satisfaction of having preserved priceless genetic resources for this and future generations, that otherwise would have disappeared for ever.

References

Allard, R.W. (1970). Population structure and sampling methods. *In* "Genetic Resources in Plants — Their Exploration and Conservation" (O.H. Frankel and E. Bennett, eds), Chapter 8. Blackwell, Oxford.

Allen, J.B. (1984). Strategies and methods for collecting *Theobroma. Plant Genetic Resources Newsletter* **57**, 8–14.

Cromarty, A.S., Ellis, R.H. and Roberts, E.H. (1982). "The Design of Seed Storage Facilities for Genetic Conservation". IBPGR, Rome.

Dickie, J.B., Linington, S. and Williams, J.T. (eds) (1984). "Seed Management Techniques for Gene Banks". IBPGR, Rome.

Ellis, R.H. (1982). Seed storage and germination of apple and pear. *Plant Genetic Resources Newsletter* **50**, 53–61.

Ellis, R.H. (1984). Revised table of seed storage characteristics. *Plant Genetic Resources Newsletter* **58**, 16–33.

Ellis, R.H. and Roberts, E.H. (1982). "Use of Deep-Freeze Chests for Medium and Long-Term Storage of Small Seed Collections". IBPGR, Rome.

Ellis, R.H. and Roberts, E.H. (1984). Procedures for monitoring the viability of accessions during storage. *In* "Crop Genetic Resources: Conservation and Evaluation" (J.H.W. Holden and J.T. Williams, eds). George Allen and Unwin, London.

Ellis, R.H., Hong, T.D. and Roberts, E.H. (1985a). "Handbook of Seed Technology for Gene Banks. Vol I. Principles and Methodology". IBPGR, Rome.

Ellis, R.H., Hong, T.D. and Roberts, E.H. (1985b). "Handbook of Seed

Technology for Gene Banks. Vol II. Compendium of Specific Germination Information and Test Recommendations". IBPGR, Rome.

Frankel, O.H. and Hawkes, J.G. (1975). "Crop Genetic Resources for Today and Tomorrow". Cambridge University Press, Cambridge.

Hanson, J. (1985). "Procedures for Handling Seeds in Gene Banks". IBPGR, Rome.

Hawkes, J.G. (1978). The taxonomists' role in the conservation of genetic diversity. *In* "Essays in Plant Taxonomy" (H.E. Street, ed.), Chapter 7. Academic Press, London and New York.

Hawkes, J.G. (1980). "Crop Genetic Resources Field Collection Manual". IBPGR and Eucarpia, Rome and Wageningen.

Hawkes, J.G. (1982). Genetic conservation of "Recalcitrant Species" — an overview. *In* "Crop Genetic Resources: the Conservation of Difficult Material" (L.A. Withers and J.T. Williams, eds). IUBS, Ser. B42 and IBPGR, Paris and Rome.

Hawkes, J.G. (1983). "The Diversity of Crop Plants". Harvard University Press, Cambridge, Mass.

Henshaw, G.G. (1975). Technical aspects of tissue culture storage for genetic conservation. *In* "Crop Genetic Resources for Today and Tomorrow" (O.H. Frankel and J.G. Hawkes, eds), Chapter 27. Cambridge University Press, Cambridge.

Holden, J.H.W. and Williams, J.T. (1984). "Crop Genetic Resources: Conservation and Evaluation". George Allen and Unwin, London.

IBPGR (1979). "Seed Technology for Gene Banks". IBPGR, Rome.

IBPGR (1984a). "The Potential for Using *In Vitro* Techniques for Germplasm Collection". IBPGR, Rome.

IBPGR (1984b). "IBPGR Advisory Committee on Seed Storage". IBPGR, Rome.

Jain, S.K. (1975). Population structure and the effects of breeding systems. *In* "Crop Genetic Resources for Today and Tomorrow" (O.H. Frankel and J.G. Hawkes, eds), Chapter 2. Cambridge University Press, Cambridge.

Marshall, D.R. and Brown, A.H.D. (1975). Optimum sampling strategies in genetic conservation. *In* "Crop Genetic Resources for Today and Tomorrow" (O.H. Frankel and J.G. Hawkes, eds), Chapter 4. Cambridge University Press, Cambridge.

Mumford, P.M. and Freire, M.S. (1984). Containers for seed storage. *In* "Seed Management Techniques for Gene Banks" (Dickie *et al.*, eds), pp. 249–267. IBPGR, Rome.

Papp, E. and Szabo, L.GY. (1984). Methods appropriate for breaking dormancy. *In* "Seed Management Techniques for Gene Banks" (J.B. Dickie *et al.* eds), pp. 219–248. IBPGR, Rome.

Porceddu, E. and Jenkins, G. (1982). "Seed Regeneration in Cross-pollinated Species". Balkema, Rotterdam.

Roberts, E.H. (1975). Problems of long-term storage of seeds and pollen for genetic resources conservation. *In* "Crop Genetic Resources for Today and Tomorrow" (O.H. Frankel and J.G. Hawkes, eds), Chapter 22. Cambridge University Press, Cambridge.

Simmons, J.B., Beyer, R.I., Brandham, P.E., Lucas, G.L. and Parry, V.T.H. (eds) (1976). "Conservation of Threatened Plants". Plenum Press, New York.

Withers, L.A. (1980). "Tissue Culture Storage for Genetic Conservation". IBPGR, Rome.

Yndgaard, F. (1983). A procedure for packing long-term storage seed. *Plant Genetic Resources Newsletter* **54**, 28–31.

A strategy for seed banking in botanic gardens: some policy considerations

CÉSAR GÓMEZ-CAMPO

Escuela Técnica Superior de Ingenieros Agrónomos, Universidad Politécnica, Madrid

Summary

Botanic gardens are the best institutions to set up wild plant seed banks with the aim of preserving rare or threatened plants, and of making available plant material for research and applied purposes. Details of how this can be done are outlined in this paper. They show that a seed bank need be neither expensive nor difficult to create, especially in the light of significant differences between wild plant seed banks and the larger and more elaborate crop plant seed banks coordinated by IBPGR. Seeds for such banks may be obtained from either the wild or from the garden, and the relative merits of each option are discussed. For wild plant conservation it is important to create a network of seed banks, as is beginning to emerge in Spain.

Resumen

Algunas consideraciones estratégicas para la creación de bancos de semillas en los jardines botánicos

Los jardines botánicos son las instituciones más adecuadas para la instalación de bancos de semillas de plantas espontàneas, con el doble fin de conservar especies raras o amenazadas y de hacer disponible dicho material para la investigación pura o aplicada. Se analizan distintas cuestiones relacionadas con estos objetivos. Un banco de semillas espontáneas no es caro ni difícil de instalar, pues ofrece importantes diferencias con los elaborados bancos de plantas cultivadas que coordina la CIRF/OAA. Las semillas de plantas silvestres

Botanic Gardens and the World Conservation Strategy

ISBN: 0-12-125462-3

pueden obtenerse de la Naturaleza o del mismo jardín, discutiéndose el valor relativo de cada una de estas opciones. En cualquier caso resultaría interesante llegar a la creación de una red de bancos de semillas de este tipo, tal y como ya ha sido iniciada en España.

Introduction

Storing seeds in conditions of low temperature and low humidity will normally result in an enormous extension of their life-span. Instead of keeping their ability to germinate for 5–25 years or so, most samples, if properly stored, should remain viable for hundreds of years.

This simple fact puts a powerful instrument in our hands to maintain large seed collections with relatively little effort. Thus it becomes much easier to improve the availability of living plant material for research or for other uses, and also to serve the conservation of species by projecting large amounts of genetic material into the future.

In the past 2 decades, a network of large seed banks devoted to cultivars of crop species has been steadily developed around the world. The International Board for Plant Genetic Resources (IBPGR) stimulates and coordinates this work. There are significant examples at Fort Collins (U.S.A.), Leningrad (U.S.S.R.), Braunschweig (West Germany), Gatersleben (East Germany), Alcalá de Henares (Spain), Bari (Italy), Izmir (Turkey) and Tsukuba-gun (Japan).

With much less financial support, the application of the same techniques to wild plants is only beginning. The efforts by our own laboratory in Madrid and by the Royal Botanic Gardens in Kew are almost 20 years old, but so far they have only been followed by a few institutions. The Botanic Gardens of Copenhagen (Olsen and Arnklit 1979), Las Palmas and Córdoba, for instance, have already taken significant steps to develop seed banks.

Without a doubt, botanic gardens are the appropriate institutions in which these activities can and should be developed (Thompson 1979). As they traditionally maintain seed collections of garden origin, they are capable of gradually transforming these collections into seed banks. This would directly result in saving much money, time and effort. Just by using the proper methods of seed conservation, the size of the collections can then be rapidly increased and their quality improved. The rôle that such collections (and botanic gardens in general) should play in conservation would then be much improved.

The aims of a seed bank

A seed bank has 2 major functions.

(1) To preserve rare or threatened plant material, thus helping to avoid possible extinctions of vulnerable or endangered species in the near future.

(2) To make plant material that would otherwise be difficult to obtain available for research or applied purposes, thus stimulating the knowledge and use of such material without damaging natural populations.

It should be stressed that protecting plant species should not only consist of keeping dry seeds in glass or plastic containers in a cold room. Conservation should preferably be done in the wild, through the protection of natural areas. By protecting the ecosystems in which the species occur, many other species will also be protected, and the ecological interactions and evolutionary mechanisms of all of them will be maintained. This is by far the cheapest and the most natural method of conservation.

However, it is also obvious that, in many countries, satisfactory protection of the areas containing endemic species may be many years in the future. Public and government awareness, money, time and effort are still needed in abundance. But, most of all, we often lack proper knowledge on how ecosystems work and how most plant species behave in their natural habitat. Many endemic species, for example, are not adapted to climatic situations but rather to ecosystems where a certain amount of exploitation — natural or artificial — is essential (Ruiz de la Torre 1985). A strict policy of non-intervention may result in a process of ecological succession and in the extinction of the species we wish to protect. On the other hand, most conservation areas (national parks, nature reserves, etc.) have usually been selected because of their landscape or fauna, but only seldom because of the presence of rare plant species.

Seed banks are therefore an important emergency method of conservation (Gómez-Campo 1979, 1985) that can help to avoid possible extinctions while other measures are fully implemented. As a reservoir of plant genetic material, they should become an important complement for protected areas, by providing seeds for the regeneration of vanishing populations or taxa.

Large amounts of plant material may also become readily available for basic or applied research. The inconveniences and limitations of the classic system of procurement and exchange of seeds through the "index semina" of botanic gardens have often been discussed and criticised (Böcher and Hjerling 1964, Heywood 1964, 1976, Gómez-Campo 1969) and botanic garden staffs are also aware of them. Most gardens, however, function on rather meagre budgets, and have a wide array of other objectives to fulfil, so that any suggestions that implies an additional activity or further expense will most likely be met with scepticism.

It should be emphasised that proper conservation of seeds is not especially expensive and that it does not require much extra work, and indeed leads to the saving of time. As well-preserved seeds can live for many years, periodic re-collection of the same taxa no longer becomes necessary — at least for extended periods. The effort saved can then be invested many times over in different activities.

To set up a seed bank is neither difficult nor expensive

Everything needed to store seeds under cold and dry conditions can be achieved through a wide array of alternative procedures with different costs, but there are cheap and reliable procedures at hand.

The erroneous idea that seed banks are always expensive installations probably comes from the lack of distinction between the crop seed banks and the seed banks for wild material. Crop seed banks deal with thousands of accessions of species whose seeds (cereals, legumes, etc.) are usually large in size. These seed banks consequently need much more space, and they also need a number of additional facilities to fulfil their objectives.

On the contrary, the seeds of wild species are usually smaller and the minimum samples needed are fewer, so that a domestic refrigerator might be enough to make a start. Drying can be done by chemical desiccants, a method that is impractical for crop seed banks. Rather than using expensive humidity-controlled chambers, it is recommended that previously desiccated seed samples are enclosed in sealed or hermetic containers. Depending on price and on the material to be stored, one can select screw-cap glass containers, glass flame-sealed capsules, metal cans or complex (aluminium/plastic) bags.

A tentative sequence of practical steps to convert a garden collection into a seed bank is given below, designed for a garden with limited funds in mind. As a first step, we might simply wish to extend the life span of all seed samples which are collected annually in the garden. This can be achieved by either low temperatures or low humidities. If we had to make a single choice, control of humidity would be the most practical. Putting the seeds in a hermetic glass or plastic chamber with a desiccant (calcium chloride or silica gel) is enough to lower seed humidity by several percent. (The desiccants can be reused many times after regeneration by heat.) Then, if subsequent storage is within complex bags, one of the sides may remain unsealed but tightly folded and fixed with a clip; this will maintain hermeticity, but allows the bag to be opened to extract samples for exchange. Though the bags can be bought, they can also be manufactured very easily and cheaply. The total cost of this method would be less than $1,500 including a roll of complex paper, a sealing tweezer, chemical desiccants and a hermetic plastic chamber. If screw-cap glass containers are preferred, the cost would not be much different. In both cases, we can expect that the life span of most of our seeds to be multiplied by a factor of 15–30 times.

As a second step we might wish to introduce low temperature storage. The price of a cold room might have deterred us at the beginning. However, a good household refrigerator can store a few hundred samples if maximum use is made of its volume; or if funds are available we could buy a larger cold chamber, of the type used to store perishable food in shops and bars. Low temperatures alone have a similar effect to desiccation, but really spectacular results are obtained when both factors are used together. To extend the life spans of seeds beyond several hundred years means that recollection or regeneration may in practice be forgotten, thus saving enormous amounts of work and time.

The time saved could then be partly invested in the special problems of those species whose seeds are recalcitrant (that is, cannot be preserved using the techniques mentioned above), or on vegetatively propagated material. It would also be possible to devote much needed attention to the proper identification and documentation of the plant material stored in the bank.

Seed collection becomes an essentially cumulative task, and our attention will be focussed on those taxa which are not yet stored. We will be free to search for the rarer and more valuable plant material, and we will realise that we have a powerful tool worth using for more ambitious purposes than just keeping the seeds collected in the garden. The importance of collecting in the wild must also be stressed.

Seeds from the garden vs seeds from the wild

When seeds are collected from plants grown in a garden, an undesirable selection occurs for those characters which are directly or indirectly related to cultivation. For instance, wild plants may exhibit seed dormancy mechanisms of several kinds, and these are gradually eliminated as we collect exclusively from those individuals grown from seeds which germinated readily in the previous season. Other characters linked, for example, to irrigation, soil composition, or to the ability for competition with other species, may be equally selected. The genetic basis present in the original sample becomes not only reduced but also modified from its former composition. In other words, we maintain the species but we do not maintain the genetic diversity.

With many plants of different kinds growing side by side, hybridisation can easily take place unless some effort is put into isolating related taxa. Isolation can be achieved either with insect-proof bags or cages, or by placing the plants sufficiently far from each other.

Identification errors in seed samples of garden origin are very frequent and they have often been criticised in the literature because they have caused many dubious results in botanical research. The errors are propagated by seed exchange, so that the same mistake is often found to occur in several seed catalogues. Although the origin of such mistakes may be diverse, the abundance of weedy species with the wrong names suggests that at least one part of them may be generated as follows: a seed sample which is dead or dormant is sown; no germination occurs but weeds grow in its place and one of these weeds, usually of the same family, is collected instead. In most cases, seed collection is routinely carried out by garden workers without adequate checking by a botanist.

Collecting exclusively in the garden means that quite often rare or endemic species growing within a few kilometres of the garden are neglected, as may be other representatives of the local flora in mountains a few tens of kilometres away. Thus, the user only has access to a limited number of species that he may also find in many other catalogues — the so-called "Botanic Garden flora" — but will find it extremely difficult to obtain seeds of valuable, rare or endemic material, which are of more interest for research and for conservation.

Collecting in the wild

Efficient collection from the wild will be the result of a balance between many factors. Planning itineraries in advance is essential, as so is a clear setting of the objectives; but too rigid schedules are not advisable since some plants searched for might not be found while other species might appear that are unexpected but interesting. Travelling by car, the number of kilometres per day should be carefully adjusted to the objectives, and balanced to leave enough time for active search and collection. Though obtaining as much data as possible on habitats and localities is important, we should not exaggerate the importance of this and risk spending our money on only a small number of seed samples, even if they are excellently documented. Planning and executing field collecting expeditions is an art which requires a special skill when time and available funds are scarce.

The general diversity of material that can be used and offered by the garden will steadily be increased by field collection, so that the garden will be in a position to assume substantial responsibilities in conservation. The enormous potential of collecting in the wild to improve the quality of a seed bank will become so obvious after the first collecting expeditions, that we might be tempted to devalue the rôle of the seeds of garden origin. However, it should be remembered that garden plants will always conserve their genetic content, even if this is somehow modified.

Living plant collections and seed banks

While a good collection of living plants can safely be derived from a seed bank, to derive a seed bank from a garden collection is subject to much inconvenience. However, it should not be interpreted that seeds of garden origin have no value at all by themselves. For "recalcitrant" species the normal method of periodic collection from growing plants should be followed — perhaps for many years — before other methods such as tissue culture or cryogeny are fully developed. Many exotics can only be provided from the garden in this way unless costly expeditions are carried out. Multiplication from tiny original samples, or special material where collection in the garden is the only option, are two additional cases to be considered. Apart from their fundamental value for display, public education, research, etc., living plant collections in the garden can also play an important rôle in the conservation and multiplication of long-lived perennials, vegetatively propagated plants, etc. They can at the same time be a valuable complement to a seed bank by serving to multiply small samples of rare species. In turn, the value of a garden collection may be greatly enhanced by the existence of an adjacent seed bank.

Thus, keeping the whole collection of seeds obtained from the garden at least in dry conditions is not only a desirable policy to start with, but a measure that will yield benefits in many specific cases. For a common weed which is abundantly collected in the wild, we need not make new collections of the same material for many years, unless we wish to sample it from different geographic origins to get some infra-specific variability. For less common plants which proved difficult to collect in the wild, multiplication may be necessary, and this can be done directly from the material displayed in the garden if it is conveniently isolated from related taxa.

For conservation purposes, samples from known wild localities will always be preferable to those of garden origin, so a gradual substitution of the latter should be the next logical step. But at this point a distinction should be made between the "base" collections, just to be stored for the future, and "active" collections that are used for either multiplication or distribution. Whatever the procedure used for active collections, be it complex bags or screw-cap containers, it is in the conservation-orientated base collections of rare, threatened or endangered taxa where we should extend our care to achieve truly long-term preservation. Here low humidity should always be combined with low temperatures, and the hermeticity of containers should be secured by any means. In this case, the use of flame-sealed glass capsules with some silica gel inside is strongly recommended.

Seed banks of wild material vs crop seed banks

Much of the experience for seed management and storage has been developed in seed banks for crop cultivars. In general, such banks are well financed, and their development has preceded that of the smaller banks for wild species. Many systems and methods from the former should be taken as valid for the operation of wild plant seed banks.

However, it would be unwise to design a seed bank of wild species just by copying the current procedures in a crop seed bank. In fact, many procedures could be done better if we take advantage of the smaller number and volume of samples to be stored. For instance, flame-sealed glass capsules are probably the most suitable container to be used, but they would be too expensive for a crop seed bank. Also the use of silica gel inside each capsule to ensure low humidity becomes prohibitive for large and numerous samples.

The differences between both types of bank are multiple and they start in the collecting phase. Collecting cultivars means visits to cultivated fields, village markets and Agricultural Stations. Collecting wild plants requires a completely different plan and itinery, usually travelling through mountains, steppes or humid zones. Only when wild relatives of crop species are collected may the objectives and procedures coincide. Proper characterisation of the material is very important for cultivars and land races. It represents an important part of the operations of a crop seed bank, and it needs much additional infrastructure (staff, fields, etc.). The equivalent operation in a seed bank of wild species coincides with the botanical research itself, and is usually undertaken by the staff of the botanic garden or by users of the material being distributed.

There are many aspects in common, so that the distinctions outlined above should not be taken too far. Crop seed banks may be used as reliable storage places for duplication of wild base collections, and they might even include a section for wild flora in general. Botanic gardens on their side might wish to specialise in some economic species, perhaps including some part of the variability of major crops. But we insist again that botanic gardens are the most suitable institutions to handle most conservation activities on wild plants, including the provision of seed banks of wild species.

Towards a network of seed banks

A further important aspect is that for rare and endemic plants, base collections need to be duplicated and sorted in different places so that the risk of destruction is minimised. This policy has been strongly recommended for crop cultivars, and it should also be applied for wild species. A high concentration of rare plant genetic material in a cold room is an enormous treasure, and it should not be put at risk from fire or accident. When IBPGR/FAO sponsors collecting expeditions of crop germplasm, it is expected that each base collection be divided into at least 3 parts and that these parts be stored in different banks; one of them should necessarily be within the country where the collection took place. Similar policies should be devised for wild material, but it seems logical that local seed banks (if there are several in a country) should play a major rôle in sharing base collections of endemic species.

A satisfactory solution in this respect might be close in Spain. Following the pioneer bank at Madrid, two more have been recently created, in Las Palmas and in Córdoba. (In addition there is a crop seed bank at Alcalá.) Already sharing of duplicates has started. The Organisation for the Taxonomic Investigation of the Mediterranean Area (OPTIMA) has repeatedly recommended the establishment of a network of similar seed banks throughout the endemic-rich Mediterranean region. The possibility of certain gardens specialising in particular plants is another attractive possibility for consideration. Many gardens already specialise in their local flora. If emphasis is placed on conservation-orientated collections, the subjects are almost unlimited — aromatic and medicinal plants, chasmophytes, hydrophytes, taxonomic groups, etc.

Seed exchange activities between botanic gardens wil be enriched as more gardens maintain conservation-orientated collections and seed banks. Seed viability, genetic purity, botanical identity and a wide range of available plant diversity will be ensured.

References

Böcher, T.W. and Hjerling, J.P. (1964). Utilisation of seeds from botanical gardens in biosystematic studies. *Taxon* **13**, 95–98.

Gómez-Campo, C. (1969). The availability of crucifer seeds from European botanic gardens. *Plant Introduction Newsl.* (FAO) **22**, 25–32.

Gómez-Campo, C. (1979). The role of seed banks in the conservation of Mediterranean flora. *Webbia* **34**, 101–107.

Gómez-Campo, C. (1985). Seed banks as an emergency conservation strategy. *In* "Plant Conservation in the Mediterranean area" (C. Gómez-Campo, ed.) pp. 237–247. Junk, Dordrecht.

Heywood, V.H. (1964). Some aspects of seed lists and taxonomy. *Taxon* **13**, 94–95.

Heywood, V.H. (1976). The role of seed lists in botanic gardens today. *In* "Conservation of Threatened Plants" (J.B. Simmons *et al.*, eds) pp. 225–231. Plenum Press, New York.

Olsen, O. and Arnklit, F. (1979). Setting up a practical small seed bank. *In* "Survival or Extinction" (H. Synge and H. Townsend, eds) pp. 185–188. Bentham-Moxon Trust, Royal Botanic Gardens, Kew.

Ruiz de la Torre, J. (1985). Conservation of plant species within their native ecosystems. *In* "Plant Conservation in the Mediterranean area" (C. Gómez-Campo, ed.) pp. 197–219. Junk, Dordrecht.

Thompson, P.A. (1979). Preservation of plant resources in gene banks within botanic gardens. *In* "Survival or Extinction" (H. Synge and H. Townsend, eds) pp. 179–184. Bentham-Moxon Trust, Royal Botanic Gardens, Kew.

Genetic conservation and the rôle of botanic gardens

J.T. WILLIAMS and J.L. CREECH

International Board for Plant Genetic Resources (IBPGR), Rome, Italy

Summary

For a decade there has been concerted international action to collect, safeguard for the future, evaluate and document the genetic resources of crop plants. This has been a result of the creation of the IBPGR, which is in many ways unique. This work has followed a number of phases, dictated by the world situation. First there was the need for widespread collection of cultivars. The second phase, just initiated, places emphasis on eco-geographic survey and expansion of the materials conserved to the genepools of wild relatives.

The work of the IBPGR is currently hampered by the slender knowledge base of the ecological preferences, patterns of variation of taxa, lack of good biological conservation methodologies, and the need, in a race against time, to target populations for sampling.

Botanic gardens — although historically of great importance in the introduction of exotic materials often linked to cash crop development — were part of the plant introduction systems out of which developed the genetic resources programmes in the past few decades. But they have been less than successful in conserving plant materials. With the rapid growth in the nature conservation movement and the economic recession limiting funds, many have realised they have changing rôles. Much of this has been sparked by dramatic appeals from botanic gardens in developed countries to help in conservation of the rapidly eroding tropical rain forest.

Current compartmentalised action needs to be coordinated, not from the viewpoint of assuming rôles and responsibilities which will only exacerbate the inactivities of many organisations over the past

Botanic Gardens and the
World Conservation Strategy

ISBN: 0-12-125462-3

couple of decades, but to produce fruitful liaison based on good science and not emotional or financial considerations.

To this end the authors outline a number of scientific imperatives which could lead to positive action and enable all organisations at this conference to assume their moral and evolutionary responsibilities. They may require radical changes in thinking (but after all evolution is based on genetic change in relation to environment). The manipulation of plants by man has led to a wealth of diversity for food, fibre, shelter and medicine. Now we have the chance to manipulate at the molecular level. We must stand ready with our material available, well documented and with clear priorities for future action.

Resumen

La conservación genética y las funciones de los jardines botánicos

Durante una década has existido una acción internacional concertada para recoger y salvaguardar para el futuro, que ha evaluado y documentado los recursos genéticos de las plantas cultivadas. Esto ha sido el resultado de la creación del IBPGR que es único en muchos aspectos. Este trabajo ha tenido diferentes fases dictadas por la situación del mundo. Primero hubo la necesidad de una extensa coleccion de cultivares. La segunda fase, ahora iniciada, pone más atención en el estudio ecogeográfico y la expansión de los materiales conservados de los "genepool" de los parientes silvestres.

El trabajo del IBPGR es corrientemente perjudicado por los escasos conocimientos básicos de la preferencias ecológicas, los modelos de variación de los taxa, falta de una buena metodología de conservación biológica y de la necesidad de una carrera contra el tiempo para identificar poblaciones para muestreo.

Los jardines botánicos — aunque históricamente de gran importancia en la introducción de materiales exóticos a menudo relacionados para hacer efectivo el desarrollo del cultivo — fueron parte del sistema de introducción de plantas desde fuera de lo cual se desarrollaron los programas de recursos genéticos en las últimas décadas. Pero han sido menos afortunados para la conservación de materiales de las plantas. Con el rápido crecimiento del movimiento de conservación de la Naturaleza y el retroceso económico limitando las disponibilidades de fondos, muchos han comprendido que ha cambiado su función. Mucho se ha clamado por las dramáticas llamadas de los jardines botánicos en paises desarrollados para ayudar a la conservación de los bósques tropicales, que están desapareciendo rápidamente.

La acción corrientemente compartida necesita ser coordinada, no desde el punto de vista de los papeles asumidos y las responsabilidades que sólo aumentarian la inactividad de muchas organizaciones a lo largo de la dos décadas pasadas, sino para producir relaciones fructíferas basadas en la buena ciencia y no en consideraciones emocionales o económicas.

A este fin los autores subrayan un número de imperativos cientificos que podrían desarrollar una acción positiva y hacer posibles todas las organizaciones en este simposio para asumir sus responsabilidades morales. Puenden exigir cambios radicales en el pensamiento (pues después de todo la evolución está basada en los cambios genéticos en relación con el medio). La manipulación de plantas por el hombre ha llevado a una gran diversidad para comidas, fibras, abrigo, medicina. Ahora tenemos la posibilidad de manipular a nivel molecular. Debemos estar preprados con nuestro material aprovechable, bien documentados y con claras prioridades para una futura acción.

In recent years we have seen genetic conservation of plants receive attention from a wide audience. Unfortunately much discussion has been on such a low scientific level that if we are not careful concepts will continue to be blurred, terminology continue to be used loosely and bandwagons will continue to roll with little effect on the grave situation at the habitat level.

Genetic conservation is aimed at the preservation of genetic diversity, and this is viewed scientifically for 2 groups of plants — firstly, the domesticates, and secondly, those in natural or semi-natural habitats and ecosystems. The domesticates are very clearly separate from wild plants because their evolution and survival depends on man-made environments (Frankel and Soulé 1981). From a biological point of view, man requires materials of the domesticates for evolutionary manipulation in a short time-frame: genetic conservation of wild species should be such that natural adaptive processes continue and at a pace not determined by man. However, the distinctions are not always clear: many useful plants such as pasture and forestry species belong in the second category and hence the application of scientific principles derived from population genetics, and the various components of fitness in time and space, can apply to both categories. Some species bridge the categories, e.g. the weedy species created by man as side products of the domestication process. We stress the need for good scientific work as guiding principles in conservation biology, because there has not been too much evidence of this in relation to the conservation movement's use of the term genetic resources in recent years. Elsewhere we have highlighted the problems which arise when a conservation topic becomes fashionable (Ingram and Williams 1984).

The maintenance of continuing adaptive evolution is the key to conservation of species in the field and this requires sufficient intraspecific genetic diversity. For the domesticates, which are mostly conserved *ex situ*, this necessitates the collection and conservation of many samples covering the range of variability of the species. Since many domesticates were derived through the interactions of several species, the conservation of sufficiently wide genepools is necessary.

A key to some recent confusion lies firstly in the word exploitation, and secondly in the alarming figures on present and projected species extinctions particularly in biomes such as the tropical rain forest. This has led to people considering value figures in economic terms — rather than practical programmes — and the fostering of political movements which have pushed science aside. The World Conservation Strategy (IUCN 1980) was promulgated to assist conservation world-wide yet it too, due to a somewhat simplistic format, has also fostered some of the misconceptions.

When the botanical community met at Kew in 1975 to discuss aspects of the role of botanic gardens in conservation (Simmons *et al.* 1976), the International Board for Plant Genetic Resources (IBPGR) had just been created. The opportunity for collaboration between IBPGR and botanic gardens simply did not exist because IBPGR was still an unknown quantity. IBPGR was created as an institute of the Consultative Group on International Agricultural Research (CGIAR) with a mandate to establish a global network of crop genetic resources centres. Its emphasis was to be on economic species, and with a limited budget it had to restrict its activities to the major food crops. During the first few years it mobilised the scientific community to provide data on collection and conservation priorities, and it mounted a major campaign to collect diversity world-wide so that breeders had the materials readily to hand. However, various evolutionary steps were taken by IBPGR, because by 1980 it was clear that genetic erosion — historically noted in the major cereals and food legumes — was also threatening vegetables, fruits and forages, and so these groups of species were taken on stage by stage. By 1981 there were over 100 species of high priority for action and many more species of lower priority, and by 1984 priorities for many genera had been assigned for forages, especially in the tropics and subtropics, arid zones and the Mediterranean. By 1983–4 it was also evident that large-scale generalised collecting was no longer appropriate, but that much more emphasis had to be placed on the associated wild species of the crop genepools and on the eco-geography of the genepools as well as subsequent work on patterns of variation (Williams 1984). Even the concepts of variation are changing rapidly; population genetics and molecular biology now remind us that genomes are in a state of flux.

Now that a decade of work has established the IBPGR as the leading international institution for the genetic conservation of crop genepools, it is in a stronger position to discuss collaboration. Through its programmes of collection, conservation, evaluation, documentation and training of nationals in the techniques necessary, IBPGR has established its credibility and, at the same time, a global network of gene banks. It currently collaborates with 106 countries, some 50 of which have established new genetic resources programmes where there were none, and of these, national committees have been created in about 25 countries. In addition, there has been a general increase by many governments in funding of genetic resources programmes. The bulk of the international funding is provided by CGIAR to IBPGR, which in 1986 will have a budget of \$4·7 million, and to the sister international agricultural research centres with which IBPGR has close working relationships (Hawkes 1985).

During the first 10 years, IBPGR underwrote over 500 plant collecting teams/expeditions utilising the expertise of scientists from many countries, established a network of gene banks which holds materials for security, and initiated research where necessary (IBPGR 1985a). The next 10 years will see a much greater emphasis on research, in many areas of which the botanic garden community can collaborate.

In as much as IBPGR is now recognised as a significant partner in the world wide effort to conserve biological resources, and has been particularly successful in defining the parameters within which we operate, it is possible that IBPGR could serve as a kind of rôle model for other similar programmes. This was recognised by the U.S. Strategy Conference on Biological Diversity in 1981 (US-AID *et al.* 1982). It was for this reason that we were invited to this meeting with the view that there might be areas of mutual interest that could be the basis of collaboration. It is an appropriate time for such discussions to occur, because IBPGR is now refining its strategy and developing programmes for the next few years. There are important shifts in our priorities, not least in the area of collecting, which require input from the academic and botanic garden communities.

The major emphasis to be placed on crop progenitors and on primitive races that occur in the remoter areas of crop cultivation will help to fill the gaps in the existing germplasm collections. Priority will still be given to those crops which are basically the most important in feeding mankind. The rationale assumes that the range of diversity available to breeders must be expanded in order to undertake crop improvement designed to enhance production in areas of stress, where the highly refined varieties are less effective and where new varieties must be geared to subsistence agriculture or to meet preference factors. At the same time, it is essential to fill the resource gaps because of the unpredictability of crop pests. Thus, IBPGR will enter a phase where the phyto-geographic basis of plant collecting will become more important than it has been in the past when collecting was mainly in the traditional centres of diversity and crop production.

It is obvious that the specific expertise which such collecting will require lies partly in the botanical institutions (although they may not always be synonymous with botanic gardens). However, many botanic gardens, particularly those associated with institutions of higher learning, will have a core of systematists and allied disciplines and also of herbaria and library facilities that make collaboration with IBPGR of mutual interest. Many botanic gardens already conduct collecting, albeit often with different objectives to the IBPGR: but the strategies are the same. Where the appropriate disciplines exist, scientists in botanic gardens might be called upon to provide background information on specific collecting localities through reviews of literature and herbarium material. Because of their special knowledge of endangered species, they could alert IBPGR to wild relatives of crops under threat. Their direct participation in the actual field operations, where plant breeders and similar crop specialists are at a disadvantage, could enhance our work. One of us (JLC) has long advocated that the ideal collecting team must include a systematist to voucher specimens and record the associated plant community. Hence a mechanism needs to be devised so that both data and specialists are readily available.

This work is not new: the IBPGR has for several years been involved in wild species collecting for *Arachis, Brassica, Pennisetum, Beta, Phaseolus, Triticum/Aegilops, Oryza, Glycine* and others, and has surveyed herbaria label data for *Mangifera*. It is to start a major new programme on the wild wheat grasses of the *Triticeae* because of their value in breeding wheat, rye, barley and forages (IBPGR 1985b). On the latter, priorities for collecting have already been agreed for over 50 species in 12 genera (see Appendix 2). For the enhanced work on wild species, in 1986 the IBPGR will start to establish a specialised computer database to include all distributional data from herbaria and publications, ecological data and data on which portion of genepools are conserved in which germplasm collections, and all keys for identification. However, this is a huge task and in the first instance priorities will be decided. Nonetheless the input from the botanical community will be necessary.

In relation to our field work, another aspect of mutual development will be where an institution or consortium of institutions undertakes work in an area of interest to IBPGR, either in relation to a specific crop genepool or to a geographic locality. The IBPGR has developed a reputation for effective networking and is willing to take a lead here. Often the IBPGR is unaware of ongoing efforts of other institutions and yet there might have been an opportunity to join collecting efforts or for IBPGR to mobilise funding. We hear of numerous collecting programmes of botanic gardens and arboreta where IBPGR would have been directly interested in considering support or where the existing knowledge and experience of IBPGR would have been useful in planning and executing the work. Many botanical institutions are deeply involved in programmes to help prevent the demise of tropical rain forests and other tropical environments that contain species of priority to the IBPGR. The findings of such scientists could very well direct attention to these areas by IBPGR. How then can we all be aware of each other's activities? If some kind of scheme could be developed to provide IBPGR with an overview of proposed useful longer-term activities — rather than the all too often short visits to areas which do little to gather representative variability — an assessment could quickly be given on the genetic resources conservation needs and all would benefit. IBPGR would remain sympathetic to proposals for support; although the IBPGR is not primarily a grant-awarding body, areas of priority for urgent work could be identified and jointly executed. Such joint proposals would help to strengthen the position of botanic gardens with their parent organisations (see Appendix 1 for areas of IBPGR research where collaboration could be effective). We must not lose sight of botanic garden tradition of "quiet, effective, scientific cooperation" (Walters 1983).

How far have botanic gardens and arboreta become involved in genetic conservation? In the past some such institutes have been famous in plant exchange and transfer of germplasm of economic species from one part of the world to another and whole industries or even economies have resulted.

Even if a few mother plants have been maintained over long periods, this hardly constitutes genetic conservation because they have been maintained largely as specimen trees or plants. More recently botanic gardens have become involved with rare species conservation through static conservation (Frankel and Soulé 1981), but this is only viable where numerous population samples of seed have been stored under suitable long-term conditions. For plants which do not produce seeds that can be stored under dry cold conditions, they are reproduced vegetatively but on the whole the number of samples in the gardens are so small that little genetic conservation results. And we must remember that botanic gardens have, over the years, not developed the best record of maintaining special purpose collections. It is hoped that this will be rectified as we all become more conservation-minded.

A great deal has been written about wild species of potential value, too much based on a range of information from tribal lore and intuition to screening programmes of industry. A great deal has been written about so-called social plants associated with primitive cultures of man and a strong case has been made for conservation activities based on vague possibilities. We have not seen any adequate, scientifically-based plans developed for these species let alone assessments of diversity in order to target viable genetic conservation. Two things are possible here. First, conservation of such wild species can occur through ecosystem conservation. Second, if there truly appears to be a potential, domestication studies could be initiated promptly by botanic gardens, for instance in the case of medicinals which are over-collected; and initiated by agronomists and foresters in the case of pasture and forest species. The rescue of a limited range of genotypes in botanic gardens will only be a useful conservation adjunct if reserves are also properly designed and properly monitored.

The World Conservation Strategy highlighted the weaknesses in conservation agencies and the need for more information. We have not advanced greatly in relation to devising practicable strategies for such minor species. Possibly the strategy sidetracked the reader too far into law and policies, and did not force the reader into basic and mission-oriented science. Here botanic gardens could well develop action. Scientifically we could even go further; it is within the bounds of current knowledge to engineer genetically some of the species, not simply for higher chemical production or whatever is needed by man, but to modify the patterns of variability, and perhaps change the genome by gene addition to stop a species well on the way to extinction from that end. We have seen no discussion of this potential in the literature and it might be far more meaningful than examples of cloning botanic garden material and reintroducing regenerants into areas where the species have become extinct. At the very least, the introduction into the genome of some hybridity from closely related taxa could also be a practical method. Action in this area will of course depend on the understanding of rarity (Harper 1981).

Movement of materials from one country to another links closely botanic gardens. Even with seed crops the IBPGR has a contract with the Royal Botanic Gardens, Kew, for its Seed Laboratory to receive samples from collecting missions, dry them, test viability and send them properly packaged to gene banks. We also recognise the important role of botanic gardens which offer facilities as a passthrough where countries cannot receive vegetative materials direct, as in the case of cacao, coffee, and rubber to name a few. The use of extra-tropical facilities, such as at Kew and the Miami Subtropical Field Station to serve as a third country propagation and clearing-house, has greatly enhanced the prospects of transporting valuable germplasm but restricting the movement of serious pests and diseases. It has greatly strengthened our ability to meet stringent quarantine regulations because movements of these important crops simply would not occur otherwise. While this appears to be a kind of service work without direct research application at first sight, it is essential and can provide meaningful research opportunities in the field of propagation, tissue culture for germplasm movement and similar studies for which botanic gardens with their special screen and glasshouse facilities are particularly appropriate.

When we examine the complementary proposals for conservation, namely *in situ* and *ex situ*, there are considerable complexities in implementing the former especially when it comes to supply of materials from reserves for breeders and more so when the materials are vegetatively propagated. In most cases such materials would have to undergo quarantine as required for that specific crop or propagation in a third country under glasshouse conditions. The IBPGR is developing methods for acquisition *in vitro* (IBPGR 1984) and the technique is already available for cocoa. If however a botanic garden was associated with specific *in situ* genetic resources reserves and had local facilities for propagation, under proper quarantine conditions, it would greatly speed up the transfer of the material to users. The same is true for material with short-lived seeds. Such linkages merit careful consideration as do the planning and running of the reserves (IBPGR 1985c). After all, if crop genetic resources are going to be used as a reason to justify *in situ* reserves, the resident curator will almost certainly be needed as well as the infrastructural back-up that we propose.

Whereas the IBPGR will not become involved very much with the practical planning of reserve areas, this being within the mandates of other organisations, it will always be ready to provide advice in the planning process. Reserve planning based on crop genetic resources considerations will flounder without both *in situ* and *ex situ* conservation being seen as part of a continuous process.

To illustrate the present shortcomings of the conservation network, the IBPGR was asked recently to look at the distribution of wild relatives of *Prunus, Allium, Malus* and some grasses in Europe, and to see how far the distribution patterns were covered by existing reserves. On the whole there are significant numbers of reserves over most of the distributional areas. What we do not have are clear inventories of species, let alone information on patterns of variation. If this is the situation in Europe, and it is estimated it will take years to get the data, how can work be planned in other parts of the world?

A small number of special purpose reserves could be set up based on existing information of threat to crop relatives; this would be more meaningful than blanket use of genetic resources of crop relatives to justify somewhat unscientific designation of reserves, or to justify actions taken in the past. Associated with those could be such parallels as the programmes conducted in recent years for the collection, quarantine, propagation and distribution of coffee germplasm using the facilities of the Coffee Rust Research centre in Portugal, the facilities of the Glenn Dale and Miami Plant Introduction Stations of USDA and various experimental stations in Latin America.

While this type of reserve with a back-up network of centres has not been planned for crops, the elaboration and implementation of such schemes calling on various disciplines (from the population geneticist and systematic botanist through to the botanic gardens and their users) makes an exciting prospect. At the same time careful field work with particular crop goals could have incidental effects on minor and unexploited species. It could even overcome such problems as a country having to conserve *in situ* something in which it has no economic interest. In another scheme it could be the user and all parties who would be partners in conservation to the benefit of all. Above all, there would be realistic rather than vague short and long-term objectives.

In this paper we have tried to relate the activities of botanic gardens and institutions to the strategies of IBPGR, and have examined some concepts which might not have been examined in the light of the rôle that botanical institutions could play in the conservation of crop genetic resources. Obviously these ideas will require considerable refinement but, with proper inputs of funds and scientific effort, they could bring the concept of *in situ* conservation into perspective. At the same time they could assist in the preservation of plants in regions and localities that might be threatened and along with them valuable genetic resources known and unknown.

References

Chapman, C.G.D. (1985). "Wheat Genetic Resources: a Review and Proposals for Future Action". Mimeographed report. IBPGR, Rome.

Frankel, O.H. and Soulé, M.E. (1981). "Conservation and Evolution". Cambridge University Press, Cambridge.

Harper, J.L. (1981). The meanings of rarity. *In* "The Biological Aspects of Rare Plant Conservation" (H. Synge, ed.) pp. 189–203. John Wiley, Chichester.

Hawkes, J.G. (1985). "Plant Genetic Resources: The Impact of the International Agricultural Research Centres". CGIAR Study Paper No. 3. CGIAR, Washington D.C.

IBPGR (1984). "The Potential for Using *in vitro* Techniques for Germplasm Collection". IBPGR, Rome.

IBPGR (1985a). "Annual Report for 1984". IBPGR, Rome.

IBPGR (1985b). "Report of a Workshop on the Triticeae held 6–8 August 1985". IBPGR (85/158), Rome.

IBPGR (1985c). "Ecogeographical Surveying and *in situ* Conservation of Crop Relatives". IBPGR, Rome.

Ingram, G.B. and Williams, J.T. (1984). *In situ* conservation of wild relatives of crops. *In* "Crop Genetic Resources: Conservation and Evaluation" (J.H.W. Holden and J.T. Williams, eds) pp. 163–179. Allen & Unwin, London.

IUCN (1980). "World Conservation Strategy". IUCN-UNEP-WWF, Gland, Switzerland.

Simmons, J.B., Beyer, R.I., Brandham, P.E., Lucas, G.L.L. and Parry, V.T.H. (eds) (1976). "Conservation of Threatened Plants". Plenum Press, New York.

US-AID *et al.* (1982). "Proceedings of the U.S. Strategy Conference on Biological Diversity". Dept. of State pub. 9262. U.S. Government Printing Office, Washington, D.C.

Walters, S.M. (1983). The role of botanic gardens in plant conservation in South-East Asia. *In* "Conservation of Tropical Plant Resources" (S.K. Jain and K.L. Mehra), pp. 230-235. Botanical Survey of India, Howrah.
Williams, J.T. (1984). A decade of crop genetic resources research. *In* "Crop Genetic Resources: Conservation and Evaluation" (J.H.W. Holden and J.T. Williams, eds.) pp. 1–17. Allen & Unwin, London.

Appendix 1

Specific programmes of the IBPGR where collaboration with botanic gardens/botanical institutes is necessary

Programmes	Potential level of collaboration
GERMPLASM ACQUISITION	
Monitoring of genetic erosion	low–high
Collecting of endangered germplasm	low–high
Selective collecting for diversity gaps	high
Facilitation of germplasm distribution	moderate
GERMPLASM CHARACTERISATION AND EVALUATION	
Data acquisition from accessions	moderate
TRAINING	
Specialised technical courses	moderate
GENETIC DIVERSITY RESEARCH	
Species mapping	high
Ecogeographical studies	high
Isozyme analysis	moderate
Wild relatives of priority crops	high

Appendix 2

IBPGR priorities for collecting Triticeae

Priority 1	**Distribution**
Agropyron fragile	central Asia to northern China
Agropyron mongolicum	north and north-west China
Hordeum intercedens	southern California (U.S.A.) into Mexico
Hordeum arizonicum	Arizona (U.S.A.) into Mexico
Thinopyrum bessarabicum	Crimea and Black Sea area
Thinopyrum elongatum (2x)	Mediterranean coast
Aegilops sp.[1]	Mediterranean and south-west Asia
Hordeum bulbosum (2x)	western and central mediterranean
Hordeum roshevitshii	southern Siberia, north and central China

Priority 2	**Distribution**
Pseudoroegneria libanotica	south-west Asia
Hordeum brachyantherum (6x)	southern California, U.S.A.
Hordeum depressum	central and southern California, U.S.A.
Hordeum erectifolium	Argentina
Hordeum chilense	Argentina and Chile
Hordeum procerum	Argentina
Thinopyrum sartorii	eastern Mediterranean
Thinopyrum curvifolium	Spain
Thinopyrum scirpeum	Mediterranean
Thinopyrum caespitosum	Crimea and Turkey
Thinopyrum intermedium	Switzerland, Austria and France
Thinopyrum ponticum	Turkey, southern U.S.S.R., Iran
Elymus sp.	eastern Asia
Leymus lanatus	central Asia
Leymus angustus	central Asia to western China
Leymus karelinii	central Asia
Leymus triticoides	western U.S.A to Canada
Secale montanum	eastern Mediterranean and south-west Asia
Psanthyrostachys juncea	central Asia to western China

Priority 3	**Distribution**
Hordeum flexuosum	Colombia, Argentina
Hordeum jubatum	
Hordeum lecheri	Argentina
Hordeum secalinum	northern Europe
Hordeum publiflorum	Argentina
Thimopyrum junceiforme	northern Europe
Thinopyrum junceum	Europe and Mediterranean
Thinopyrum runemarkii	Mediterranean
Thinopyrum corsicum	Corsica
Thinopyrum nodosum	Crimea and Black Sea
Thinopyrum scythicum	
Thinopyrum gentryi	Iran
Elymus bakeri	
Elymus psammophilus	Great Lakes, U.S.A.
Elymus californicus	California, U.S.A.
Elymus sp.	south-west Asia
Leymus pacificus	California, U.S.A.
Leymus multiflorus	California, U.S.A.
Leymus Condensatus	California, U.S.A.
Henrardia sp.	south-west Asia
Crithopsis sp.	south-west Asia

[1]Already covered by IBPGR as near relatives of wheat (Chapman 1985).

The rôle of the Jardín Botánico Canario "Viera y Clavijo" in the conservation of endangered Canarian endemics

DAVID BRAMWELL

Jardín Botánico Canario "Viera y Clavijo", Las Palmas de Gran Canaria, Canary Islands

Summary

The World Conservation Strategy has helped to define the rôle of botanic gardens such as the "Viera y Clavijo" in conservation of endemic species. The work of the garden is described paying particular attention to research and rescue. Projects in the fields of seed banking, micropropagation, natural resource assessment, etc., are outlined and the scientific value and potential resources of the flora considered. The usefulness of *ex situ* cultivation as a method of conservation is also discussed.

Resumen

El papel del Jardín Botánico Canario "Viera y Clavijo" en la conservación de endemismos canarios amenazados

La Estrategia Mundial para la Conservación ha ayundado a definir el papel de los jardínes botánicos como el "Viera y Clavijo" en la conservación de especies endémicas. La labor del jardín es prestar particular atención a la investigación y rescate de plantas amenazadas. Se destacan los proyectos en el campo de banco de semillas, micropropagacion, asentamiento de recursos náturales, etc. y se considera el valor científico y los recursos potenciales de la flora. Se discute la utilidad del cultivo *ex situ* como método de conservación.

Botanic Gardens and the World Conservation Strategy

ISBN: 0-12-125462-3

Introduction

The publication in 1980, by the International Union for Conservation of Nature and Natural Resources (IUCN) and the World Wide Fund For Nature (WWF), of the World Conservation Strategy (WCS) has given conservationists all over the world a basic document with which to build their conservation programmes. Over the last 5 years on the island of Gran Canaria, the Island Council (Cabildo Insular de Gran Canaria) has started to implement the Strategy through its "Reverdecer Gran Canaria" programme. The Jardín Botánico Canario "Viera y Clavijo" plays an important part in this programme with its leading role in the preparation of the "Plan Especial de Protección de Espacios Naturales" (PEPEN), which proposes the protection of large areas of natural vegetation and the recuperation of degraded areas of special interest. The garden also has a very extensive environmental education project, which is operated by means of an agreement with the Autonomous Government Education Commission, who provide teaching staff; it involves over 20,000 schoolchildren in visits to the garden every year.

In this paper, however, I will limit myself to recounting briefly how the garden functions in the field of research and rescue, in trying to save the many endangered Canarian endemic plants and discovering how they can be propagated and maintained *ex situ* with the long-term aim of reintroducing them into natural reserves or restored natural habitats.

Ex situ conservation is not the ideal means for saving species. Biologists generally agree that *in situ* conservation is the better alternative, but in difficult "fire brigade" situations conservationists have to be practical as well as idealistic. *Ex situ* conservation is probably the only alternative we have at this moment for saving many thousands of species on a world-wide scale. My personal view of *ex situ* conservation has, of course, been formed as an island biologist dealing with tiny but reasonably well-known populations of highly endangered endemics; it is probably very different from that of a tropical rain-forest specialist faced with an overwhelming mass of almost unknown species with virtually unresearched population structures, and covering vast tracts of the Earth's surface. Indeed our approach to the subject of conservation will not and should not be the same and I can sympathise deeply with my tropical colleagues when they view *ex situ* conservation in botanic gardens with certain practical and philosophical scepticism. The fact is that we are facing different problems under entirely different circumstances and a perfectly valid solution to one situation may be of very little help to another.

Some of the arguments against *ex situ* conservation are of a rather heuristic nature and can be resolved by careful planning and common sense. These include the problems of genetic impoverishment and the creation of "botanic garden ecotypes". As a fail-safe strategy to ensure the survival of a considerable proportion of our plant genetic resources for future generations, I can see few viable alternatives to the combination of botanic garden and seed bank. It is not my intention to present here a case for the defence of *ex situ* conservation but rather to put forward an indictment of those who are not yet convinced of its value.

The effects of genetic erosion and the alteration of selection pressures on populations held in botanic gardens probably do not differ very much from those on tiny natural populations occurring in habitats which are extremely vulnerable to the activities of man, overgrazing, erosion and so on. Even if we establish habitat reserves and have to manage, "garden" or otherwise manipulate such reserves then we are also changing the natural selection factors, and in some ways such small populations both in reserves and in gardens will behave as founder populations with all the implications that this has for genetic change. To present *in situ* reserves as the means of ensuring continuity of natural selection is not, in such cases, as valid as it may seem at first.

For a single, very rare species in a highly endangered habitat, a typical island conservation situation, and in a habitat which is of no other particular natural value than that of being host to an endangered species, a combination of botanic garden and seed bank can conserve a wider range of genetic potential more cost-effectively than creating and maintaining a reserve. In the Canary Islands the *ex situ* conservation of some of our endangered species is a practical and achievable target and will also provide the basic material for the necessary background studies needed to aid both *ex* and *in situ* conservation. In this context Frankel and Soulé's conclusion is very relevant: ultimately the effectiveness of *ex situ* conservation will depend on the interest, perseverence and dedication of individuals in addition to traditions and obligations, and that "in the long term all preservation is a social as much as a biological issue" (Frankel and Soulé 1981).

The Canarian flora as a natural resource

The flora of the Canary Islands is of considerable scientific interest. In *Aeonium, Sonchus, Echium* and *Argyranthemum* we have some of the most complete examples of insular plant adaptive radiation on a world scale. *Aeonium* with over 30 species has few parallels as a practical example of adaptive radiation even in the animal kingdom, but several species of the genus are already under very severe pressure in their natural habitats although, fortunately, none are yet extinct.

The pine and laurel forests are "living fossils" of the warm temperate and subtropical forests which covered the northern shores of the Tethys Sea in the mid and late Tertiary Period (Sunding 1979). Many species of these ecosystems, which are locally becoming rarer by the day, are known only from the Canary Islands and as fossils from Miocene and Pliocene deposits from the Mediterranean region.

The succulent and dry zone floras of the islands with representatives of *Euphorbia, Kleinia, Aeonium, Ceropegia* and *Caralluma* are related to other relict semi-arid floras from southern and eastern Africa and appear to form part of an old African Tertiary Flora (Bramwell 1986).

The Canarian Flora is also rich in potential genetic resources. Among the most popular warm temperate and subtropical ornamentals are a number of native Canarian plants, such as the Canary Palm (*Phoenix canariensis*), the Canary pine (*Pinus canariensis*), the Dragon's Blood Tree (*Dracaena draco*), the Florist's Cineraria (*Senecio (Pericallis) hybrida*) and the Paris Daisy (*Argyranthemum* spp.). There are many other less well known but equally valuable species making up an important subtropical horticultural resource.

The flora is also a potential source of useful crop-plant genes with endemic species of *Brassica, Beta, Dactylis, Avena, Olea, Phoenix, Solanum, Crambe, Persea,* etc. In *Crambe*, where *C.abyssinica* is being studied as an important potential oil-seed crop, the largest seeded species in the genus is the highly endangered *Crambe sventenii* endemic to a couple of cliff faces on the island of Fuerteventura.

The pharmacological value of a number of Canarian species is well known thanks mainly to the research carried out by Professor Antonio González and his co-workers at the University of La Laguna, Tenerife. The *Digitalis* relative *Isoplexis* is a potentially important source of digitalis. Other plants such as *Tamus edulis, Maytenus canariensis* and species of *Sideritis* contain steroid precursors and are also potentially valuable. In the Canarian flora, therefore, we have a rich heritage to conserve for the future — a heritage of world importance.

Conservation in practise

At the Jardín Botánico "Viera y Clavijo" we have a substantial programme for "research and rescue" of the endangered Canarian flora. The rescue of critically endangered species is carried out by means of three approaches.

The seed bank contains almost 500 accessions, over 90% of which are field-collected and which include a large number of the endangered species.

The seed bank, as well as being a valuable conservation tool, is also the source of material for many other studies on endangered species, for example cytology, breeding systems, seedling behaviour, germination and establishment, etc. The seed bank records are also a valuable source of field data on population size, natural seed production, etc., and of course, by seed exchange within the international botanic garden system, it is a means of getting endangered species into cultivation in other gardens for safe-keeping.

Tissue culture

A number of species present serious problems for seed banking. They may have recalcitrant seeds, have low fertility or low seed production in the field, or simply not have enough individuals in the natural populations to give a sufficiently large sample for seed banking. In some local cases such as *Senecio hadrosomus* from Gran Canaria and *Lotus berthelotii* from Tenerife other problems such as parasitisation of seeds and high self-incompatibility add further to the problems. In such cases we have been successful in propagating by means of tissue culture to bulk-up material for seed production and possible reintroduction into the natural habitat.

With tissue culture, as with any other form of vegetative propagation, there is the problem of the limited number of clones propagated and consequent genetic erosion. In cases such as *Senecio hadrosomus* with a known wild population of under 20 plants and *Lotus bethelotii* with even fewer, it is possible to sample the wild population adequately as the source for culture material. In such cases it is possible to bring into cultivation via tissue culture the whole of the wild genepool. In the case of *Senecio hadrosomus*, where the natural habitat is owned by a public body, its protection in the wild is ensured and a reintroduction programme has already begun.

Tissue culture presents many problems especially as many of the species to be propagated are extremely rare and from families on which little or no previous research on micropropagation has been carried out (e.g. Globulariaceae, Santalaceae). In such cases very basic research is necessary on media, nutrients, pH, etc., and many of the data generated may then be useful at a later stage as bases for understanding cultivation requirements in the nursery, garden and even in natural habitat reserves.

Conventional garden cultivation

This botanic garden, like all others relies enormously on the acquired skills of nursery and garden staff to propagate and cultivate plants by conventional methods to provide material for display and education purposes, to provide research material, to bulk-up seed for banking and to maintain a general reserve collection as a long-term project with the objective of reintroduction and re-establishment of rare species in the wild.

Conservation-orientated research

In the garden laboratories we have a number of conservation-orientated research projects underway. Our research on potential natural resources involves studies of plants such as *Tanacetum ptarmaciflorum*, *Gonospermum* and *Lugoa* species and *Limonium*, which all have ornamental possibilities. The research, at the moment, is concentrated on cytology and genetic diversity to provide information which may be used later as the basic data for plant-breeding projects. Similar work is being carried out on local *Beta* species, which are resistant to nematode infections, and on *Sideritis*, which contains steroid precursors of pharmacological value and has several extremely endangered species. It is a cytologically difficult genus and was, until recently, one of the major gaps in our knowledge of chromosome numbers of Canarian plants. One of our research workers has now studied virtually all the species to karyotype level (Marrero 1986) and we are now involved in interpreting the cytology of a genus showing adaptive radiation and a wide range of chromosome numbers — a unique situation in the Canarian flora.

The genus *Lotus* with a large number of Macaronesian endemic species in the Section *Pedrosia* has also been surveyed cytologically and biochemically. We are now expanding this to a study of the forage potential of the most interesting perennial species. In the field of natural resource studies we are also working in collaboration with the University of Wageningen on the latex of the endemic *Euphorbia* species.

Among the local endemic flora we have a number of endemic grass species, several of which are potentially important forage species but are extremely rare and usually confined to almost inaccessible cliff-faces. This is probably due to overgrazing by goats and rabbits in historical times. A member of our staff is now working on a general survey and taxonomic study of the Canarian grass flora, as well as an evaluation of the potential productivity of some of the species including *Dactylis smithii* and *Brachypodium arbuscula*.

Other projects involving endangered Canarian species include a pollen survey of the endemic flora and the application of palynological data in biogeographical studies of the relationships between the Canaries and Africa. Palynology helps in the confirmation of taxonomic relationships, monophyly in disjunct genera and to identify near relatives. It has helped to elucidate the relict nature of the Canarian flora and thus its scientific value as a living fossil.

The garden has 2 further projects of interest to conservation; one is on breeding systems and dispersal, involving both botanists and ornithologists. The other is a group working on marine algae both from a taxonomic and floristic point of view, and on algae as bio-indicators of pollution and of environmental changes due to the effects of the development of the coasts for tourism, fishing ports and marinas, industrialisation, etc.

Obviously in the course of this work we generate considerable amounts of data and we have our own computerised monitoring unit and data bank (using dBase III) which provides back-up for research, documentation of living collections and seed bank, state of natural populations (Red Data sheets), preparation of species lists, check-lists, bibliography, etc.

Support for conservation

The information presented above is a brief summary of some of the garden's activities in the field of environmental conservation. The work is supported principally by the Island Council of Gran Canaria (Cabildo Insular de Gran Canaria), who are the owners of the garden.

Other organisations, however, also provide financial assistance for projects; these include locally the Dirección General de Medio Ambiente del Gobierno de Canarias; nationally the Dirección General de Medio Ambiente del Ministerio de Obras Publicas y Urbanismo; and internationally IUCN and WWF.

For small islands such as the Canaries conservation is not a luxury nor is it the domain of a "lunatic fringe". It is a basic necessity of life and the basis for future development, and so must be in the forefront of government policy and financing now so that we can pass on this important heritage as intact as possible to future generations. This is also not a luxury but an absolute obligation that we have to our own children.

References

Bramwell, D. (1986). *Bótanica Macaronésica* **14**, 3–34.

Frankel, O.H. & Soulé, M.E. (1981). "Conservation and Evolution". Cambridge University Press, Cambridge.

IUCN/UNEP/WWF (1980). "World Conservation Strategy". IUCN, Gland, Switzerland.

Marrero Rodriguez, A. (1986). *Botánica Macaronésica* **14**, 35–58.

Sunding, P. (1979) Origins of the Macaronesian flora. *In* "Plants and Islands" (D. Bramwell, ed.) pp. 13–40. Academic Press, London and New York.

Woody endemic species of the wet lowlands of Sri Lanka and their conservation in botanic gardens

C.V.S. GUNATILLEKE and I.A.U.N. GUNATILLEKE

Department of Botany, University of Peradeniya, Sri Lanka,

B. SUMITHRAARACHCHI

Botanic Gardens, Peradeniya, Sri Lanka

Summary

In Sri Lanka, which shares a geological history of Gondwana origin with Peninsular India, 23% of the 3,360 indigenous flowering plant species and an equally high proportion of its lower plant species are endemic to the island. Most of these endemics are restricted to the tropical rain forests of the south-western region of which only about 9% now remain. Except for a few protected reserves, most of them are degraded and fragmented.

Re-examination of data gathered from phytosociological surveys of the woody vegetation in 9 different lowland wet zone forests in a total sample of 44 ha shows that 68% of the identified species are endemics and comprise 81% of all the endemic tree species recorded from the entire lowland wet zone. Of the endemics enumerated, 52% represented the canopy/subcanopy and 48% understorey tree strata. While the proportion of those restricted to any one of the 9 sites sampled was 25%, a similar proportion had densities of less than 0·5 individuals/species/ha.

Based on distribution and population densities, around 90% of the 187 endemics encountered in these surveys belong to the Rare,

Botanic Gardens and the World Conservation Strategy

ISBN: 0-12-125462-3

Vulnerable and Endangered categories of the IUCN Red Data Book. The Sinharaja Biosphere Reserve, the largest least-disturbed lowland rain forest, harbours over 70% of the endemics enumerated; the remaining 30%, recorded mostly from lower altitude forests outside Sinharaja, are possibly endangered as they are not adequately protected. At least 14 species of these lowland wet zone forests have not been recollected during the last 60 years and so may be considered extinct in the wild. The need for arboreta of adequate size to conserve endemic plant species restricted to these lower altitude rain forests (below 250 m) is emphasised.

Among the 3 botanic gardens in Sri Lanka, only Peradeniya Gardens has a collection of lowland endemic trees, in this case of 71 species. Fifty of them were encountered in the phytosociological surveys. For a few of the remaining 22, the botanic garden is the last refugium. Here too, however, most are represented by only one individual, thus critically limiting their genetic diversity. A greater participation of botanic gardens in *ex situ* conservation is therefore advocated.

Resumen

Especies endémicas leñosas de las tierras bajas húmedas de Sri Lanka y su conservación en jardines botánicos

En Sri Lanka, que comparte con la India Peninsular la historia geológica del origen de Gondwana, el 23% de las 3.360 especies de plantas con flores y una igualmente alta proporción de sus especies de plantas de rango inferior son endémicas de la isla. Más de estos endemismos están restringidos a los bosques tropicales de la región sudoccidental de los que sólo el 9% subsisten actualmente. Excepto unas pocas reservas protegidas, muchas de ellas están degradadas y fragmentadas.

Un nuevo estudio de los datos recogidos en investigaciones fitosociológicas de la vegetación leñosa en 9 diferentes zonas forestales de tierras bajas húmedas realizado sobre una zona de 44 ha. muestra que el 58% de las especies constituyentes son endémicas y corresponden a 81% de todas las especies de árboles endémicos registrados en todas las zonas de las tierras bajas húmedas. De los endemismos enumerados, el 52% representaba la cúpula/subcúpula y el 48% el estrato bajo el nivel arbóreo. Mientras que la proporción en algunos de los 9 lugares de los que se tomaron muestras fue el 25%, una proporción similar tenía densidades de menos del 0.5 individuos/especies/ha.

Basado en la distribución y en la densidad de población, por encima del 90% de los 187 endemismos encontrados en estas investigaciones pertenecen a la categoría de los Raros/Vulnerables/Extinguidos del Libro Rojo de Datos de la UICN. La Reserva de la Biosfera de Sinharaja, bosque tropical de zona baja menos alterada, alberga más del 70% de los endemismos enumerados; el 30% restante, recogidos en su mayor parte en los bosques de más baja altitud de las afueras de Sinharaja, están posiblemente en peligro pues no están adecuadamente protegidos. Al menos 14 especies de los árboles y arbustos dominantes de estas zonas de bosques de zonas bajas húmedas, no han sido recogidos durante los últimos 60 años y por tanto pueden considerarse como extinguidas en forma silvestre. La necesidad de arboretas de un tamaño adecuado para conservar las especies de plantas endémicas restringidas en estos bosques tropicales de más baja altitud (250 m.) es bastante acentuada.

De los tres jardines botánicos de Sri Lanka, sólo el de Peradeniya tiene una colección de árboles endémicos de zonas bajas, en este caso, de 72 especies. 50 de ellas fueron encontradas en las investigaciones fitosociológicas. Unas pocas de las restantes 22 tienen como último refugio los jardines botánicos. Aquí también, por otra parte, muchas están representadas por un solo individuo, hasta el punto que limitan críticamente su diversidad genética. Se aboga por una gran participación de los jardines botánicos para la conservación *ex situ*.

Introduction

Sri Lanka, which shares the same tectonic plate with Peninsular India and formed part of the southern Gondwana continent in ancient times, has over 3,360 species of indigenous flowering plants representing about 1,070 genera and 180 families. While none of these families are endemic to the island, 26 genera and 830 species (23% of the flora) are endemic (Peeris 1975). Among the lower plant groups studied, 57 of the 314 indigenous pteridophyte flora (18%) (Sledge 1982) and 39 of the 110 species (35%) of the lichen family Thelotremataceae (Hale 1981) are reported endemic to the island.

Geographically, most of these endemics are concentrated in the rain forests of the ever wet perhumid south-western part of the island, known as the "Wet Zone" encompassing 23% (about 14,850 km^2) of the island's land area (Table 1). However, even within this zone the endemics are highly localised or restricted in their distribution, and sadly the area now under natural vegetation has been recently estimated at only 9%. Of this too, large parts are quite degraded, and the extent of relatively undisturbed primary forest, based on our field experience, is less than 1% and 3% of the land area in the wet lowland and montane zones respectively (Fig. 1). These remaining forests are quite fragmented ranging in size between 23 ha and 8,800 ha (Perera 1972), with the exception of a few large blocks that enjoy full protection as Biosphere Reserves, Wild Life Sanctuaries or Strict Natural Reserves. They are the Sinharaja MAB Reserve (8,800 ha, about 50% undisturbed), Horton Plains Strict Natural Reserve (3,162 ha, 50%

Table 1
Distribution of endemic species in different climatic zones of Sri Lanka according to their life forms after Peeris (1975) where data has been gathered from Trimen (1893–1900)

Climatic Zones / Life form	Montane	Montane and wet lowlands	Wet lowlands	Wet lowlands and dry	Dry, intermediate and common to both	Total in each and all life forms
Trees	72	46*	156*	15*	17	306
Shrubs	84	32	82	15	17	230
Herbs	128	51	88	9	18	294
Total in each and all climatic zones	284	129	326	39	52	830

* Based on the Revised Handbook to the Flora of Ceylon (Dassanayake and Fosberg 1980–4) there are 14 additional tree species in the three categories (marked with an asterisk, considered in this paper making a total of 231.

montane forest and 50% montane grassland), Peak Wilderness Sanctuary (22,400 ha) and Hakgala Strict Natural Reserve (1,142 ha).

Phytosociology of the tree and liana species above 30 cm gbh in 9 different lowland wet zone sites (Fig. 1) have been studied enumerating over 29,500 individuals (Peeris 1975, Gunatilleke and Gunatilleke 1984, 1985) in 176 plots, each 0·25 ha in size. They were identified as far as possible to their respective species by referring to the standard floras (Trimen 1893–1900, Dassanayake and Fosberg 1980–4). These studies provide valuable information on geographical distribution and population sizes of adult individuals of the species enumerated.

The objective of this paper is to assess the extent of *in situ* and *ex situ* conservation given to the endemic woody flora of the lowland wet zone by examining the following, using the phytosociological data available from 9 sampling sites (Table 2).

(1) The Red Data Book (RDB) category (Synge 1981) of each endemic species recorded in the 9 sites.
(2) The population sizes of adult individuals of species in each RDB category.
(3) How restricted or widely distributed each species is, to decide which areas should be given, or maintained for, *in situ* conservation.

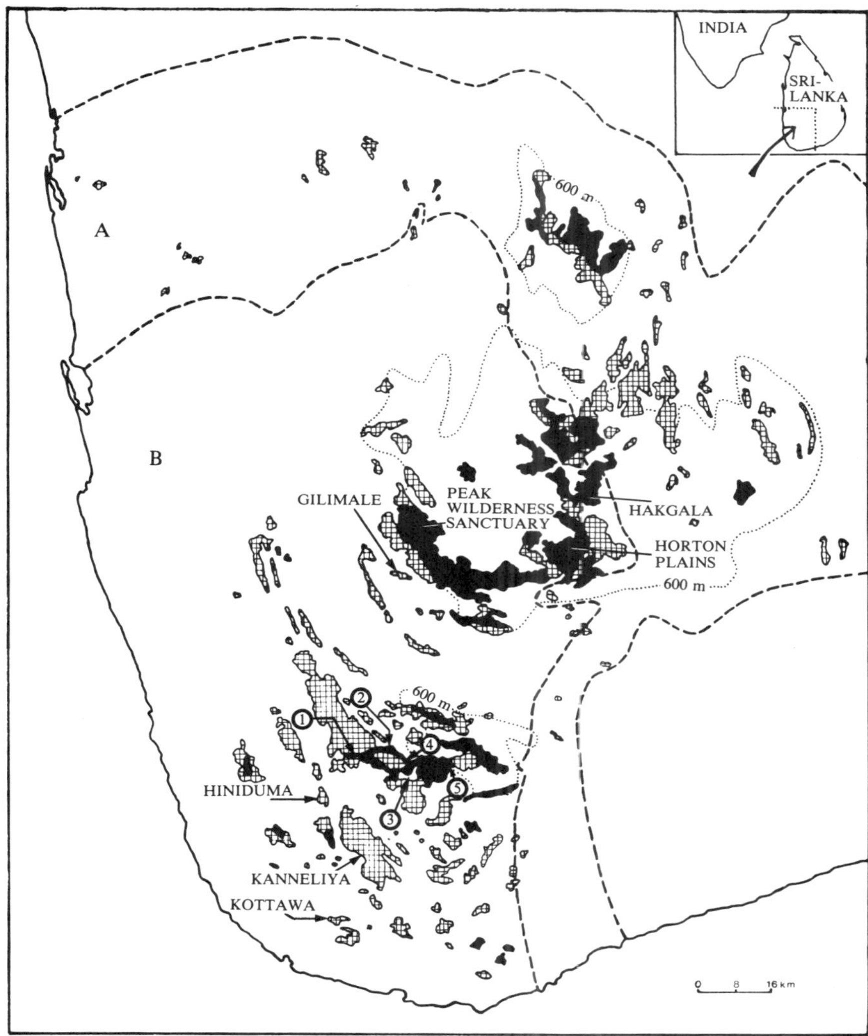

Fig. 1. *South-western part of Sri Lanka showing the relatively undisturbed (shaded) and disturbed (square hatched) natural forest extents in the intermediate* (A) *and wet zone* (B). *Arrows indicate study sites of the phytosociological surveys of the lowland wet zone. 1–5 are the Sinharaja MAB reserve study sites.*

Table 2
Total number of taxa in the 9 lowland wet zone sites sampled and the proportion of endemic species in each site given as a percentage of the total number of species in it.

Sites sampled	Families	Genera	Species		Percentage endemic species
			Identified	Unidentified	
Kottawa	31	65	101		79*
Kanneliya	34	77	125		76*
Hiniduma	39	92	140		75
Sinharaja					
1	34	76	121	44	73*
2	38	73	115		72*
3	39	87	144		72
4	35	75	125		75
5	39	88	140		62*
Gilimale	41	88	136		66*
All sites	48	146	273	44	68

* The difference in percentage endemicity values given here and those in Gunatilleke and Gunatilleke (1985) is due to updating of the endemic status of some species using the Revised Handbook to the flora of Ceylon (Dassanayake and Fosberg 1980–4).

(4) The number of endemic tree species planted in botanic gardens to assess how many remain to be salvaged for *ex situ* conservation or considered for *in situ* conservation.

Results

The woody flora of the 9 sites in the lowland wet zone sampled contains 48 families, 146 genera and 273 identified species and 44 unidentified taxa in the vegetation above 30 cm gbh (Table 2). Sixty-eight per cent (187 species) of those identified species are endemic to the island and they constitute 81% of the total 231 endemics recorded from the lowland wet zone (Table 1).

The identified species were grouped into canopy (30–40 m), subcanopy (15–30 m) and understorey (5–15 m) tree strata according to the position occupied by mature individuals of each species in the vertical structure of the forest. If data in our sample alone are considered, some of the understorey tree species whose reproductively mature individuals are better represented below the arbitrary minimum girth size (30 cm) used in sampling would appear to have very small population densities. Vegetation below 30 cm gbh would have to be studied to obtain realistic population sizes of reproductively mature individuals of these understorey species. Of the 187 endemic species in the sample, 99 represented subcanopy/canopy species and 88 were typical of the understorey tree stratum. Based on whether a species was (1) recorded at a few or many sites within the wet zone (Table 3); (2) confined to the lowland wet zone or much more widespread; and (3) its population size (Table 4), each species, as far as possible, was identified to its appropriate RDB category. Consequently it was found that 98% of the subcanopy/canopy species and 85% of the understorey species belonged to the Rare, Vulnerable and Endangered categories. The distributional range of species among the 9 sites, based upon the phytosociological data, show that 16 subcanopy/canopy and 31 understorey species are restricted to any one of the 9 sites sampled, and together constitute 25% of the endemic flora identified in study sites (Table 3).

Table 3
Number of endemic species of the canopy/subcanopy and understorey strata enumerated in 9 sites sampled in the lowland wet zone according to their range of distribution among sites and RDB categories.

Stratum and RDB category	Distributional range of species in sites sampled						
	Any 1 site	Any 2 sites	Any 3–4 sites	Any 5–6 sites	Any 7–8 sites	9 sites	1–9 sites
Canopy/sub-canopy							
Endangered	11	5	5	1	1	1	24
Vulnerable	3	1	8	13	7	1	33
Rare	1	—	2	6	17	14	40
Insufficiently known	1	—	—	—	—	—	1
Out of danger	—	—	—	—	—	1	1
Subtotal	16	6	15	20	25	17	99
Understorey							
Endangered	12	4	1	1	—	—	18
Vulnerable	10	5	10	3	3	—	31
Rare	2	3	5	5	8	3	26
Insufficiently known	7	—	2	1	—	—	10
Out of danger	—	—	—	—	1	2	3
Subtotal	31	12	18	10	12	5	88
Total in each distributional range	47	18	33	30	37	22	187

Table 4
Distribution of endemic species in relation to their respective population densities and IUCN RDB categories

Population densities / IUCN Red Data Book	1 individual only in 44 ha	Individuals/ha					Total in each category
		0·05–0·5	0·5–1	1–2	2–5	>5	
Canopy/subcanopy							
Endangered	1	13	8	2	—	—	24
Vulnerable	—	5	10	10	8	—	33
Rare	1	—	2	1	13	23	40
Insufficiently known	—	1	—	—	—	—	01
Out of danger	—	—	—	—	—	1	01
Subtotal	2	19	20	13	21	24	99
Understorey							
Endangered	5	13	—	—	—	—	18
Vulnerable	5	16	3	4	3	—	31
Rare	—	9	1	3	10	3	26
Insufficiently known	5	3	2	—	—	—	10
Out of danger	—	—	—	—	1	2	3
Subtotal	15	41	6	7	14	5	88
Total in each density class	17	60	26	20	35	29	187

The population densities of species in each RDB category examined show that each of 56 understorey and 21 subcanopy/canopy species has less than 0·5 individuals per ha (Table 4). In contrast, among those with more than 1 individual per ha, there are fewer understorey species (26) as compared to the subcanopy/canopy species (58). Among the 42 Endangered species enumerated, there are also two which are found in all 8 or 9 sites (44 ha) but having low densities — *Garcinia terpnophylla* (49 individuals, i.e. less than 2 individuals/ha) and the variegated timber species *Diospyros quaesita* (67 individuals, i.e. lower than 1·5 individuals/ha). Other species worthy of note are *Dipterocarpus insignis, Shorea lissophylla, Shorea ovalifolia, Shorea pallescens,* each restricted to one of the 9 sites, and *Shorea dyeri*, restricted to 2 sites, each with a population density of less than 1 individual/ha.

The suitability of each sampled site for *in situ* conservation of endemic species was evaluated by examining its size, the total number of species belonging to each RDB category and the human impact on each of them. Of the 187 endemic species in the total sample, the number of species at individual sites varies from between 80 at Kottawa to 105 at Hiniduma (Table 5). Of the 42 Endangered species, 23 are restricted to any one of 9 sites (Table 3). Kanneliya (17) and Hiniduma (16) records the highest number of Endangered species. The 5 Sinharaja sites together contain 139 endemics. The remaining 48 are distributed among the other sites, which have the following shortcomings as conservation areas. Kottawa comprises only 4 ha of natural forest; — this extent may not hold viable populations for most of its constituent species. Kanneliya, which comprises 5,737 ha, has been selectively logged many times and even the 40 ha once demarcated as a MAB area and where phytosociological studies were carried out, has subsequently been selectively logged. Hiniduma and Gilimale-Eratne comprising 360 ha and 4,975 ha respectively are not adequately policed and are subject to rampant illicit felling. Thus, at present Sinharaja alone remains adequately protected and consequently the sole refugium for most wet lowland endemics. Although several appeals have been made to designate representative samples of other lowland forests particularly those at lower altitudes (<250m) as conservation areas and to manage them for *in situ* conservation, the overall dearth of timber in the country and lack of field staff prevent them from being effectively managed.

Table 5
Species richness of the woody endemics (over 30cm gbh) in sites enumerated and their distribution according to IUCN Red Data Book categories

	E	V	R	K	O	Total
Kottawa	11	27	38	1	3	80
Kanneliya	17	27	46	1	3	95
Hiniduma	16	28	53	4	4	105
Sinharaja 1	8	29	47	1	3	88
Sinharaja 2	6	22	51	1	3	83
Sinharaja 3	11	29	57	3	4	104
Sinharaja 4	5	29	54	2	4	94
Sinharaja 5	5	28	49	2	3	87
Gilimale	11	31	42	2	3	89
All sites	42	63	67	11	4	187

E = Endangered; V = Vulnerable; R = Rare; K = Insufficiently known; O = Out of danger

These results certainly confirm that endemic species are quite localised even within the small area of the wet zone and that most of them have very low population densities. Both these factors suggest that more areas outside

Sinharaja should be set apart for *in situ* conservation, e.g. Hinidum Kanda, Kanneliya, etc. Considering the importance of the local flora, particularly that of the south-west wet lowlands, as one that may be traced back to the Gondwana flora, it is not too much to ask that more areas of the wet lowlands be set aside for *in situ* conservation, particularly sizeable forest extents of the coastal lowlands below 250 m altitude, whose flora is relatively different to that above 250 m elevation represented by the Sinharaja sites. Although, 26,462 ha of lowland wet zone reserves have been declared as Man and the Biosphere reserves by the MAB Committee of Sri Lanka (Sri Bharathie 1979), the extents of relatively undisturbed natural forest in them include only a fraction of the total demarcated area and even these are not properly conserved.

The herbarium collections made during the extensive botanical surveys carried out for the revision of the Flora of Ceylon show that at least 14 species belonging to 4 plant families of the lowland wet zone have not been collected again during the past 60 years and are considered to be extinct in the wild (Table 6). Corresponding information for other plant families is not available at present and awaits revision of the remaining families.

Table 6
Some plant species not collected during recent botanical surveys

Taxa	Recorded locality and zone	
MYRTACEAE		
Eugenia floccifera	Raigam Korale	(LWZ)
Eugenia fulva	Morawak Korale	(LWZ)
Eugenia glabra	Galle and Ambagamuwa	(LWZ)
Eugenia haeckeliana	Weligama	(LWZ)
DIPTEROCARPACEAE		
Stemonoporus lancifolius	Hewessa	(LWZ)
Stemonoporus marginalis	Hewessa	(LWZ)
Stemonoporus nitidus	Hinidum Korale	(LWZ)
MELASTOMACEAE		
Memecylon orbiculare	Kalubowitiyana Kanda	(LWZ)
ANNONACEAE		
Uvaria cordata	Matara	(LWZ)
Alphonsea hortensis	Raigam Korale	(LWZ)
Polyalthea moonii	Kalutara/Raigam Korale	(LWZ)
Orophea zeylanica	Hantana, Raxawa	(LWZ)
Milliusa zeylanica	Raigam Korale	(LWZ)
Anaxagorea luzorensis	Maskeliya	(SMZ)

LWZ = Lowland Wet Zone; SMZ = Submontane Zone

Table 7
Number of endemic species cultivated in the Botanic Gardens at Peradeniya, giving their respective climatic zones and IUCN RDB category

Climatic zone / IUCN Red Data Book category	Montane	Dry and intermediate	Wet lowlands including those common to other areas	Total
Endangered	—	02	18	20
Vulnerable and Rare	01	03	49	53
Out of danger	01	01	04	06
Total	02	06	71	79

Finally, the contribution of the Botanic Gardens at Peradeniya to *ex situ* conservation of the endemic tree species may be examined. At present, 71 lowland wet zone, 2 montane and 6 intermediate/dry zone endemic tree species are cultivated in the Botanic Gardens. While 49 of these lowland endemics have been enumerated in the phytosociological survey, 22 have not been recorded (Table 8). This leaves 137 endemics recorded in the phytosociological survey. Of these, 58% are in the Endangered and Vulnerable categories, and a further 22, which are very rare according to Trimen (1895–1900) and possibly in the Endangered category of IUCN, are not yet covered by *ex situ* conservation. In the Botanic Gardens most of the

Table 8
Comparison between the total number of endemic species in the lowland wet zone, those enumerated in the phytosociological surveys and the Botanic Gardens, Peradeniya

ICUN Red Data Book category	E	V	R	K	O	Total
All endemics in lowland wet zone	72	141	—	11	7	231
Endemics recorded in phytosociological survey	42	63	67	11	4	187
Endemics in Botanic Gardens, Peradeniya and common to phytosociological survey	10	17	21	—	1	49
Endemics only in Botanic Gardens not in phytosociological survey	8	11		—	3	22
Endemics neither represented in Botanic Gardens nor in phytosociological survey	22	—		—	—	22

E = Endangered; V = Vulnerable; R = Rare; K = Insufficiently known; O = Out of danger endemics are each represented by one individual. Thus the genetic diversity of the species is extremely narrow. Bearing in mind this shortcoming, it would be more desirable to include (or even affiliate) suitable patches of forests, which have a wider spectrum of species and a greater genetic heterogeneity per species than in botanic gardens or as satellite sites to them. This situation prevails in one of the three botanic gardens — the Hakgala Botanic Gardens in the montane zone, where the Hakgala Strict Natural Reserve administered by the Wild Life Department abuts onto the garden.

Acknowledgements

For introducing us to tropical forest ecology and for continued interest and enthusiasm imparted to us in related work ever since, we extend our appreciation to Professor P.S Ashton, Arnold Arboretum of Harvard University, U.S.A. The phytosociological surveys carried out over the years were funded by the International Foundation for Science, Sweden, and the Natural Resources, Energy and Science Authority of Sri Lanka. Funds to participate at the symposium were provided by the Organising Committee, and the International Foundation of Science, Sweden. To all of them we extend our sincere thanks.

References

Ashton, P.S. and Gunatilleke, C.V.S. (in press). New light on the plant geography of Ceylon I, Historical plant geography. *J. Biogeography*

Dassanayake, M.D. and Fosberg, F.R. (1980–4). "A revised handbook to the Flora of Ceylon". 5 vols. Amarind Publishers, New Delhi.

Gunatilleke, I.A.U.N. and Gunatilleke, C.V.S. (1984). Distribution of endemics in the tree flora of a lowland hill forest in Sri Lanka. *Biol. Conserv.* **28**, 275–85.

Gunatilleke, C.V.S. and Gunatilleke, I.A.U.N. (1985). Phytosociology of Sinharaja — A contribution to rain forest conservation in Sri Lanka. *Biol. Conserv.* **29**, 21–40.

Hale, M.E., Jr (1981). A revision of the lichen family Thelotremataceae in Sri Lanka. *Bull. Br.Mus. Nat.Hist.(Bot.)* **8**(3), 227–332.

Peeris, C.V.S. (1975). "The ecology of endemic tree species of Sri Lanka in relation to their conservation". Ph.D. Thesis, University of Aberdeen, U.K.

Perera, W.R.H. (1972). A study of the protective benefits of the wet zone forestry reserve of Sri Lanka. *Sri Lanka Forester* **10**, 87–102.

Sledge, W.A. (1982). An annotated checklist of the Pteridophyta of Ceylon. *Bot. J. Linn. Soc.* **84**, 1–30.

Sri Bharathie, K.P. (1979). Man and Biosphere Reserves in Sri Lanka. *Sri Lanka Forester* **14**, 37–40.

Synge, H. (Ed.) (1981). "The Biological Aspects of Rare Plant Conservation". John Wiley, Chichester.

Trimen, H. (1893–1900). "A Handbook to the Flora of Ceylon". Dular, London.

4

Botanic Gardens for Sustainable Development

A botanic garden in the Indian context: a case study

T.N. KHOSHOO

Tata Energy Research Institute, New Delhi, India

Summary

In 1948 the Government of Uttar Pradesh set up the National Botanic Gardens (NBG) at Lucknow. Although envisaged as a botanic garden and applied research laboratory, constraints of the site led to an emphasis on the latter rôle and the garden was renamed the National Botanical Research Institute (NBRI) in 1978. The aims and objectives of the NBRI are outlined in the context of the country's needs, with details of the institute's facilities, organisation, activities and achievements. Emphasis is placed on applied research on economic plants, particularly non-agricultural/non-traditional plants and ornamentals. To achieve its objectives, the institute's niche in the developmental and botanical community, nationally and internationally, is discussed and outlined, from which conclusions are drawn about possible objectives for other botanic gardens in tropical countries.

Resumen

Un jardín botánico dentro de un contexto Indú: un ejemplo estudiado

En 1948, el Gobierno de Uttar Pradesh puso en funcionamiento los Jardines Botánicos Nacionales (NBG) en Lucknow. Aunque concevido como jardín botánico y laboratorio de investigación aplicada, las limitaciones del sitio dieron lugar a un énfasis en esta última activadad y en 1978, el jardín botánico paso a llamarse Instituto Nacional de Investigación Botánica (NBRI). Las metas y objetivos del NBRI se

Botanic Gardens and the World Conservation Strategy

ISBN: 0-12-125462-3

encuadran en el contexto de las necesidades del país; así mismo se detallan aspectos de la organización, activadades, logros y las facilidades ofrecidas por el Instituto. Se muestra especial atención en la investigación aplicada de plantas de importancia económica, en particular en aquellas especies no utilizadas tradicionalmente en agricultura y especies ornamentales. Para lograr sus objetivos, se discute y subraya el papel del Instituto en la comunidad botánica desarrollada, nacional e internacionalmente, del que se obtienen conclusiones sobre los posibles objetivos de otros jardines botánicos en países tropicales.

Introduction

An anonymous author has defined a botanic garden in the following words:

> A botanic garden is a hybrid type of organisation combining some of the functions of a university, a museum and an experimental station, with the informal recreational aspects of a park system. The tools of a botanic garden, its plant collections, are so employed that they exhibit great aesthetic appeal, along with instructional and inspirational values as well as existing for their primary scientific purposes.

In line with this definition, the Government of Uttar Pradesh in 1948 set up the National Botanic Gardens (NBG) by converting a neglected public park of historical importance. It was, however, taken over by the Council of Scientific and Industrial Research (CSIR), Government of India, on 13 April 1953.

The NBG is located at Lucknow, the capital of the State of Uttar Pradesh, in the heart of the town. The garden part of the Institute is the old historical "Sikander Bagh", which was laid out around 1800 AD by the nawabs of Lucknow, as a royal garden. It was named "Sikander Bagh" by Nawab Wajid Ali Shah, the last nawab of Lucknow, after the name of Sikander Mahal Begum, one of his favourite queens. It lies along the western bank of the river Gomati and is bounded on 2 sides by 2 arterial roads, the Rana Pratap Marg and the Ashok Marg. The main laboratory complex lies outside the garden, across the Ashok Marg.

From the very beginning, the National Botanic Gardens (NBG) was envisaged to be a combination of a botanic garden and an applied botanical research laboratory. However, due to its location along the bank of a flood-prone river, the area of the garden instead of extending, over the years, went on shrinking, making it impossible to develop into a truly national botanic garden, and, like its sister laboratories in the CSIR family, the National Botanic Gardens went on laying increasing emphasis on its applied and developmental research functions in keeping with the national needs and priorities, as spelt out, from time to time, by the Government. In recent years, particular emphasis has been laid on the discovery, domestication, detailed agro-botanical study and utilisation of new vegetable raw materials of non-agricultural/non-traditional economic plants, including ornamentals, a segment of India's rich plant wealth that has not received the attention it is due. The resources of the NBG were mainly canalised towards multi-disciplinary research and development work, by building up a wide spectrum of expertise in basic and applied botany, together with very good laboratory, field and library facilities.

Consequently, it was felt that the name National Botanic Gardens had become a misnomer for the organisation as it no longer projected the correct nature, extent and scope of its functions and activities. Accordingly, the Governing Body of the CSIR decided to change the name to The National Botanical Research Institute (NBRI) with effect from 25 October 1978. The new name well reflects the distinctive character and the research and development activities of this leading applied botanical laboratory cum botanic garden.

A UNESCO-sponsored symposium on "The Role and Goals of Tropical Botanic Gardens", held in 1974 at Kuala Lumpur, Malaysia, highlighted an enlightening but paradoxical fact that though 90% of the world's total plant wealth lies in the tropics and subtropics, these are among the most underdeveloped areas of the world. Another disquieting feature of this situation is that very little systematic research effort has been directed in these countries, of which India is one, towards a scientific study and proper utilisation of the renewable natural resources of the vast vegetational wealth, although there have been some excellent historical precedents in this direction. As far as India's flowering plant wealth is concerned, we are the second richest country in the world, possessing about 15,000 plant species. In a predominantly agricultural country like India, therefore, where dependence of the population on plant resources is complete, planned conservation, multiplication and utilisation of her abundant floristic wealth is an area in which any amount of research activity will not be excessive to meet the country's needs.

That is the reason why increasing productivity of agricultural crops and livestock has always been among the prime concerns of the central as well as the state agricultural/horticultural research institutes and universities. Crop species are not the only important plants, however. There is a whole range of subsidiary agricultural economic plants and plant products, such as medicinal, aromatic and essential oil plants, indigenous herbal drug plants, non-edible oil seeds, seed gums and mucilages, hydrocarbon or petro-plants, ornamental plants, etc., which have an annual turnover of several hundred crores of rupees, but until recently, had not attracted the attention of plant scientists. For instance, the annual turnover of betelvine (*Piper betel*) trade is over Rupees 700 crores with an area of 30,000 ha under its cultivation. No scientific input worth the name had gone into its production and processing, and there was a serious decline in its cultivation. Based on the painstaking basic and applied studies at NBRI, an All-India Coordinated Study has been planned to take care of the research and development needs of this important non-agricultural plant. This is a crop which involves millions of very poor people. The NBRI undertakes research work on such neglected but important plants.

Aims and Objectives

The research and development effort of the NBRI is geared to the introduction, conservation, propagation, protection, genetic upgrading and utilisation of native as well as exotic plant wealth of India, particularly the non-agricultural/non-traditional plants of economic importance, and ornamentals, leading to identification and development of production technologies for new plant resources of commercial importance.

The main aims and objectives of the Institute are:

Basic and applied botanical and related phytochemical researches, aimed at utilisation of plans and plant products and development of production technologies for new plant sources of commercial importance;

Building-up of a National Germplasm Bank of the afore-mentioned plants and maintenance of special collections;

Providing necessary expertise and assistance for identification, supply and exchange of plants and propagules, garden layout and landscaping, and organisation of flower shows, exhibitions and training courses in garden technology and systematic botany, particularly of cultivated plants;

Dissemination of information on the research and development activities of the institute, through publication of scientific and popular literature for scientists and amateur and professional gardeners and planters.

Areas of research/Activities

All the research and development activities of the NBRI fall into the following 11 multidisciplinary areas:

Introduction, Conservation and Documentation of Germplasm
Non-agricultural Industrial Seed Resources
  Seed Mucilages and Gums
  Protein and Lipid-rich Seeds
Ornamental Plants
Pharmaceutical Plants
  Indigenous Herbal Drugs
  Medicinal Plants
  Aromatic Plants
Other Economic Plants
  Subsidiary Food and Fruit Plants
  Sewage-grown Algae and Edible Mushrooms
  Petro-crops
  Betel-vine
Ethnobotany
Environmental Sciences
  Aerobiology
  Environmental Pollution
Utilisation of Alkaline Soil
  "Usar" Land Ecosystems (Woody Biomass/Energy Plantations)
  Pilot-scale Cultivation
Other Areas of Research Work
  Taxonomy and Morphology
Information Services
Extension Services (including Teaching Aids and Materials).

Organisation

The NBRI comprises a well laid-out botanic garden, a rich herbarium, a voluminous library, well-equipped and staffed research laboratories, a modern auditorium and one field research station. The research and development activities of the institute are carried on by the following 13 Disciplines:

Plant Introduction and Acclimatisation
Garden
Taxonomy and Herbarium
Morphology and Palynology
Plant Pathology and Protection
Tissue Culture and Plant Physiology
Genetics and Plant Breeding
Horticulture
Phytochemistry

Environmental Sciences
Information, Publications, Planning and Liaison
Research Stations
Extension Services
Essential Services

The garden is spread over an area of 25 ha and has over 2,000 taxa. It has an arboretum, rosarium, fern house, conservatory, cactus house, palm house, vinetum, large varietal collections of *Bougainvillea, Chrysanthemum, Ixora, Hibiscus, Canna, Nymphaea* and many bulbous plants, experimental plots, a mist propagation facility and glass houses.

The herbarium has about 90,000 plant specimens, properly identified, classified and incorporated.

The Library houses 40,000 volumes and reprints and subscribes to 450 research journals.

The Banthra Research Station is a field-station with about 85 ha of alkaline land located near Banthra, a village about 23 km away from Lucknow, on the Lucknow–Kanpur Road.

Technical aid and advice

The NBRI provides technical aid and advice on:

The occurrence, distribution, identification and economic utility of plants

Garden layout, landscaping and gardening

Cultivation of ornamental, subsidiary food, aromatic, essential oil and medicinal plants together with other non-agricultural economic plants;

Extraction of essential oils

Attends to scientific and technical enquiries.

Flower and plant shows and exhibitions

The Institute organises the following 5 flower and plant shows every year:

Annual Rose and Gladiolus Show, during the month of January
Annual Bougainvillea Show during the month of March/April
Annual House Plants Show during the month of October
Annual Chrysanthemum and Coleus Show during the first or second week of December.

Apart from flower shows, as and when needed, it also organises exhibitions like:

Flowers-in-season
Open days and Science Exhibition
Indigenous Herbal Drugs
Science and Art of Floriculture

There is also a very well laid out permanent exposition of the institute's activities which is very popular with students and the public.

Exchange

The NBRI carries on a regular exchange of information, literature, plants and seeds with over 250 major botanic gardens and industries of the world and other research and teaching organisations.

Public facilities

The garden is open to the public all year round, from sunrise to sunset. Conducted tours of the institute, in particular to the exposition on NBRI activities, are arranged at request. There is a spacious and unique open-air theatre for cultural performances and sale of authentic nursery stock, flower seeds and planting material of non-agricultural economic plants. Help is also given to horticultural societies and clubs.

Economic botany information service

An Economic Botany Information Service (EBIS) to cater to the information needs of the scientist, the industrialist and the planner, covering the research activities of the NBRI, has been set up at the institute. The EBIS is associated with the National Information System for Science and Technology, Government of India, and fulfils a long-felt need in the country for an information agency in the field of economic botany, particularly the non-agricultural economic plants like indigenous herbal drugs, gums and mucilages, non-edible oils and fats, ornamentals, etc.

Scientific expertise and research activities

There is very good scientific expertise at the institute for advanced research and development work in the fields of plant taxonomy, especially of cultivated plants, and plant morphology, seed morphology, palynology, plant tissue culture, indigenous herbal drugs, cytogenetics and plant breeding, mutation breeding, floriculture, plant chemistry, with special reference to seed mucilages and gums, non-edible protein- and lipid-rich seeds and aromatic and essential oil plants, mushroom cultivation, plant virology, and utilisation of alkali soils.

The total staff strength stands at around 600, out of which the scientific and technical personnel engaged directly in research and development activities number about 125 excluding over 30 research scholars working for their doctorates. The NBRI is recognised by a large number of Indian universities as a centre for post-graduate research and advanced studies, leading to the award of PhD degree in various branches of botany, agriculture, horticulture and plant chemistry.

The research activities are organised under 11 research disciplines and 36 projects grouped under 7 areas of work. The individual or discipline-oriented approach prevailing before July 1973, when each laboratory was manned on average by 2·9 and each project by hardly 1·2 scientific workers, was replaced by a collective and institutional approach with over 10 and 5 scientific workers per Discipline and Project respectively. It was not uncommon to find Junior Scientific/Technical Assistants as heads of the Laboratories/Disciplines in July 1973. However, the present Disciplines are headed by Deputy or Assistant Directors. There is now a greater emphasis on the areas of work rather than the Disciplines. The relationship between the 2 is that while scientific and technical workers sit in discipline-oriented laboratories, they are seconded to different multidisciplinary areas of work and Projects.

For historical reasons, the Disciplines of Morphology and Taxonomy have the largest manpower. The transfer of these workers to other Disciplines would not have been a proper solution. However, the research teams under these Disciplines were assigned work on applied morphology, e.g. botanical authentication of indigenous herbal drugs and their differentiation from adulterants; development of easy-to-carry-out morpho-anatomical field tests of important indigenous drugs; correlation of morphological markers and chemical parameters in medicinal-aromatic plants; seed technological studies; seed anatomy, development and histochemistry of non-agricultural industrial seeds; epidermal structure as bio-indicators of the extent and nature of air pollution, etc.

In 1971–2 only 20·86% of funds were available for research purposes while 79·14% was spent on salaries and allowances. Earlier (up to 1971–2), there was steady decline in the expenditure on chemicals, glassware, apparatus and equipment and surrender of funds under these headings was not uncommon. Obviously, one of the very first tasks was to step up the tempo of research. This resulted in a continuous increase in our expenditure on research facilities, so that during 1975–6 nearly 31·10% of the budget was utilised for research work. Soon the figure rose to 40% which under Indian conditions is an ideal balance. Today the NBRI is among the best equipped plant-based institutes in India.

Functioning

In order to discharge its functions effectively and with a sense of urgency, the activities and functioning of NBRI were reorganised in 1973 along the following 3 directions:

Scientific-technical programming;
Laboratory organisation and managerial aspects; and
Liaison with other organisations.

After making a critical appraisal of the past work in 1973, future objectives were clearly enunciated; utility-oriented research and development projects were formulated and work initiated. This resulted in almost total elimination of open-ended and freelance research, and better accountability on the part of workers. Soon the results of oriented-basic and/or applied value began to flow. Simultaneously, managerial aspects were examined in depth and a system was evolved for greater participation by scientists at all levels in decision-making, resulting in better motivation and involvement in work. An atmosphere conducive to collaborative team work was generated in the institute, which is now poised for doing very good and useful work. Essentially, a system was created in NBRI which could plan, organise, execute, communicate, motivate, monitor and evaluate the entire research and development programme. The last two aspects needed to be put on a firmer basis.

Regarding the third direction, while the image of NBRI improved considerably as a result of the reasonably good liaison with other organisations, much remained to be done to develop and strengthen suitable linkages because, in the ultimate analysis, the work of any institute is judged by the quality of research and its socioeconomic impact. It may be pointed out that NBRI, by the very nature of its character, is unique in the CSIR family in as much as it has to play a far greater and more meaningful rôle than many other laboratories in liaison with other organisations, and more so with the public at large. More effective communication and interaction between NBRI and the research and development and teaching institutions, plant industry and public was called for so as to identify the actual problems of the end-users on one hand, and transfer to them the solutions developed by NBRI on the other. Our efforts needed to be canalised in the following directions:

National Level:
- research and development and teaching institutions;
- plant-based industry;
- public.

International Level.

National level

Research and development and teaching institutions

As an applied botanic laboratory-cum-garden, NBRI has research and educational functions and, therefore, it is necessary to interact with botanical/agri-horticultural teaching/research institutions by participating in multi-institutional projects and multi-location trials relevant to NBRI's objectives. It is also necessary to help to improve botanical teaching by providing authentic teaching materials to students and research workers, conduct educational-cum-training courses and field excursions, and answer a large number of enquiries regarding authentication of taxonomic identification, etc. Such an interaction was built up with other laboratories of CSIR, the Indian Council of Agricultural Research, the Central Council of Research in Indigenous Medicine, the Indian Council of Medical Research, U.P. Agro-Industrial Corporation, Forest Departments, Development Authorities, and other agencies, and teaching institutions in botany like the universities, colleges and schools.

Plant-based industry

In order to understand the actual problems faced by the plant industry and to ensure rapid transfer of the results of the work of the institute, get-togethers, discussion-meetings and joint forums are arranged with nurserymen and seedsmen, florists; betel-vine growers, indigenous drug dealers and Ayurvedic and Unani pharmacies; indigenous perfumers; industries utilising pharmaceutical plants, seed mucilages and gums; architects; interior decorators; and greeting card designers and manufacturers.

Public

Providing recreational and public amenities is an important aspect of NBRI's activities. The number of flower shows and festivals was increased and more emphasis was given to organising courses in cultivation and training of ornamentals, horticultural practices, bioaesthetics (flower arrangements; dried flowers, cereal stalks, and drift wood decorations), utilisation of garden waste, landscaping, and garden lay-out. Sale of authentic seeds, seedlings, planting materials, plants, etc., was promoted. Women's Welfare Organisations and Social Welfare Boards were involved in several developmental activities of the garden and in popularising the cultivation of mushrooms. The importance of a botanic garden in an environmental context through educating the public in setting right the ecological imbalance is also being highlighted.

The three-way interaction of NBRI is diagrammatically represented in Fig.1.

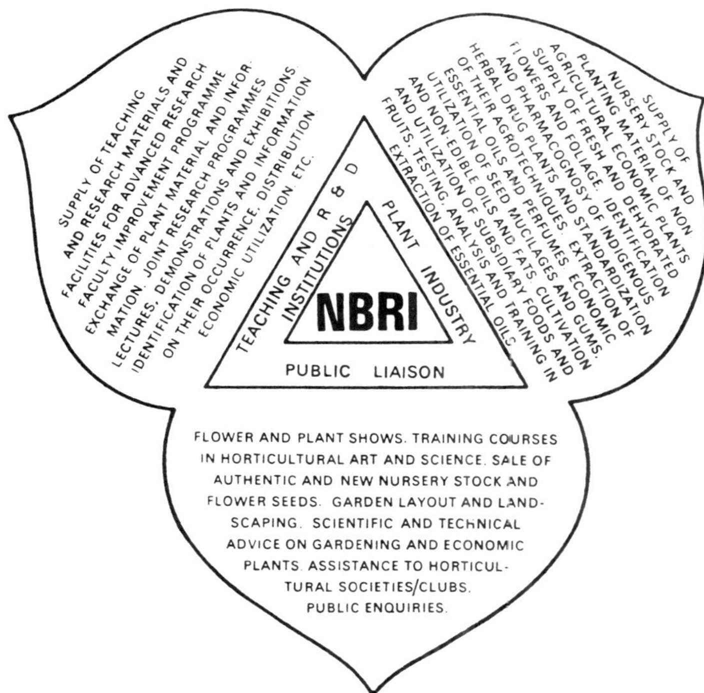

Fig. 1. Three-way interaction of NBRI.

International level

Another aspect progressively emphasised was the rôle that NBRI could play as a tropical botany research and training centre. This was brought out by the author at the UNESCO-sponsored symposium held at Kuala Lumpur. It may be pointed out that at present there is no organisation comparable to NBRI in the region and this institute has the potential to become a leading centre for research and training in tropical botany if, in the next 2–3 years, it is possible to strengthen disciplines of Economic Ecology, Plant Physiology, Agronomy and Germplasm Bank (Seed, Pollen and Tissue Banks), create an Information and Documentation Centre for tropical economic botany, and start training courses in the management of botanic gardens and in plant taxonomy based on the use of computers.

A seed exchange service was established with 190 organisations throughout the world, principally those located in south-east Asia, South and Central Americas, the Middle East, Africa and southern Europe. An *Index Seminum* has been issued and NBRI has been put on the international map of cooperation for seed exchange. In the course of time, seed exchange was expanded to seed and plant exchange. A number of national, regional and international symposia and workshops were also held on important subjects.

In conclusion it may be stated that with the completion of the reorganisation of the research and development programming and managerial aspects, it was possible to highlight the rôle and goals of NBRI in the national and international context.

Banthra Research Station

Due to paucity of arable land, wastelands have become relevant to meet the escalating demand for diverse land-uses. However, utilisation of such lands poses a major research and development challenge. At NBRI Banthra Research Station, experiments for the utilisation of alkaline (*usar*) land through ecodevelopment were conducted successfully. The approach was holistic and involved ecologically clean, low-energy as also low-engineering inputs and utilisation of human power on a voluntary basis. Essentially the work is multi-sectoral in character, and is aimed to help the small farmer through intensification and diversification of biomass production for food, fodder, fuel, fibre, fertiliser, small timber, aquaculture, medicare, and small village-level biomass-based mini-industry. A by-product of such an approach is an improvement in bioaesthetics and micro-meteorology. Initially different trees, shrubs and annuals tolerant to *usar* conditions were used. In turn, these gave place to less tolerant species because of a perceptible amelioration in the quality of the land itself. Cultivation of many species was also cost-effective. Accordingly, a Biomass Research Centre especially for firewood was established at Banthra with the help of the Department of Science and Technology of the Government of India.

The transformation of barren wasteland devoid of any worthwhile vegetation into a green area was possible through community action involving local people. This brought about environmental, social and economic rehabilitation of the area as a whole. Today, the challenge lies in converting such a micro-level success into macro-level one. This model of ecodevelopment has 5 components — cropland, grassland, woodland, aquaculture and biomass-based vocations (Fig.2).

Research highlights

Results of the scientific investigations carried out in the institute have been published in more than 1,400 research papers and review articles, 10 books, 300 popular articles and 163 bulletins. Research projects sponsored by various agencies in and outside the country have been under investigation.

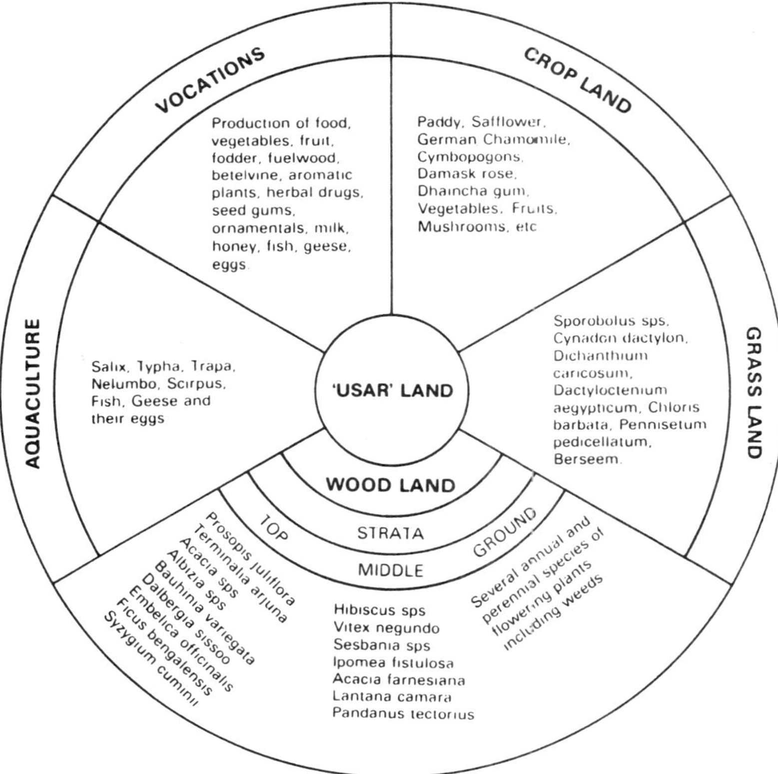

Fig. 2. Model of ecodevelopment.

Collaborative research programmes with sister CSIR laboratories and other agencies are also in progress. The NBRI is a centre of advanced study leading to the award of PhD degrees of a large number of universities. Over 80 members of the scientific staff and research fellows have so far obtained their PhD degrees for research work done at NBRI in various subjects. The chief research achievements are listed below:

Introduction of nearly 250 species of ornamental, aromatic and medicinal plants with particular reference to trees and shrubs.
Pharmacognostic studies on the indigenous herbal drugs, their adulterants and substitutes with a view to standardising their botanical identity.
Upgrading phytochemical processing of aromatic plants like damask rose, jasmines and tuberose.
Genetic upgrading of medicinal plants like opium poppy and *Dioscorea floribunda.*
Mass multiplication through tissue culture of medicinal, aromatic, forestry and ornamental plants.
Cytogenetical studies in relation to evolution and improvement of ornamentals, subsidiary foods (e.g. amaranths) and forest tree species together with related biosystematical and experimental evolutionary studies.
Circumscription of the genepools together with extent and nature of gene exchange between members of a genepool.
Programme of precision blooming in chrysanthemum through photperiodic control.
Control of virus diseases in some aromatic, medicinal and aromatic plants.
Botany and phytochemistry of gum-yielding seed legumes, non-edible oil and protein rich seeds.
Biotechnology of sewage grown *Spirulina.*
Improvement in agronomic practices and shelf-life, control of diseases and phytochemical analysis of betel-vine varieties in India.
Microbiological production of gibberellins.
Breeding biology and assessment of natural variability in Himalayan conifers.
Introduction and acclimatisation of grapes in the Gangetic plains.
Economic utilisation of alkaline land using biological methods.
Ethnobiological studies on tribals of Uttar Pradesh.
Palynological and aerobiological studies in relation to taxonomy and allergy.
Morphology and taxonomy of pteridophytes and lichens.

Seed anatomy of angiosperms.
Cuticular and epidermal studies as an aid to taxonomy and in relation to air pollution.
Preparation of district Floras.
Phytochemical screening of over 400 species of Indian angiosperms.

Some of the problems pursued by cytogenetics, plant breeding and tissue culture teams are indeed basic-oriented and also support applied work in progress in different areas. The basic-oriented work of the cytogenetics team can be considerably widened if a multi-disciplinary crop evolution team is organised. This is essential because today biosystematics and experimental evolutionary studies have become out of fashion with plant breeders and biotechnologists. However, the importance of these subjects has not decreased, particularly the utilisation and conservation of wild stocks. Furthermore, conventional genetics and plant breeding will continue to remain relevant for the foreseeable future. As a result, centres of diversity, the extent and nature of gene exchange within the genepools and circumscription of the limits of genepools, etc., are also equally relevant. However, in today's contex, basic botany deals with physiological, genetical and developmental aspects at cellular/molecular level. Judged against this background, most of the so-called basic botany at the NBRI is actually classical in character. However, even classical information is not available for tropical plants. The importance of tropical botany has been stressed time and again ever since the Fairchild Report was submitted in 1960. The tropics harbour the world's most underdeveloped countries, and also about 90% of the world's plant wealth, but so far only 10% of the total research effort has been put in this area. In the tropics, plant resources provide the basic and primary means of humankind's subsistence and, therefore, on them depends improved human welfare. Centres of origin/diversity of many of our important cultivated plants (about 152 species in India alone) are located in the tropics. Therefore, for the advancement of knowledge and for obtaining a total picture of the science of botany, classical botanical information on tropical plants is of vital importance, and would be a prelude for working out fully the economic potentialities of tropical plants. In India, and perhaps even in the region, NBRI may be the only place where such work has been taken up in an integrated fashion, and the first step in this direction was to organise an advanced centre for management of botanic gardens and training in plant taxonomy. Furthermore, while the Botanical Survey of India is the repository for flowering plants including their type specimens, there is no such organisation for the lower plants, and NBRI can take up this work.

Developmental work in connection with proper economic utilisation of non-agricultural economic plants is composed of two different but interrelated components, i.e. pilot scale cultivation and processing of the actually utilisable chemical product. We had not paid attention to both the components of this chain. Although, over the years, we have been taking up small-scale cultivation of an unusually large number of items, our actual performance was poor for want of proper priorities, time-phasing and a statistical approach, with hardly any appreciation for and emphasis on such work. In fact, this had been one of the serious lacunae in our functioning. However, since 1973, due importance was given to this aspect, and, what is most important, efforts have been made to establish a firm linkage between the agronomical and chemical components of the technology so as to produce a single coherent package, from cultivation to production, processing and utilisation. It is only recently that we have tried to expose these packages to cost-benefit analyses so as to have an idea if they are commercially viable. Some of the successful packages are: cultivation of ornamentals for cut-flower and propagule production; aromatic plants like *Abelmoschus moschatus*, Damask rose, matricaria, tuberose, etc. Other such cost-effective packages are: Dhaincha gum, *Dioscorea floribunda*, betel-vine, firewood production using shrubs and trees, and year-round mushroom cultivation.

For the developmental work relating to chemical processing of the pharmaceutical plants and non-agricultural industrial seed resources, NBRI co-ordinates its efforts with and utilises the pilot plant facilities and expertise already available with sister institutes.

Products and technologies developed

From the research and development work briefly summarised above, the following products and technologies have emanated:

(1) Over 75 new cultivars of 11 well-known ornamental plants, viz. amaranths, antirrhinum, bougainvillea, chrysanthemum, erythrina, gladiolus, marigold, *Nyctanthes arbor-tristis* (Har Singar), portulaca, rose and tuberose (*Polyanthes tuberosa*). Many of these have gone into the nursery trade.

(2) Chrysanthemum varieties evolved for every month in the year.

(3) Photoperiodic control of blooming in specific chrysanthemum cultivars, making the plants bloom for 6 months instead of the usual 6 weeks.

(4) Technology for raising F1 hybrid seed of triploid marigold,

produced by crossing male sterile African tall and male fertile French dwarf varieties.

(5) Rapid mass-propagation of some difficult-to-grow orchids through tissue culture.

(6) Formulations for prolongation of vase-life of bougainvillea, gladiolus, tuberose and chrysanthemum.

(7) Dehydration of fresh flowers and foliage, and their utilisation.

(8) Introduction of unusual economic plants, such as Jojoba (*Simmondsia chinensis*) — a source of a superior lubricant; *Kallstroemia pubescens* — a new source of diosgenin; *Euphorbia tirucalli* — a possible petro-crop; Guayule (*Parthenium argentatum*) — a rubber plant, etc..

(9) Rapid mass propagation of *Dioscorea floribunda*, the pill-plant, through tissue culture, at a nominal cost.

(10) Introduction into north India of Winged Bean (*Psophocarpus tetragonolobus*), a multipurpose legume with high content of protein and oil in seed and starch in tuberous roots.

(11) Technology for cottage-scale cultivation of edible mushrooms like: button mushroom (*Agaricus bisporus*), paddy-straw mushroom (*Volvariella volvacea*), ''Dhingri'' (*Pleurotus sajor-caju*), and *Pleurotus cystidiosus*.

(12) Aerobiological survey of Indian cities and its correlation with allergenic diseases.

(13) Introduction of grapes to north India and forecasting productivity, preponing ripening and curing mineral deficiencies of grapes.

(14) Cost-effective cultivation of shrub and tree biomass on alkaline land for firewood.

(15) Extraction of industrial gum from ''Dhaincha'' (*Sesbania bispinosa*) seeds.

(16) Production of virus-free propagating material of chrysanthemum and citrus by tissue culture.

(17) A process and field distillation unit for Rose Oil distillation — Indian Patent No. 117510.

(18) A new apparatus for distillation of Perfumed Waters and Attars — Indian Patent No. 130396.

(19) A process for preparation of raisins from ''Pusa Seedless'' variety of grapes — Indian Patent No. 118187.

(20) A process for making substitutes for Canada Balsam — Indian Patent No. 118017.

(21) A process for separation and isolation of aromatic principles of *Abelmoschus moschatus* seed — Indian Patent No. 72408.
(22) A new process for preparation of frozen mango pulp from "Dashehri" and "Chausa" varieties — Indian Patent No. 124082.
(23) A new process for the improvement of Tamarind Kernel Powder from seeds of *Tamarindus indica* — Indian Patent No. 130997.
(24) Production, in sewage, of *Spirulina platensis*, a protein-rich alga, for animal feed.
(25) Eradication of noxious algae in garden tanks and ponds.
(26) Pollen-spore Teaching Aid Kit.

Concluding remarks

Because of its concern with botany, whose direct industrial/economic implications were regarded as limited, NBRI was at one time considered a misfit in the Council of Scientific and Industrial Research (CSIR) of the Government of India. However, with time, this very weakness became NBRI's principal strength due to the realisation that plant wealth is a renewable resource and there is a world-wide awareness about its conservation and planned utilisation.

In India, there is a well organised Botanical Survey of India (BSI) whose sole mandate is to prepare the Flora of India. BSI is backed by a strong university set-up. We also have a very strong agricultural set-up in the form of agricultural universities and the Indian Council of Agricultural Research maintaining a number of agricultural research institutes. The connecting link between the BSI-University System and ICAR-Agricultural University System is NBRI. In this context, the chief function of NBRI is, therefore, to discover and/or introduce new economic plants as sources of commercial raw materials and to develop suitable agro-botanical and processing technologies for them. Besides new food and ornamental plants, the types of commercial raw materials that are being sought are seed-oils, proteins and gums, steroids, alkaloids, glycosides, tannins, hydrocarbons, fibrous cellulose, biodegradable pesticides, insect repellents, etc.

By the very nature of its charter, NBRI is indeed unique as far as its potentialities of interaction with industry, teaching and research institutions and the public are concerned. The machinery for this purpose had, however, to be strengthened with appropriately qualified staff.

When planning a National Botanical Garden in any country in the Commonwealth, one immediately thinks of the Royal Botanic Gardens, Kew, as the model, often forgetting that Kew Gardens is over 200 years old and that whole of the Commonwealth has contributed materially towards building it. Due to ever-increasing recurring maintenance costs, Kew itself may need to have second thoughts about maintaining world collections. During discussions with many colleagues at Kew and elsewhere, it emerged that for the developing countries, the NBRI concept of having a botanic garden-cum-economic botany laboratory may offer an alternative to the Kew model. Furthermore, for a variety of reasons, the Kew concept is difficult to follow by any country, more so a developing south-east Asian one. For instance, to maintain tropical collections under temperate conditions by central heating is far cheaper and easier than to maintain temperate collections in tropical conditions by air-conditioning. Fortunately India, because of the great diversity in climates, can achieve the same results by resorting to substations in different agro-climates. This would turn out cheaper for us and if we are able to create a National Grid or Consortium of all our botanic gardens, preferably under one central authority, then maintenance of germplasm collections would be still cheaper and the system would function far more efficiently.

Several botanists all over the world have expressed appreciation and NBRI is now considered among the first 5 botanic gardens of the Orient (cf. *Missouri Botanic Garden Bulletin* **53**(2), 9–12). NBRI is regarded as a new, evolving and dynamic concept of a botanic garden. That botanists/agriculturists/horticulturists and decision-makers in India even today insist on the Kew model is because of the British influence and also because it is the only concept known to them. Such a lack of appreciation was, in fact, the reason the clock was reversed during 1965–71, and due to over-enthusiasm to fall in line with CSIR's industrial culture, without proper scientific back-up support, attempts were made to convert NBG into a very routine horticultural institution. This demolished the very edifice on which NBG-NBRI stood. Such thinking was entirely erroneous because an industrial culture cannot exist without proper scientific support. It can now be stated with confidence that with a healthy mix of scientific and industrial work, NBRI has turned the corner and is in a creative phase. It has earned a definite place for itself in Indian botany and in the tropics. It is necessary to build upon the concept of tropical economic botany and many of the facts stated in this paper are pointers towards such a national, regional and international rôle for NBRI.

Botanic gardens in the process of national development: the case of Indonesia

S. SASTRAPRADJA, DIDIN S. SASTRAPRADJA and USEP SOETISNA

Puslitbang Bioteknologi, Bogor, Indonesia

Summary

Long before Indonesia proclaimed her independence, 2 botanic gardens had been established by the Dutch, namely the Bogor Garden (1817) and the Cibodas Garden (1874). At the end of the Dutch Administration another garden, Purwodadi (1941), came into being, soon to be followed by Bedugul, Bali (1959). Each garden has its own characteristics and their beauty reflects the richness of Indonesia in plant resources.

In a developing country like Indonesia it is not easy to develop and maintain a botanic garden. Scientific merit and conservation alone would not be sufficient to attract the government's support. Ways and means have been put forward to convince the policy-makers that a botanic garden performs useful rôles in national development, given a chance. Another important factor which ensures the sustenance of the botanic gardens is the involvement of the public at large.

In this paper, the individual botanic gardens are described and the most recent development of such a garden in Indonesia, Serpong (1984), is outlined with special reference to its challenging future.

Botanic Gardens and the
World Conservation Strategy

ISBN: 0-12-125462-3

Resumen

Jardines botánicos en el proceso de desarrollo nacional: el caso de Indonesia

Mucho antes de que Indonesia proclamase su independencia, los holandeses habían establecido dos jardines botánicos, denominados el Jardín de Bogor (1817) y el Jardín de Cibodas (1874). Al término de la administración holandesa se constituyó otro Jardín, el de Purwowadi (1941) que pronto sería seguido por el de Eka Karya Bali (1959). Cada jardín muestra sus características y belleza siendo el reflejo de la riqueza de Indonesia en recursos de plantas.

En un país en vías de desarrollo como es Indonesia, no es fácil desarrollar y mantener un jardín botánico. El mérito científico y la conservación únicamente, no serían suficientes para atraer el apoyo del Gobierno. Se han propuesto vías y argumentos para convencer a los políticos para que den una oportunidad a los jardines botánicos a fin de que puedan desempeñar útiles papeles en el desarrollo nacional. Otro importante factor que asegura el sostenimiento de los jardines botánicos es el de implicarlos grandemente con el público.

En esta ponencia se somete a discusión el desarrollo más reciente de un jardín en Indonesia, el de Serpong (1984) haciendo especial referencia a su desafiante futuro.

Introduction

Established in 1817, the Bogor Garden (Kebun Raya) is the first and the oldest botanic garden in Indonesia. Therefore, compared to the other Indonesian botanic gardens, Bogor offers the full beauty of mature trees interspersed with lawns and ponds. Bogor also has the highest number of species of any garden in Indonesia.

Following the Bogor Garden, the mountain garden at Cibodas was established in 1874, on the slope of Mount Gedeh, about 45 km south of Bogor. Not until 70 years later then, in 1941, was Purwodadi chosen as the place for a lowland dry garden. Later, in 1959, another mountain garden was established at Bedugul, Bali (1959), complementing the Cibodas Garden.

In 1976 a centre for science and technology was developed in Serpong, which is about 20 km south of Jakarta. This complex occupies about 300 ha of land, of which 150 ha is allotted for buildings. The National Biological Institute, to which the botanic gardens belong, was asked to design the open space for the plant collections. The planning was carried out in 1978, but the actual planting was not done until 1982. Concurrent with the inauguration of the whole complex as a scientific and technological centre, the Serpong Garden was officially established on 18 December 1984. How the garden differs from the sister gardens and how these gardens fit into the programme of the national development are discussed below.

The history of the gardens

As a tropical country, Indonesia is rich in plant resources. Many of the plants are economically important. Therefore, in the early days, the Europeans were attracted by this potential. Spices were the main reason for them to come to Indonesia. Realising the importance of plant resources, in 1815 an institute of plant research was established to exploit the natural resources, in particular plants. This institute also studied the adaptation of the introduced crops with the aim of developing them into cash crops.

As a consequence, a garden was developed in the grounds of the Governor's Palace occupying 87 ha, and the Bogor Garden was founded in 1817 by Reinwardt. It was soon placed in the charge of Teysman and it was he who organised the garden in such a way that the plants are grouped according to family. Since then, a combination of science and landscape architecture has been an essential in developing new gardens.

A research laboratory for biology was developed in 1884, complementing the garden and the herbarium (1844). This laboratory accommodates visiting botanists who intend to do research on tropical flora. This laboratory was named after Treub, one of the famous directors, and is at present known as Pusat Penelitian Botani. Soon, in 1894, a sister laboratory emphasising zoology was developed, which is known even today as the Museum Zoologicum Bogoriense. The 4 organisations, namely the botanic gardens, the herbarium, the plant laboratory and the zoological museum, are the daughter institutes of the National Biological Institute.

To appreciate the individual character, as well as the beauty of the botanic gardens of Indonesia, brief accounts of each of them follow.

The Bogor Garden (87 ha)

Located about 60 km south of Jakarta, the capital city of Indonesia, the Bogor Garden was established on the slope of Mount Salak. Compared to Jakarta, Bogor has a cooler climate, averaging 26°C throughout the year. Bogor is also known as the Rain City, because of its high annual rainfall of about 2,500 mm.

There are now about 12,000 registered accessions comprising more than 5,000 species, excluding annuals, of which the tree species predominate. In this garden, the oldest oil palm of Asia (1848) is still in existence, from which all oil palms cultivated in south-east Asia are derived. Other economic palm species grown in the garden include sago (*Metroxylon sago*), aren (*Arenga pinnata*) and salak (*Salacca edulis*). The garden also has an excellent general palm collection.

Another group worthy of mention are the bamboos; the giant bamboo (*Dendrocalamus giganteus*), which was uprooted by the heavy storm in 1971, dominates the group. The giant arum (*Amorphophallus titanum*) attracts visitors to the garden when it is in flower. Native to Sumatra, this species is closely related to the elephant yam (*A. campanulatus*). The corm, which may attain a diameter of 1 m, is likely to be exploited in the future for its manna. The current producer of commercial manna is *A. blumei*, a wild species common in East Java.

There are many introduced species in the Bogor Garden. The famous cannon ball tree (*Couroupita guianensis*), the giant water lily (*Victoria amazonica*, Syn. *V. regia*) and the cacao (*Theobroma cacao*) are just a few examples. This reflects its early function as a tropical botanic garden that put emphasis on the botanical exploration of the region and on the introduction of economic plants from elsewhere regardless of their origin.

The Cibodas Garden (82 ha)

Beyond Bogor, on the way to Bandung, the capital city of West Java Province, there are two volcanoes, Gedeh and Pangrango. Located 45 km south of Bogor, on the slope of Mt Gedeh, the Cibodas Garden was established in 1874 at an altitude of about 1,400 m. The average temperature is 17°C during the day and 10°C during the night, while the annual rainfall is 4,000 mm, distributed almost evenly throughout the year.

The Cibodas Garden is noted for its large collection of conifers. *Araucaria, Cupressus* and *Podocarpus* are commonly planted, giving an impression of a European landscape. Eucalypts are another group that dominate the plantings in parts of the garden.

The garden is celebrated for its cacti and succulent collection. These groups are, of course, of foreign origin, but they are well adapted to the Cibodas climate. Together with the gloxinias, they are permanently exhibited in the glasshouses.

The original site of the garden occupied 54 ha of land. In 1976, an area of 27 ha was developed as an annex of the old garden. This new area has been designed in such a way that scenic beauty is given much more emphasis. The idea behind it is to ease the number of visitors to the old establishment, because on Sundays and holidays they make the garden overcrowded. The main reason why the average person visits the garden is for recreation rather than for any scientific purpose. However, a plan to develop a new area for maintaining collections is on the way.

As part of the effort to conserve natural beauty and plant resources, Indonesia is also developing a system of national parks all over the country. Gedeh-Pangrango National Park is just behind the garden toward the crater of Mt Gedeh. This particular park is also designated as one of the 5 biosphere reserves of the UNESCO/MAB Programme.

A laboratory for biological research was donated by UNESCO in early 1953. When a College of Biology was established in 1955, part of the laboratory was turned into a dormitory and a classroom. Soon, the college was moved to Bogor and the space was occasionally used by the groups of students who wanted to stay in the garden for short visits. Unfortunately, in early 1984, a strong storm hit the garden, leaving damage in several areas including part of the laboratory, restoration of which is now underway.

Realising the potential of this area for tourism, the District Government allotted 25 ha of land beside the garden for a camping site. With the 3 points of interest — the national park, the botanic garden and the camping site — on weekends and on holidays the garden can attract as many visitors as does the Bogor Garden.

The Purwodadi Garden (90 ha)

Compared with the 2 gardens described above, this garden is relatively young. Located between 2 major cities in the Province of East Java — Surabaya, the capital of the province, and Malang, a resort city — Purwodadi is at an altitude of about 200 m. Unlike Bogor or Cibodas, the annual rainfall amounts to only 1,000 mm and the rainy season is limited to approximately 6 months between November and April. The daily temperature is higher than Bogor with an average of 30°C.

During the dry season, water supply is a problem. Though a water system has been developed to assure the availability of water all the year round, the capacity of the creek on which the garden depends is small. Because of this, the garden looks different in the dry season from in the rainy season.

It was mentioned earlier that the development of Purwodadi Garden was first started in 1941. Not until recently, however, did the garden concentrate its efforts on the quality of the collection. Attempts have been made in the past several years to focus collecting on the lowland dry region in East Java, Bali, Nusa Tenggara Barat and Nusa Tenggara Timur. Tree species are the main target.

The collection of *Ficus* species is quite outstanding. The garden has made use of the services of Mr Mudjiono, a retired gardener, for the purpose of local exploration. The genus *Ficus* is well represented in eastern Indonesia; in Nusa Tenggara Timur, for example, one can easily collect many different species. For those who wish to study the development and the establishment of fig trees, this area is certainly an open laboratory.

Eastern Indonesia also offers a great diversity of native *Croton* species. Therefore, the garden is proud of having a great variety of crotons showing differences in leaf morphology and combination of colours. At least 50 different varieties are in the collection.

In future the Purwodadi Garden will specialise in tree legumes. It is already beginning to accumulate species of *Albizia, Acacia* and *Dalbergia*. Many of these species are appreciated for their timber and foliage. Among other tree legumes which are attractive in the dry season are *Erythrina* and *Cassia* species.

From the point of view of economic plants, the garden maintains a collection of races of three species, namely sugar cane, banana and mango. The IBPGR twice sponsored the sugar cane collection in Indonesia and a set of samples have been maintained by the Sugar Institute in Parasuan which, for the time being, is kept in the botanic garden. As for banana, both wild and cultivated samples are in the collection, which is supported financially by IBPGR.

The Bali Garden (129 ha)

To complement the Cibodas Garden, the late Professor Kusnoto, the first Indonesian director of the chain of botanic gardens, decided to develop a mountain garden in Bali, which was officially opened in 1959. This garden is located between the 2 major cities in Bali — Denpaser, the capital city of the province, and Singaraja, a large city in the north of the island. Located at 1,500 m, the Bali Garden has a cool climate, averaging 18°C in temperature with an annual rainfall of 3,000 mm, which is not distributed evenly over the year.

The garden originally occupied 57 ha of forest land of which only 30 ha were developed into a garden in early 1960. Ten years after the development, another 72 ha were added so that the garden now occupies 129 ha of land.

In terms of the collections, the systematic exploration of the mountain region in east Indonesia has been organised over the past few years. The most significant results are the orchid collections. These are now maintained outdoors in contrast to those of the Bogor collection, which are kept in the glasshouses.

Among the tree species collected are *Manglietia glauca, Podocarpus javanicus, Bischoffia* sp., *Ficus* spp., *Alstonia* spp., *Pometia* sp., *Flacourtia* sp. and *Elaeocarpus* sp. However, they are still in the nursery, because they are not yet ready to be planted in the garden.

Funds have been allocated by the Central Government to finance the physical development of the garden. A large scale nursery, a two-storey office building, concrete paths and water reservoirs are now under construction. By next year, all the physical development is expected to be complete.

Like the Cibodas Garden, the Bali garden is adjacent to a protected area, Batu Kau Nature Reserve. The population of cemara pandak (*Podocarpus javanicus*) is a prominent feature of that mountain. The garden overlooks the beautiful Beratan Lake, which is the main water reservoir for west Bali.

The Serpong Garden (about 150 ha)

Since 1976 a science and technology research and development centre has been developed in Serpong, south of Jakarta. This occupies 300 ha of land hosting several modern technological laboratories from aircraft development to chemical engineering.

Of the 300 ha of land, approximately 150 ha will be used for building sites and residential areas, leaving the 150 ha for open space. This complex is under the jurisdiction of the Minister of Research and Technology who, from the very beginning of this central development, has expressed the desire to turn the open area into a sort of botanic garden.

To convert his wish into a reality, in 1978 the National Biological Institute was asked to formulate the development of the Serpong Garden. While the planning took place, the planting was started immediately in 1979. Unfortunately, however, the project was terminated due to lack of funds. In January 1982, the new Minister of Research and Technology was so impressed by the Bogor garden that he renewed the commitment of the National Biological Institute to develop the garden in Serpong.

Facing up to the many challenges, the National Biological Institute did not hesitate in accepting the task. In September 1982, the land that was ready to be cleared was prepared and planted. The National Biological Institute has asked the Minister for at least a 10-year commitment to lay out the foundations of the whole complex.

Unlike the previous gardens, Serpong Garden has a unique status, both from a philosophical as well as from a managerial point of view. While the scientific background is still maintained, the type of collection is planned differently. In the botanic gardens, the core of the collections is species diversity, but in Serpong variation of cultivated species will also be covered, especially species of selected economic value. The latter have been deliberately introduced in Serpong to ensure an annual income for the maintenance of the collection.

Though Serpong Garden is not normally open to the public, unlike the rest of the botanic gardens, schoolchildren and group visitors are welcome. In the garden, all 27 provinces of Indonesia are represented through species which have been selected from each province. The selection of the species was based mainly on their ethnobotanical value, rather than for their aesthetic appearance. Therefore, central Java, for example, is represented by *Michelia alba*, Nusa Tenggara Timur by *Santalum album* and Sulawesi Tengah by *Pigafetta filaris*. The wayside trees are species such as *Agathis damara, Podocarpus fimbricata, Pigafetta filaris, Palaquium* sp. or the famous *Mimusops elengii.*

Introduced species are also used in Serpong. Most of the borders were planted with *Canna indica* and *Bougainvillea*. For lawns, the National Biological Institute selected *Stenotaphrum dimidiatum*, a grass introduced in 1974 by a Mr Webster for trial in Bogor and subsequently found to be very fast-growing and adapted to various soil conditions.

The Bogor Garden and national development

During its early history, the Bogor Botanic Garden was very active in plant introduction. Crop species of foreign origin such as oil palm, rubber, cacao and coffee were introduced and evaluated. Soon it was discovered that the wide range of climatic regions in Indonesia offered a good opportunity for these plants to grow and produce whatever was expected of them. Since then they have been among the cash crops of the country. However, plant introduction is now the responsibility of the agricultural institutes.

As one of the daughter institutes of the National Biological Institute, the garden, together with the herbarium, is responsible for plant exploration and collection. For the living collections priority has been given to those indigenous species with economic potential and scientific importance. Moreover, those species which are subjected to genetic erosion receive the highest priority.

Research activity is focussed on the evaluation of the biological characters, which in turn are employed for determining taxonomic status, estimating economic value, and preparing for the mode of conservation. This kind of activity is executed with the other daughter institutes of the NBI (Treub Laboratory, the Herbarium and the Zoological Museum). Every 5 years, the objectives of the research programme are adjusted according to the 5-year national development plan, and the activities are linked with those of agriculture and forestry. The programme of 1984–9, for example, is focussing on the legumes, especially the genera *Acacia* and *Albizia*. Many of these species are valuable as fodder, timber, fuelwood or even for honey-bee farming, especially in the dry regions of Indonesia.

In the field of conservation, the botanic garden performs its function as a place of *ex situ* conservation, complementary to *in situ* conservation, together with the collections of other gardens, arboreta and the provincial gardens. The networking of conservation efforts is coordinated through the National Committee on Plant Genetic Resources Conservation, which was established in 1976. With the networking, there is division of labour among the institutes handling the plant resources conservation: the botanic gardens are responsible for botanical species with their variation, while the agricultural institutes are responsible for crop variation. In addition to this, the gardens are also active in *in situ* conservation through the MAB Programme together with the Directorate General of Forest Protection and Nature Conservation. The *in situ* conservation areas are under this Directorate General, which is also the CITES Management Authority, while the botanic gardens are the CITES Scientific Authority.

Indonesia is active in dealing with environmental problems. As in conservation, the gardens are active in joining the environmental research programme through, among others, the programmes of the Indonesian National MAB Committee. The gardens provide planting materials for landscape purposes and give advice on landscape architecture to other institutes that need it. Taking advantage of this opportunity, the gardens select the indigenous species to be used in developing a garden. In this way, a contribution is made to the conservation of the species.

One of the functions of botanic gardens is public education. In this particular field, the Indonesian gardens offer a lot. The display of the collections readily demonstrates the rich diversity of the country's flora together with the richness of the tropics. Each plant is provided with a label showing the place of origin so that one can easily learn where the specimen has come from. Information on the properties of species are given to the public through the garden's semi-popular bulletin. At every opportunity, the garden exhibits the uniqueness of Indonesian species, such as *Amorphophallus titanum* and *Rafflesia arnoldii*.

Being a government institute, the garden has to be open to the public as a place of recreation. To avoid excessive visitor numbers the garden charges a higher fee during weekdays, but allows free entry for those who visit the garden for educational or scientific purposes.

The garden is considered as one of the main tourist attractions. Because of the many visitors, the economy based on the garden is developing as well. Many food stalls, vegetable stands and souvenir trays are offered around the gardens. However the negative impact of the public visits is greatly felt. They leave rubbish everywhere, pluck the flowers, even though this is prohibited, change the labels, etc. All of these problems will continue to occur in the years to come, in addition to the inadequacy of the maintenance budget and insufficient qualified staff. Despite this the garden should and will survive. For almost 170 years the garden has overcome difficulties. Given the same opportunity, the chain of botanic gardens will be in existence indefinitely.

80,000 plants in South America: the case for creating more botanic gardens

E. FORERO

Missouri Botanical Garden, St Louis, U.S.A.

Summary

Interest in tropical and subtropical South American botanic gardens has increased over the last few years. Efforts are being made by various people and institutions to gather information about them. Governmental agencies as well as non-governmental organisations in some countries have shown their willingness to create new botanic gardens and to support those already in existence.

Readily available published information on the botanic gardens in the region is scarce. Nevertheless, this paper aims at presenting an overview of existing South American botanic gardens covering their location, administrative organisation, size, history, present programme and future goals.

The rôle of botanic gardens in tropical and subtropical South America is discussed. Suggestions are made as to what these institutions might do in order to have a greater impact on local and international programmes to conserve natural resources through education, basic and applied research, data-gathering, and so on.

The need for closer cooperation among botanic gardens at the national and South American levels is pointed out, and the convenience of establishing sound collaborative programmes with sister institutions in Europe and the United States is strongly emphasised.

Botanic Gardens and the World Conservation Strategy

ISBN: 0-12-125462-3

Resumen

80.000 plantas en Sudamérica; el caso para los jardines botánicos

En los últimos años ha crecido el interés acerca de los jardines botánicos tropicales y subtropicales de Sudamérica. Se han realizado esfuerzos por parte de diversas personas e instituciones para recopilar información sobre ellos. Los organismos gubernamentales así como las organizaciones no gubernamentales han mostrado en varios países su buena voluntad para crear nuevos jardines botánicos y sostener aquellos ya existentes.

Actualmente resulta escasa la información publicada disponible sobre los jardines botánicos que existen en la región. No obstante esta ponencia está encaminada a presentar un vistazo sobre los jardines botánicos sudamericanos, abarcando su localización, organización administrativa, dimensiones, historia, programa presente y futuro éxitos o realizaciones.

Se somete a discusión el papel de los jardines botánicos sudamericanos de la zona tropical y subtropical. Se formulan sugerencias acerca de lo que estas instituciones pueden realizar para tener todavía un mayor impacto en programas locales e internacionales para conservar los recursos naturales, por medio de la educación, la investigación básica y aplicada, la recopilación de datos y así sucesivamente.

Interest in the botanic gardens of tropical and subtropical South America has, in recent years, increased considerably, and various people and organisations have been making efforts to collect together the maximum possible amount of information about them. In some countries, governmental and non-governmental agencies have demonstrated interest in creating new botanic gardens or in helping the development of those that already exist.

In this paper, I shall give some examples of the work that is currently going on in this field in various parts of South America. These initiatives are more or less isolated, but serve to demonstrate the interest and dedication of our botanists in all aspects of plant conservation.

It has repeatedly been said that the flora of Latin America is, in terms of species numbers, the richest in the world. The figure which appears in the title of this contribution "80,000 species of plants" is a reasonable estimate although the real figure covered could be as high as 90,000. At present it is not possible to give a precise figure because so much of the region is inadequately explored botanically and because modern Floras have been written for only a tiny percentage of the species and areas. Every new collecting trip may find new species in abundance, a situation that does seem very different from all other regions of the world. Raven (this symposium) quotes figures of 35,000 species for tropical and subtropical

Africa and 50,000 for tropical and subtropical Asia. Whatever the precise figures, the flora of South America is certainly very rich in absolute terms and certainly richer than the flora of any other tropical region. Yet the dangers to this wealth of plant life are very grave as forests continue to be cut down and degraded. The challenge before us is very great.

Functions of botanic gardens

Let us consider briefly some of the basic functions of botanic gardens. Most of the active botanic gardens in South America carry out research, teaching and conservation, etc., although in different combinations and with varying emphases on each. Many are actively studying the local flora. Among these we can list those of Medellín, Cartagena, Tuluá and Bogotá in Columbia; Coro in Venezuela; Rio de Janeiro in Brazil; and Punta Arenas in Chile.

One of the largest herbaria of Latin America is the Jardim Botânico do Rio de Janeiro with almost 300,000 specimens. Many botanic gardens have their own herbarium or are in some way associated with universities or other institutions which have them. These two examples serve to emphasise 2 important points: firstly, we cannot conserve that which we do not know and, therefore, floristic studies are very important. Secondly, even now it is still necessary to awaken the conservationist conscience of many taxonomists, especially because there is frequently no direct relationship between work in the herbarium and the living plants of the surrounding area or associated botanic garden.

It is necessary in botanic gardens to promote conservation-related research on aspects such as seed production, dispersion, pollination mechanisms, etc. In the Jardín Botánica "José Celestino Mutis" in Bogotá, studies on the phenology of native Andean plants have been in progress over the last 10 years; the aim of these studies has been to provide information that will be of value for reafforestation and for establishing these plants as ornamentals. Also, in the Jardim Botânico de São Paulo, studies are going on into ornamental plants, principally orchids.

Botanic gardens should serve as showcases for botany, and for this should organise their exhibitions principally using local native plants. They should arrange the plants in an attractive way to demonstrate among other things the fundamental principles of ecology and the dependence of man on plants.

Living collections are organised in distinct ways in different gardens. So-called "Systematic Gardens" exist in almost all of them. In the Instituto Nacional de Tecnología Agropecuaria, Castelar, Argentina, the plants are also organised by photogeographical regions; in the Jardín Botánico de

Valdivia, Chile, they have been placed according to the classification systems of Cronquist and Takhtajan, but there are also displays of aquatic plants, cultivated plants, edible species, epiphytes, plants of the Northern Hemisphere, of central Chile, of the Andes, etc.

Two of the most important functions of botanic gardens are, necessarily, education, and in agreement with the theme of this conference, conservation. Education should be carried out at all levels — with children, with their teachers, with the general public and with the politicians and in the universities. One of the objectives of the Jardín Botánico de Cartagena is "to educate teachers and students at all levels". The Jardim Botânico de São Paulo has organised numerous courses in environmental education. In Tuluá, in Colombia, courses are given on horticulture and gardening. There are similar activities in other gardens.

The botanic garden should be a "living laboratory" and the objectives of the Museu Goeldi de Belem, in Brazil, are defined in this way. The Jardín Botánico "Lucien Hauman", Buenos Aires, is designed, among other things, to "provide materials for the teaching of botany".

The work of botanic gardens is known to lesser or greater extent through periodical publications, illustrative leaflets, guide books, catalogues, etc.

As far as conservation is concerned, I will begin by indicating that the objectives of the Jardín Botánico de Cartagena are "to advance studies on the protection and conservation of nature in the north of Colombia and carry out cultural, scientific and educative work related to these objectives".

Botanic gardens, as has been said many times, should be involved in both *ex situ* and *in situ* conservation. For many plants, cultivation in a botanic garden may be the only option for their survival. As it is impossible to cultivate them all, it may be necessary to give priority to plants with economic potential or with scientific value. One of the principal objectives of the Jardín Botánico de Tuluá in Colombia, is the conservation of potentially valuable species from the economic point of view. We should not forget that in the tropics we are involved in a race against time as a third of all tropical species will possibly be extinct by the end of this century.

Botanic gardens should establish nature reserves or cooperate with other organisations so that reserves are established. These reserves serve as living laboratories and could be the best way of saving many species. Here are some examples of how this can be done: the Jardín Botánico de Bogotá has started a study of one of the few forests which still exist near the city, and to suggest the creation of reserves. The Jardim Botânico São Paulo in Brazil has under its care the Biological Reserves of Paranapiacaba in the Serra do Mar and of Moigiguacu in the region with "cerrado" vegetation in the State of São Paulo.

These and other functions make more sense and are more effective if there is good coordination between botanic gardens. To achieve this, it is convenient to organise networks of botanic gardens at national and international level, with the aim of:

Co-ordinating action programmes;
Avoiding duplication of effort;
Encouraging specialisation;
Stimulating the creation of new botanic gardens where necessary;
Exchanging information;
Obtaining the support of both the public and governmental and non-governmental organisations.

Another way of promoting botanic gardens in developing countries, at least in South America, is by means of establishing bi- or multilateral agreements with botanic gardens of other regions of the world for training of personnel, exchange of information, support for research, etc.

Conclusions

The tropical and subtropical botanic gardens of South America should be important centres for research and education, and should play a very important rôle in the exploration and search for promising plants, in the recognition and conservation of species in danger of extinction and in the establishment of reserves and protected areas.

Great potential exists in our region, and in effect, some of our botanic gardens are already achieving some of these aims. But there is still much to do.

This conference on Botanic Gardens and the World Conservation Strategy constitutes an important step towards stimulating coordination and cooperation in this work. The expectations and hopes are great and I am certain that the results of the conference will provide a positive response to these expectations and hopes.

Note: This is a preliminary list from the IUCN database to which the author has contributed. It is in the process of development for the next edition of the IUCN/IABG Directory of Botanic Gardens. Corrections for that Directory would be greatly welcomed.

Appendix 1
List of botanic gardens of Latin America

ARGENTINA

Departamento de Botánica Agricola
INTA (Instituto Nacional de Tecnología Agropecuaria), 1712 Castelar, Provincia Buenos Aires.

Jardín Agrobotánico "Prof. Ing. Enrique C. Clos"
Facultad de Agronomía, Universidad Nacional de La Plata, Casilla de Correo 31, 1900 La Plata.

Jardín Agrobotánico de Santa Catalina
Instituto Fítotécnico de Santa Catalina, Llavallol, FNGR.

Jardín Botánico de la Universidad Nacional de Rio
Facultad de Agronomía, Enlace Rutas 8 y 36 — km 603, 5800 Rio Cuarto, Córdoba.

Jardín Botánico
Facultad de Agronomía y Veterinaria, Universidad de Buenos Aires, Avenida San Martín 4453, 1417 Buenos Aires.

Jardín Botánico "Carlos Thays"
Instituto Municipal de Botánica, Avenida Santa Fe 3951, 1425 Buenos Aires.

Relaciones Fitológicas Intercontinentales
Burela 3519 — Villa Urquiza, Capital Federal, Buenos Aires.

"Chacras de Coria" Garden
National University of Cuyo, Almirante Brown 500, 5505 Chacras de Coria, Mendoza.

"L.R.Parodi" Garden
Santa Fe.

BOLIVIA

Jardín Botánico de Santa Cruz de la Sierra
Casilla 123, Santa Cruz.

Jardín Botánico "Martín Cárdenas"
Casilla 538, Cochabamba.

BRAZIL

Jardim Botânico
Divisão de Botânica do Museu Nacional, Quinta da Boa Vista, Rio de Janeiro, RJ.

Jardim Botânico de Botucatu
Instituto Básico de Biologia Medicinal e Agrícola, Caixa Postal 526, 18.610 Botucatu, São Paulo.

Jardim Botânico de Porto Alegre
Fundaçao Zoobotânica do Rioã Grande do Sul (FZB), Caixa Postal 1188, 90.000 Porto Alegre, RS.

Jardim Botânico de São Paulo
Instituto de Botânica, Caixa Postal 4005, 01.000 São Paulo.

Jardim Botânico dã Rio de Janeiro
Rua Jardim Botânico 1008, 22.460 Rio de Janeiro, RJ.

Museu de História Natural
Rua Gustavo da Silveira 1035, Horto, Belo Horizonte, Minas Gerais.

Museu Paraense "Emilio Goeldi"
Caixa Postal 399, Av. Magalhaes Barata 376, 66.000 Belem, Para.

Parque Botânico do Morro Bau
Aven. Coronel Marcos Konder 800, 88.300 Itajai, Santa Catarina.

Reserva Ecologia do IBGE
Laboratório de Ecologia, Edifício Venancio II — 2 Andar, 70.302 Brasilia — DF.

CHILE

Arboretum Antumapu
Departmento de Silvicultura, Facultad de Ciencias Forestales, Casilla 9206, Santiago.

Arboretum Frutillar
Departamento de Silvicultura, Facultad de Ciencias Forestales, Casilla 9206, Santiago.

Arboretum
Instituto de Silvicultura, Universidad Austral de Chile, Casilla 853, Valdivia.

Jardín Botánico Hualpen
Departamento de Botánica, Universidad de Concepción, Casilla 1367, Concepción.

Jardín Botánico Nacional
Casilla 683, Viña del Mar.

Jardín Botánico
Instituto de Botánica, Universidad Austral de Chile, Casilla 567, Valdivia.

Jardín Botánico "Carl Skottsberg"
Instituto de la Patagonia, Casilla 102 — D, Punta Arenas.

COLOMBIA

CORPONARIÑO
Calle 18 # 19 - 64, Pasto.

Jardín Botánico Eloy Valenzuela

Corporación de Defensa de la Meseta
Calle 34 # 17 - 20, Bucaramanga.

ICECOL
Apartado Aereo 050326, Barranquilla.

Jardín Botánico de Santa Marta
Apartado Aéreo 1017, Santa Marta.

Jardín Botánico del Quindio
Apartado Aéreo 123, Armenia.

Jardín Botánico "Alejandro von Humboldt"
Universidad de Tolima, Ibagué.

Jardín Botánico "Joaquin Antonio Uribe"
Carrera 52 No. 73 – 298, Apto Aéreo 51 – 407, Medelín, Antioquia.

Jardín Botánico "Guillermo Piñeres"
Apdo Aéreo 5456, Cartagena.

Jardín Botánico "José Celestino Mutis"
Carrera 66 – A No. 56 – 84, Bogotá.

Jardín Botánico "Juan María Céspedes"
Tuluá, c/o Instituto Vallecaucano de Investigaciones Científicas, Apto Aéreo 5660, Cali, Valle.

Jardín de la Facultad de Agronomía
Universidad de Caldas, Apto Aéreo 275, Manizales, Caldas.

COSTA RICA

Lankester Botanical Garden
Escuela de Biología Universidad de Costa Rica, Ciudad Universitaria "Rodrigo Facio", San José.

Las Cruces Tropical Botanical Garden
San Vito de Coto Brus.

EL SALVADOR

Jardín Botánico La Laguna
Urbanización Industrial La Laguna, Apartado Postal 2260 CG, San Salvador.

ECUADOR

Tropical Botanical Garden
Universidad Técnica "Luis Vargas Torres", Esmeraldas.

FRENCH GUIANA

Institut de Botanique
ORSTOM, Route de Montabo, B.P. 165, 97323 Cayenne Cedex.

Jardin Botanique Municipal
97300 Cayenne.

GUYANA

Botanical Gardens
Guyana Forestry Commission, 1 Water Street, Georgetown.

Georgetown Botanical Gardens
Ministry of Agriculture, Turkeyen, Greater Georgetown.

GUATEMALA

Jardín Botánico
Departamento de Biología, Universidad de San Carlos de Guatemala, Avenida de la Reforma 0 – 43 Zona 10, Cuidad de Guatemala.

HONDURAS

Wilson Popenoe Multiple Use Botanical Garden
Apartado Postal 49, Tela, Atlántida.

MEXICO

Asociación Mexicana de Orquideología A.C.
Apartado Postal 53 - 123, México 11320, D.F.

Estación de los Tuxtlas
Instituto de Biología — UNAM, Apartado Postal 94, San Andrés Tuxtla, Veracruz.

Jardín Botánico
Departmento de Difusión y Enseñanza, Universidad Nacional Autónoma de México (UNAM), Cd. Universitaria, Delegación Coyoacán, 04510 Mexico D.F.

Jardín Botánico
Centro de Investigaciónes de Quintana Roo, Apto Postal No. 886, Cancún, Quintana Roo.

Jardín Botánico
Universidad Autónoma de Chapingo, Chapingo, C.P. 56230.

Jardín Botánico
Escuela Nacional de Medicina y Homeopatía, Instituto Politécnico Nacional, Arroyo de Guadalupe 239, Col. Ticomán, Fraccionamiento la Escalera, C.P. 07270.

Jardín Botánico
Centro de Investigación Científica de Yucatán, Apartado Postal 924, Merida, C.P. 97000, Yucatán.

Jardín Botánico
Instituto de Silvicultura y Manejo de Recursos Naturales, Ex-Hacienda Guadalupe, Apartado Postal 104, Monterrey, Nuevo León.

Jardín Botánico
Instituto Tecnológico de Oaxaca, Apartado Postal 893, Oaxaca, C.P. 68000.

Jardín Botánico
Colegio Superior de Agricultura Tropical, Apartado Postal 24, H. Cárdenas, Tabasco, C.P. 86500.

Jardín Botánico Acuario de Mazatlán
Av. de Los Deportes No. 11, Apartado Postal 770, Mazatlan, Sinaloa, C.P. 82000.

Jardín Botánico
Escuela Nacional de Estudios Profesionales "Iztacala", Los Reyes Tlanepantla, C.P. 54090.

Jardín Botánico de la Universidad
Absolo 33, Chilpancingo, Guerrero, C.P. 39000.

Jardín Botánico de Instituto de Historia Natural
Apartado Postal 6, Tuxtla Gutiérrez, Chiapas, C.P. 29000.

Jardín Botánico de Tecolutla
Administración del Jardín Botánico, Apartado Postal 11, Gutiérrez Zamora, Veracruz, C.P. 93550.

Jardín Botánico "Jorge Victor Eller T"
Guadalajara, Av. Patria 1201, Guadalajara, Jalisco, C.P. 44100.

Jardín Botánico Didáctico de Cactáceas
Dirección General de Reforestación y Manejo de Suelos Forestales, Netzahualcoyol 109 - 90, Piso, 06080 México 1 D.F.

Jardín Botánico Escuela de Ciencias
Universidad del Noreste, Apartado Postal 14 - 79, Tampico, Tamaulipas, C.P. 89000.

Jardín Botánico Escuela Secundaria No.1
Apartado Postal 423, Oaxaca, C.P. 68000.

Jardín Botánico Faustino Miranda
Calzada de los Hombres de La Revolución, Tuxtla Gutiérrez, Chiapas, C.P. 29000.

Jardín Botánico Fco. J. Clavijero
INIREB, Apartado Postal 63, Km 2·5 Antigua Carretera a Coatepec, 91000 Xalapa, Veracruz.

Jardín Botánico Forestal
Inst. Nacional de Investigaciónes Forestales, Av. Progreso 5, Coyoacán, México D.F., C.P. 04110.

Jardín Botánico Gustavo Aguirre Benavides
Universidad Agraria — Antonio Narro, Apartado Postal 554, Saltillo, Coahuila, C.P. 25000.

Jardín Botánico Municipio de Ahome
Ayuntamiento de los Mochis, Ahome, C.P. 81310.

Jardín Botánico San Cristóbal de las Casas
Instituto Nacional de Investigaciones Sobre Recursos Bióticos (INIREB), Apartado Postal 63, Xalapa, Veracruz, C.P. 91000.

Jardín Botánico
Universidad Autónoma de Morelos, Avenida Universidad 1001, Col. Chamilpa, Cuernavaca, Morelos, C.P. 62201.

Jardín Botánico
Estudio Florístico Ecología del Estado de Morelos, Unidad Biomédica, Avenida Universidad 1001, Col. Chamilpa, Cuernavaca, Morelos, C.P. 6200.

Museo de Herbolaria y Medicinales
Instituto Nacional de Antropología e Historia, Córdoba 45, Col. Roma, C.P. 06700.

PARAGUAY

Jardín Botánico y Zoológico
Parque y Museo de Historia Natural, Asunción.

PERU

Jardín Botánico
Universidad Nacional de Huanuco, "Hermilio Valdizan", Cuidad Universitaria, Cayhuayna.

Jardín Botánico
Universidad Nacional Mayor de San Marcos, Jiron Puno 1002, Lima.

Jardín Botánico
Universidad Nacional Agraria, Aptdo 456, La Molina, Lima.

Museo de Historia Natural "Javier Prado"
Avenida Arenales 1256, Lima.

SURINAM

Agricultural Experimental Station
Paramaribo, Surinam.

URUGUAY

Jardín Botánico
c/o Casilla Correo 1013, Montevideo.

VENEZUELA

Jardín Botánico
Instituto Botánico, Apto 2156, Caracas.

Jardín Botánico de Barinas
Universidad Nacional Experimental de los Llanos Ezequial Zamora, Apartado 19, Barinas.

Jardín Botánico
Universidad Central, Caracas.

Jardín Botánico de Maracaibo
Autopista al Aeropuerto "La Chinita", Apto. Postal 10.123, Maracaibo.

Jardín Botánico San Juan de Lagunillas
Facultad de Ciencias Forestales, Departamento de Botánica, Universidad de los Andes, Mérida 5101.

Jardín Botánico Xerofítico de Coro
Coro, Falcon.

Universidad Central
Facultad de Agronomía, Apartado 4579, El Limón, Maracay, C.P. 2101, Edo. Aragua.

The work of Xishuangbanna Tropical Botanical Garden in conserving the threatened plants of the Yunnan tropics

XU ZAIFU

Yunnan Institute of Tropical Botany, Academia Sinica, China

Summary

With about 7,500 species of vascular plants the Yunnan tropics are exceptionally rich in plant species. This paper presents some background to the region and on Xishuangbanna Tropical Botanical Garden.

Located on the northern edge of tropical south-east Asia, Yunnan's climate is influenced by monsoons. Mountains, whose ecosystems are fragile, occupy about 90% of the region. For many reasons, about half the forests have disappeared during the last 35 years; this has caused serious water and soil erosion and has widely threatened the native flora.

Even if establishing large nature reserves in various parts of the region would appear to be the most promising way to conserve rare and endangered plants, the cultivation of threatened plants in botanic gardens and other habitats is available as a complementary measure of species conservation.

This paper introduces the contributions that the Xishuangbanna Tropical Botanical Garden has made so far in conserving the threatened plants of Yunnan. Examples include the establishment of a living collection of native threatened species; research into the cultivation of wild economic plants; investigation of the nature reserves of the region; research in ethnobotany connected with environmental protection; research on integrating plant conservation into regional development; and environmental education for local people. It is a

Botanic Gardens and the World Conservation Strategy

ISBN: 0-12-125462-3

good start for the botanic garden to work in these fields; much however still remains to be done.

Resumen

La Flora del Jardín Botánico Tropical Xishuangbanna en conservación y tratamiento de las plantas del Yunnan tropical

Con aproximadamente 7.500 especies de plantas vasculares, el Yunnan tropical es extraordinariamente rico en especies. El presente trabajo muestra una panorámica de la región y del Jardín Botánico Tropical de Xishuangbanna.

Localizado en el extremo norte del sudeste de Asia tropical, el clima del Yunnan está influenciado por los monzones. Montañas con ecosistemas frágilesque ocupan aproximadamente la mitad de los bosques han desaparecido durante los últimos 35 años; esto ha ocasionado graves problemas de erosión, suelo y agua y ha amenazado ampliamente a la flora nativa.

También si se establecieran grandes reservas naturales en varias partes de la región sería la mejor forma de asegurar y conservar las plantas escasas y amenazadas; el cultivo de plantas en peligro en los jardines botánicos u otros hábitats es útil como medida auxiliar de la conservación de especies.

Este trabajo presenta las contribuciones que el Jardín Botánico de Xishuangbanna ha hecho hasta ahora en la conservación de las plantas amenazadas de Yunnan. Incluye ejemplos del establecimiento de colecciones vivas de especies nativas amenazadas; investigación en el cultivo de posibles plantas económicas; investigación de reservas naturales de la región; investigación en etnobotánica conectada con protección ambiental; investigación en la integración de la conservación de plantas en el desarrollo regional y educación ambiental para gente del lugar. Para el jardín botánico trabajar en estas cuestiones es un buen destino, sin embargo, todavía queda mucho por hacer.

Introduction

Bordered by Vietnam, Laos and Burma, the Yunnan tropics are situated in the southern part of the Yunnan Province of China (Fig. 1). The tropical part of Yunnan is about 80,000 km^2 in size, forming 20% of the total area of the province, and 90% of it is covered by mountains. The region has a large variety of land-forms ranging from high mountains consisting of a series of parallel and longitudinal folds which extend through the region into adjacent countries, to a particular kind of basin surrounded by mountains. The major rivers draining the region are the Red, Mekong and Salween Rivers, which rise in the north and north-west parts of the Yunnan Province and extend through the region into adjacent countries.

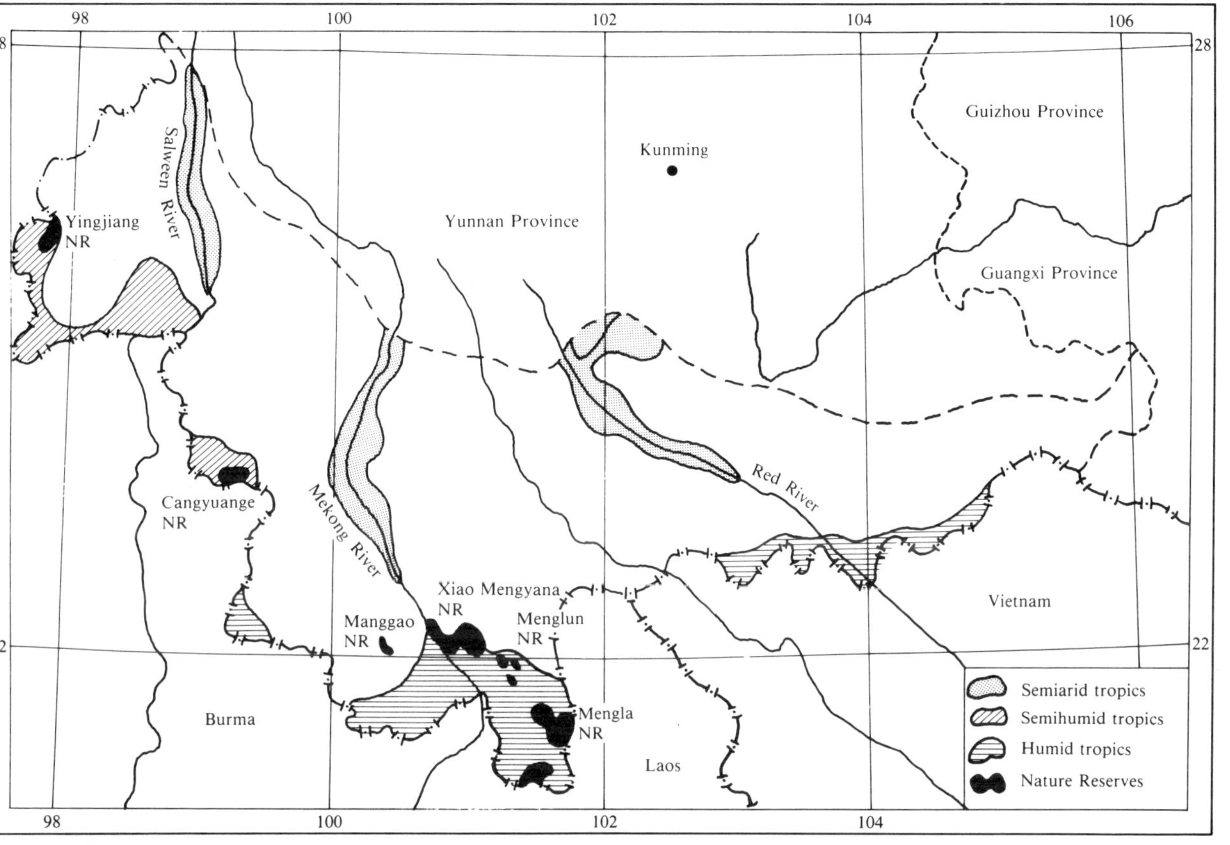

Fig. 1. The tropics of Yunnan Province and their nature reserves.

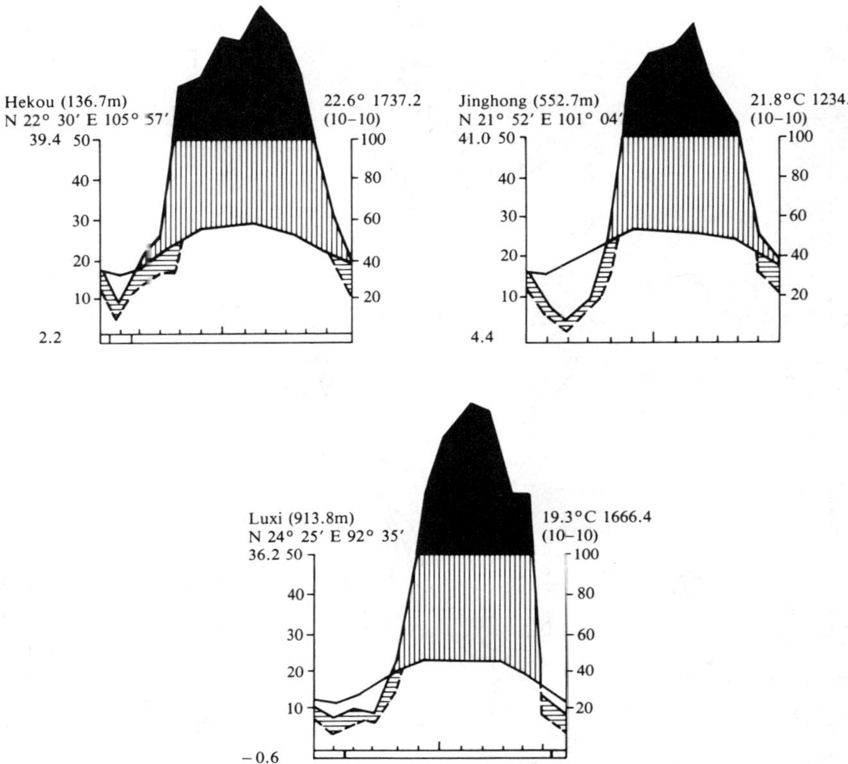

Fig. 2. The climatic diagrams of Yunnan Tropics (from Wu Cheng-yih et al. *1980).*

The climate of the region is influenced by the south-east monsoon from the Pacific in the east and by the south-west monsoon from the Indian Ocean in the middle and the west. The annual precipitation ranges from 600 to 2,000 mm, of which 80–90% is concentrated in the rainy season from May to October. The dry season is from November to April. Climatic diagrams of the tropical zone are given in Fig. 2. According to the aridity index, the Yunnan tropics can be divided into 3 types of climate — humid, semi-humid and semi-arid regions, as shown in Fig. 1.

About 18% of the area is covered by natural forest. The tropical vegetation can be divided into 3 major types — tropical evergreen forests, tropical monsoon forests and tropical savanna woody formations. The vegetation of the Yunnan tropics is, therefore, very complex.

Some 15,000 species of vascular plants occur in the province of Yunnan. Of these about 7,500 species occur in the tropics. The region is, therefore, extraordinarily rich in plant species. According to Wu Cheng-yih (1979), the

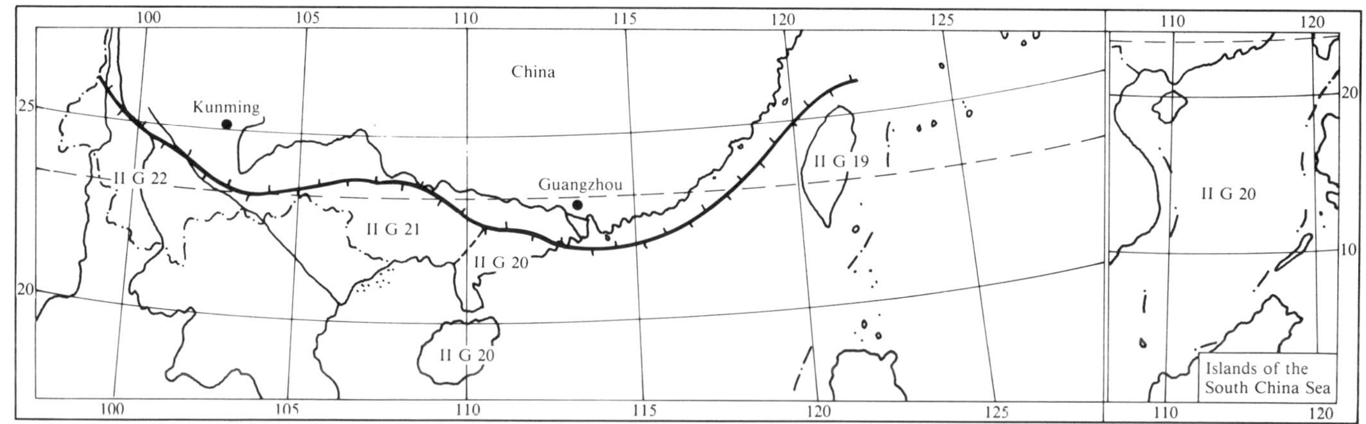

II G 19 — Taiwan region, II G 20 — South China Sea region, II G 21 — Tongking Gulf region
II G 22 — Yunnan, Burma, Thailand region

Fig. 3. The region of the paleotropic kingdom of China (from Wu Cheng-yih, 1980).

tropical flora of China belongs to the Malaysian Sub-kingdom of the Paleotropic Kingdom and can be divided into the following 4 regions — the Taiwan region, the South China Sea region, the Tongking Gulf region and the Yunnan–Burma–Thailand region (Fig. 3).

According to our studies (Xu Zaifu *et al.* 1982), it is quite evident that over 1,000 species of plants in Xishuangbanna, a region of the Yunnan tropics, have a high economic potential which can be readily exploited. For example, more than 500 species have long been used by local people in medicinal care; some 150 species are oil plants with seeds containing over 15% of oil; a further 150 species are fast-growing and valuable timber tree species, including bamboo; others are sources of essential oils, fibres, resins, gums, starches, vegetables and fruits, etc. Many species have been collected by local minority ethnic groups and used to sustain their daily life, for example as medicines, playing an important rôle in the development of traditional cultures based upon plants.

Our research has also shown that there are about 120 species of wild types or close relatives of cultivated crops distributed in the Yunnan tropics. According to N.I. Vavilov, this number is similar to that of the Indian, Central Asiatic or North American Centres. This germplasm is very important for studying the origin and development of cultivated plants and is very useful in plant breeding and the selection of genetic stock to improve our domestic plants.

Since only incomplete information is so far available on how many endemic species there are in the Yunnan tropics, the author has completed from recent publications a list of about 500 endemic species.

Due to its extraordinary richness in plant species, its high economic usage and potential, its genetic diversity and its degree of endemism, the flora of the Yunnan tropics is of exceptional interest in China. Located on the northern edge of tropical south-east Asia, with the climate influenced by monsoons, and 90% of the tropics in the montane area, the ecosystems of the region are fragile. In order to meet the need for species and ecosystems conservation, 6 nature reserves have been set up in the tropical zone in recent years. The total area of these reserves is 260,000 ha or 3·2% of the total tropical area; their distribution is shown in Fig. 1.

Although the establishment of large nature reserves in various parts of the region would appear to be the most promising way to conserve the rare and threatened plants, the cultivation of these plants in botanic gardens and other habitats has also served as a complementary measure of species conservation. Located in the humid tropics, Xishuangbanna Tropical Botanical Garden is the only botanic garden in the tropical zone of China. With an area of 1,000 ha, surrounded by a large area of tropical and subtropical forest and by 3 blocks of the Menglun Nature Reserve, it has shown since it was established that the botanic garden is an ideal place for protecting the rare and threatened plants of the Yunnan tropics *ex situ*. The following account outlines the contributions made by the garden since 1959 to conserving the rare and threatened plants of the region.

The study and assessment of the regional threatened plants and their conservation priorities

Recent estimates suggest that out of the 250,000 species of flowering plants, as many as 60,000, that is an average of 1 in 4, could become extinct by 2050 if present trends continue (Anon. 1986, Raven, this volume). However, the detailed study of rare and threatened species of vascular plants has not been completed for the flora of the Yunnan tropics.

In recent years, the study of the rare and threatened plants of the Xishuangbanna region, where about 4,000 species of native vascular plants are distributed in a total area of 19,000 km^2, has just been carried out (Yunnan Institute of Tropical Botany 1984). Botanists have proposed a list of 345 species of rare and threatened plants from Xishuangbanna, to be considered for protection (Li Yan Hui 1984). Among these, some 53 species were listed as requiring urgent protection in China overall. The species of rare and threatened plants may constitute more than 10% of the regional flora, because we estimate that roughly one half of the tropical forests have disappeared through deforestation since the 1950s.

In order to provide a basis for conserving the dangerously rare and severely threatened species *in situ* and *ex situ* in the Xishuangbanna region, we have worked out a method of using the IUCN Red Data Book categories to indicate the degree of threat to individual species and the priorities for their conservation in specific regions.

Even though the concepts of rarity and threat have been defined and the categories of threat designated (Lucas and Synge 1978, IUCN/UNEP/WWF 1980, Reveal 1981), we find them somewhat subjective and difficult to apply. We have, therefore, sought for some years a quantitative method for determining degrees of threat and for the establishment of conservation priorities. To assess the environmental or indicator value of a threatened species, we emphasise ecological factors, such as the following.

(1) Individual species ecology. Species of narrow ecological amplitude possess lower ecological equivalence and are merely adapting to the limiting changes of the environment. They are, therefore, more likely to be threatened than species with a wider ecological amplitude (Dajoz 1972).

(2) Population dynamics. The relative abundance, dominance and frequency of plants in the community reflect their competitiveness in threatened situations. The age classes of plants in the community reflect their regenerative ability under environmental stress.
(3) Community ecosystems. Even though characteristic species may have more individuals than preferential or indifferent species, their ecological equivalence may be lower with their more limited range (Dajoz 1972).

The following environmental indicators are suggested.

(1) The patterns of range.
(2) The regional distribution.
(3) The limitation of the community.
(4) The importance in the community.
(5) The population structure in the community.

The direction and strength of the impact on plants differ according to the different species involved, different environmental changes and different degrees of exploitation of the plant populations. The effects may be assessed in quantitative terms on a scale of 1 to 5. A mean coefficient for the degree of threat can then be worked out and used to place the threatened species in the categories of the IUCN Red Data Book.

As mentioned above, we must determine an order of priority for the conservation of threatened species in order to apply our limited resources to the most threatened species and to the species with very important scientific values and economic potentials. One of the most important aims is to protect germplasm from erosion. In the field of genetic diversity, the World Conservation Strategy (IUCN/UNEP/WWF 1980) indicates that we must pay particular attention to higher taxa (families and genera). It suggests that "Priority should be given to species that are endangered throughout their range and to species that are the sole representatives of their family or genus, according to the following formulation: the greater the potential genetic loss the less imminent that loss need be to justify preventative action". The formula is illustrated in Fig. 4. From this, we can see that 1, 2 and 3 have the highest priority and can be considered as the first level for conservation; 4, 5 and 6 are of intermediate priority (second rank); and 7, 8 and 9 have low priority (third rank).

Even though the formula used by the World Conservation Strategy takes into consideration the scientific interest covered by the term "size of loss", we consider that the emphasis given to species of important scientific interest, economic value, genetic potential and ecological potential, etc., is not high enough. In these cases we may introduce the term "species of

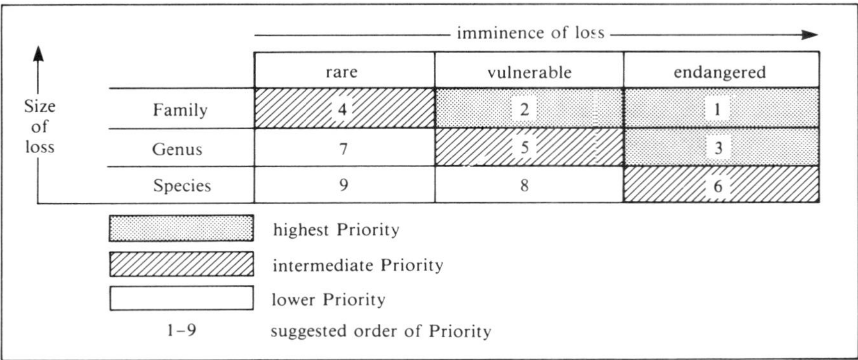

Fig. 4. Formula for determining priorities amongst threatened species.

special concern'' to adjust the calculated coefficient for the degree of threat and the suggested order of priority, according to the formula of the World Conservation Strategy, in the direction of those species that have special interest or value to man.

Using this method, 49 species of plants, which are distributed in the Xishuangbanna region and indicated in the lists for requiring urgent protection in China, have been studied. We feel that the method is easily manageable and that the results give a more objective reflection of the real situation of plants growing under the influences of natural selection and human action. We are, therefore, going to use the method to determine the regionally threatened species and their conservation priorities in the coming years.

The establishment of a living collection of native species

The theory of island biogeography suggests that if 10% of an ecosystem type is preserved, no more than 50% of the species dependent upon it can be preserved in the long run (MacArthur and Wilson 1967). Even if 4 nature reserves were to be set up in Xishuangbanna, with 200,000 ha in the reserves or 11% of the total area of the region, there would still be some 1,500 species of plants, or one third of the species of the local flora, outside the reserves; moreover, the fragmentary ecosystems of the reserves could hardly preserve all the species dependent on them in the long run. Therefore, if the conservation of a species in its natural habitat proves to be impossible in the long term, cultivation in botanic gardens or in other habitats will need to serve as an additional measure. In recent years, the emphasis on the conservation of rare and threatened species has been built into the long-term planning of Xishuangbanna Tropical Botanical Garden, due to the great increase in deforestation and to the large number of threatened species in the Yunnan tropics.

Various living collections have been established in the garden, which now serves as a centre for especially rare and threatened species of native tropical plants and so meets the requirements of protection and scientific research. Several special living collections are being established in the garden, including a bamboo collection, a palm collection, a dipterocarp collection, a shade plant collection, a tropical conifer collection, a collection of important economic plants and a collection of rare and threatened plants of Xishuangbanna. About 1,700 native species of the Yunnan tropics have been involved in these collections, especially in the last collection, which covers 50 ha and was set up where the remains of rain forest and monsoon forest are located. There are about 1,300 species of Xishuangbanna native plants in this collection and 30 of them are the species listed for requiring urgent protection in China.

It is impossible to collect all the native species of plants for *ex situ* conservation in the botanic garden; therefore special attention has been paid to the protection of endemic species (avoiding overlap with other botanic gardens), native rare and threatened species that urgently require saving, species with genetic value for agriculture and forestry, which are important to both development and science, and species of economic potential, which are easily threatened by unwise utilisation. It is also impossible for a botanic garden to cultivate many populations of each species; therefore the cultivated population for each species has been limited according to the diversity of vegetative forms, but includes as many ecotypes as possible using the method of multiple genepool sampling. It is also impossible for us to create habitats for collected species as naturally diverse as the ecosystems they came from; therefore the unfavourable conditions of living collection cultivation have been remedied by the careful selection of microhabitats as near as possible to the natural conditions.

In order to manage introduction techniques and to improve the management of *ex situ* conservation, research on seed storage, seed germination, seedling growth and plant developmental phases have been carried out over several years. We have also studied the growth and adaptation of some native plants of the region under cultivation (Xu Zaifu *et al.* 1985), from which we have shown that the developmental rhythm of plants has been changed under cultivated conditions owing to the change in microhabitat. The changed rhythm is in direct proportion to the magnitude of difference of environmental conditions between the natural habitats and the environment of cultivation. For instance, in the case of the seasonal rain forest species in cultivation, the change of developmental rhythm in the upper-storey element is less than that in the under-storey element. Generally cultivated conditions can promote the growth and development of cultivated plants, in direct proportion to the difference in limiting factors between nature and cultivation. However, there are some ill-effects of cultivation due to critical changes both in autecology and synecology of individual species and populations. For example, the seedlings of seasonal rain forest species are easily burned by sunlight in open field conditions, some species have a reduced reproductive capacity due to loss of their pollinators in cultivated conditions, and some species are seriously damaged by insects under intensive cultivation due to the loss of the natural enemies of their pests. This information has helped us to set up appropriate living collections under favourable conditions and to improve their maintenance in an efficient manner.

Integration of conservation and development

The newly developed theory of island biogeography has demonstrated that a limited area, and the fragmentary ecosystems of nature reserves and botanic gardens, cannot protect most species from danger and/or extinction in the long term. Therefore, we not only have to see these problems from the species protection angle itself, but also to keep regional development planning in mind in order to improve the regional environment available for the survival of species. As the World Conservation Strategy mentioned, the vicious circle by which poverty leads to ecological degradation which in turn leads to more poverty can be broken only by development. Conservation must, therefore, be combined with measures to meet short-term economic needs (IUCN/UNEP/WWF 1980). Based on these concepts, we have, since the establishment of the garden, encouraged our scientists to participate in the development process of the region. Our aims have been to find some measures of development suitable to the regional conditions in order to reduce pressure on and depletion of the living resources essential for human survival and sustainable development. Our contribution to the development of the region is as follows.

The development of agroforestry ecosystems

Studies have already shown that shifting cultivation and monoculture are not suitable systems for tropical land-use and for species protection alongside an increasing human population. The most radical approach in recent years to solve the urgent problem of optimising the productivity of ecosystems has been achieved by agroforestry. In the Yunnan tropics, agroforestry has been practised by the native people for generations. For instance, in the third year of shifting cultivation of upland rice, the seeds of *Cassia siamea*, cultivated by the Dai people as a traditional fuel wood for at least 400 years (Pei Sheng-ji 1983), are spread with the rice seeds, the trees grow quickly above the crops and have a deep root system which avoids root damage by the cultivators. Three years after sowing, the trees are ready to be cut and the stocks then continue to grow by sprouting. Planting tea, *Camellia sinensis* var. *assamica* under the Yunnan camphor tree (*Cinnamomum glanduliferum*) increases the productivity of tea and camphor from the same plot of land, and has made the Yunnan tropics the home of the world-renowned Puer Tea. Planting shade-requiring crops like tea, *Amomum villosum* (Chinese traditional medicine) and *Baphicacanthus cusia* (for blue dye used by local people) under tropical forests, by means of clearing the understorey, is another kind of agroforestry practised by the local people.

The development of agroforestry ecosystems has been studied in the botanic garden for 25 years, and is based on both the traditional experiences accumulated by local people and the studies of tropical forest ecosystems. The following patterns have been very successful:

rubber (*Hevea brasiliensis*) + tea
rubber + amomum (*A. villosum*)
rubber + cinchona (*Cinchona ledgeriana*)
rubber + *Rauvolfia* (*R. vomitoria*) + *Homalomena* (*H. occulta*)

The pattern of rubber + tea is the most successful one to have been developed and now extends over 10,000 ha in the tropics of China. The increased food, wood and other produce from the same plots of land in the stable agroforestry ecosystems, the decreased water and soil erosion, and the improvement of the microclimate will relieve natural ecosystems from exploitative pressure and will improve the regional environment, thus making conservation more possible than it would otherwise be.

Rural energy development

About 90% of the wood production is used for fuel wood consumption in the Yunnan tropics, and this has been a main reason for the deforestation of the region. There is lack of fossil fuel in the region whereas water power is plentiful, so that the government has paid special attention to the construction of hydro-electric power stations in order to meet the need for rural energy. However, the development of fuel wood production is still the basic measure used to meet the energy supply. *Cassia siamea* is one of the best fuel wood tree species that has been planted by the Dai people of Xishuangbanna for hundreds of years, and they have accumulated considerable experience of how to cultivate and maintain it. Besides studying *Cassia siamea*, the botanic garden has tested some 30 species of fuel wood and fast-growing trees introduced from home and abroad, of which *Cassia spectabilis* has proved to be more productive and to yield a higher quality of fuel wood than *Cassia siamea*. We are sure that only by solving the rural energy consumption problems can we relieve the pressure of deforestation and save more species of rare and threatened plants from extinction.

The cultivation of wild, potentially economic plants

Since ancient times the local people have regularly collected hundreds of wild plants for their shelters, vegetables, medical care and other uses. They did this on a small scale, using simple tools without any harmful effect on the survival of species (Pei Sheng-ji 1982). On the other hand, the local people have also cultivated some species of wild plants to meet their daily needs. From our investigations in the Xishuangbanna region (Xu Zaifu and Yu Pinhua 1984, Yu Pinhua *et al.* 1985), there are 69 native species which have been cultivated by the Dai people in their home gardens or plantations. Even though they sometimes sell surplus products in the local markets, most of them have still not become cash crops except a few species such as tea and Yunnan camphor.

Today, however, the situation has been changed by deforestation on a great scale. Many species have declined considerably, especially the rare species and the species with strict habitat requirements for their regeneration. These will become even more rare or threatened as more populations and habitats are lost. On the other hand, many species of wild plants have been found to be of value in recent years for new drugs, for industrial uses or for export. Considerable damage is done to these plants by collecting according to the philosophy of ''draining the pond to get all the fish''. For example, *Calamus gracilis* has become a vulnerable species due to uncontrolled collecting for furniture making. *Rauvolfia yunnanensis, Homalomena occulta, Maytenus hookeri* and *Gloriosa superba* are important medicinal plants but are becoming rare or endangered due to over-collection as cash plants.

Unfortunately, there are very few people willing to cultivate these wild cash plants before the species become depleted. The reason why people do not want to cultivate them is not only because they can easily collect them in the wild, but also because they know that wild plants are not always easy to cultivate on a large scale. In order to solve these problems, a project for domesticating native wild economic plants has been carried out in the Xishuangbanna Tropical Botanical Garden for over 25 years and from which a few species may be expected to become new cultivated crops, for instance, *Rauvolfia yunnanensis, Maytenus hookeri* and *Gloriosa superba*. Cultivating such economic plants in great quantity can not only help to discourage the collecting of them from the wild, but can also contribute directly to the preservation of the species by spreading them widely in cultivation.

With about 7,500 species of vascular plants in an area of some 80,000 km^2, the Yunnan tropics are no doubt the "plant treasury of China". Unfortunately, about half the forests have disappeared in the region during the last 35 years due to many reasons. This has caused serious ecological problems, especially to the survival of wildlife. Therefore special attention must be given to forest and plant (and animal) conservation in the region.

In recent years, the Xishuangbanna Tropical Botanical Garden has modified its traditional rôles and begun to play a vital part in conservation of rare and threatened plants of the region. The contributions made by the garden so far extend beyond those described above to research on the nature reserves of the region; research on the ethnobotany associated with environmental protection (Pei Sheng-ji 1983) and the education of local people in environmental protection. The garden has made an important start in working in these fields, but much more still remains to be done. For instance, the reintroduction of rare and threatened plants into the wild is the ultimate aim of the botanic garden. However, it still has a long way to go because reintroduction in tropical ecosystems may be much more difficult than in temperate ecosystems. In order to provide us with a wider range of habitats for different rare and threatened plants for *ex situ* conservation, the establishment of a series of small satellite gardens in different vegetation and climate zones of the tropics should be considered as part of long-term planning if possible.

Acknowledgements

My sincere thanks are given to my colleagues at Xishuangbanna Tropical Botanical Garden for their contributions to the garden in conserving the rare and threatened plants of the Yunnan tropics. The author also wishes to thank Professor Pei Sheng-ji, the Director of the Yunnan Institute of Tropical Botany, Academia Sinica, for reading the manuscript.

References

Anon. (1986). 60,000 plants under threat. *Threatened Plants Newsletter* **16**, 2.

Dajoz, R. (1972). "Précis d'Ecologie", Deuxieme Edition. Paris. (Translated into Chinese by Zhang Shen *et al.* 1982. Gansu Province Peoples Press. pp. 16–17, 281–284.)

Francisco *et al.* (1976). Thoughts on the Biosphere Reserve Concept and Its Implementation, Selection, Management and Utilization of Biosphere Reserves, pp. 173–146. Proceedings of the US-USSR Symposium on Biosphere Reserves.

IUCN/UNEP/WWF (1980). "World Conservation Strategy". IUCN, Gland, Switzerland.

Lucas, G. and Synge, H. (1978). "The IUCN Plant Red Data Book". IUCN, Switzerland.

Li Yan Hui (1984). Rare and Precious Plants in Nature Reserves of Xishuangbanna. *Journal of Tropical Plants* **25**, 1–22.

MacArthur, R.H. and Wilson, E.O. (1967). "The Theory of Island Biogeography". Princeton University Press, Princeton, N.J.

Pei Sheng-ji (1982). "A Preliminary Study of the Ethnobotany in Xishuangbanna", pp. 16–32. Collected Research Papers on Tropical Botany, Yunnan People Press, Kunming, China.

Pei Sheng-ji (1983). Some Effects of the Dai People's Cultural Beliefs and Practices upon the Plant Environment of Xishuangbanna, Yunnan Province, Southwest China. Paper prepared for the Conference on Cultural Values and Tropical Ecology, Eastwest Environment and Policy Institute. Honolulu, U.S.A.

Reveal, J.L. (1981). The concept of rarity and population threats in plant communities. *In* "Rare Plant Conservation: Geographical Data Organization" (L.E. Morse and M.S. Henifin, eds) pp. 41–47. The New York Botanical Garden, New York.

Wu Cheng-yih (1979). The regionalization of Chinese Flora. *Acta Botanica Yunnanica* **1**(1), 1–22.

Wu Cheng-yih *et al.* (1980). "The Vegetation of China", pp. 43–44. Sciences Press, Peking.

Xu Zaifu and Yu Pinghua (1984). Present conditions and potentiality of the tropical crops of south Yunnan. *Tropical Geography* **4**(1), 14–20.

Xu Zaifu *et al.* (1982). "Research on Reservation of Germplasm Resources of Tropical Plants in Xishuangbanna", pp. 7–15. Collected Research Papers on Tropical Botany, Yunnan People Press, Kunming, China.

Xu Zaifu *et al.* (1985). Growth and adaptation of native plants of southern Yunnan under cultivated situations. *Plant Introduction and Acclimatization* **4**, 15–21.

Yu Pinghua *et al.* (1985). The study on traditional cultivated plants in Tai villages of Xishuangbanna. *Acta Botanica Yunnanica* **7**(2), 169–86.

Yunnan Institute of Tropical Botany, Academia Sinica (1984). "List of Plants in Xishuangbanna", p. 509. Yunnan People Press, Kunming, Yunnan.

Investigation and introduction of some rare and endangered species in Nanjing Botanical Garden, Mem. Sun Yat-Sen

HE SHAN-AN, YANG ZHI-BING, WANG MING-JUN, ZHONG SHI-XIAN, SHEN JIA-YU and TAO JIN-CHUAN

Jiangsu Institute of Botany, Nanjing Botanical Garden, Mem. Sun Yat-Sen, Nanjing, China

Summary

The authors provide an outline of the work of the Nanjing Botanical Garden in conserving rare and threatened species of Jiangsu, Zhejiang and Anhui Provinces of the People's Republic of China. The affinities of the flora are discussed briefly, and 137 rare and threatened plant taxa are listed.

Resumen

Investigación e introducción de algunas especies raras o en peligro en el Jardín Botánico de Nanjing, Mem. Sun Yat-Sen

Los autores dan a conocer la línea de trabajo adoptada por el Jardín Botánico de Nanjing para la conservación de especies raras o amenazadas de las provincias de Jiangsu, Zhejiang y Anhui de la Republica Popular China. Se discuten brevemente las afinidades de la flora, enumerándose 137 táxones vegetales raros o en peligro.

Botanic Gardens and the World Conservation Strategy

ISBN: 0-12-125462-3

Investigation of rare and endangered species in Jiangsu, Zhejiang and Anhui Provinces

Jiangsu, Zhejiang and Anhui provinces are located in eastern China, between latitudes 27° and 35°06′N. and longitudes 114°50′ and 123°E. They include warm temperate, north subtropical and central subtropical climate zones. The total area is 343,900 km^2 with an elevation from 500 to 1,000 m. Most of the region is mountainous with summits reaching 1,900 m, for example Mt Fongyang and Mt Tianmo in Zhejiang Province, and Mt Huangshan and Mt Dabiaoshan in Anhui Province. Temperature and precipitation increase from north to south. The mean annual temperature is 13°–19°C and the mean annual precipitation is 700–1700 mm.

Because of the complexity of habitats in mountainous regions, there is a diverse flora, which includes endemic species restricted to eastern China, some elements of the flora of central and south-western China, and elements of south and north China floras. The vegetation of these mountainous areas was greatly modified by the glacial periods of the Quaternary, but those species which survived have since recovered.

The flora has close affinities to that of Japan, for example in species such as *Cephalotaxus sinensis, Cercidiphyllum japonicum, Kirengoshema palmata* and *Rhodotypos scandens*. There is also a close relationship with the North American flora, shown in species such as *Liriodendron chinense, Pseudotaxus gaussenii, Sassafras tzumu, Nyssa sinensis, Carya cathayensis, Calycanthus chinensis* and *Halesia macgregorii*.

Most of the rare and endangered species in these 3 provinces are found in very remote regions. Nevertheless, habitat destruction has taken place gradually over the years, resulting in serious decline of the populations of some species. Investigation has resulted in a list of 137 rare and threatened taxa, 60 of which are included in the first volume of the Chinese Red Data Book (see Appendix 1). Many reserves have been established by local government in order to protect rare and endangered species.

The growth and development of rare and endangered species in *ex situ* cultivation in Nanjing Botanical Garden

The conservation of plant germplasm, especially that of rare and endangered species, is one of the main objectives of botanic gardens throughout the world. Investigation of rare and endangered species native to Jiangsu, Zhejiang and Anhui Provinces has been carried out at Nanjing Botanical Garden over a number of years. Observations on the biological characteristics of introduced species have also been conducted in the garden.

Since 1954, there has been a programme to introduce rare and threatened species to the garden. Today, the garden has a collection of 67 species classed as rare or threatened, representing 34 families and 58 genera, including 15 monotypic genera.

Two sections of the garden are concerned with rare and threatened species, namely the living collection and the natural vegetation area for reintroductions. Studies have been made on the growth, oversummering and overwintering of 62 species. Altogether, 34 species are growing well and producing flowers and fruits; however, the fruits of some species, such as *Keteleeria fortunei* and *Liriodendron chinense*, abort or do not develop fully. This research may indicate one possible reason why these particular species are rare.

Experiments have been undertaken on propagation and cultivation of certain species which show promise for sylviculture, nut growing, landscape and other decorative uses, such as *Pseudolarix amabilis, Nyssa sinensis, Carya cathayensis, Torreya grandis, Tapiscia sinensis, Taxus chinensis, Sinojackia xylocarpa* and *Magnolia cylindrica*. In this way, plant conservation has a direct link to sustainable economic development.

The garden is located at *c.* 32°07′N., with mean annual temperature of 15·4°C (max. 43°C, min. −13·1°C), mean annual precipitation of *c.* 1,000 mm and 2,057·6 hours of sunshine per annum.

Acknowledgements

The authors extend their sincere acknowledgements to the Honorable Directors of the Institute and Botanical Garden, Professors Shan Ren-hua and Sheng Cheng-gui for their advice, and also to the assistants Mr Chen Yong-hui and Wu Shou-peng.

Appendix 1
Rare and endangered taxa in Jiangsu, Zhejiang and Anhui Provinces

Js: Jiangsu Province; Zj: Zhejiang Province; Ah: Anhui Province.

* Included in Volume 1 of the Plant Red Data Book for China.
† Introduced to Nanjing Botanical Garden.

Name	Distribution
* *Abies beshanzuensis* M.H. Wu	Zj
Acer anhweiense Fang & Fang f.	Ah
Acer nikoense Maxim.	Zj, Ah
† *Acer sinopurpurascens* Cheng	Zj
**Acer yangjuechi* Fang & P.L. Chiu	Zj
† *Alniphyllum fortunei* Park.	Zj, Ah
† *Altingia gracilipes* Hemsley	Zj
Brachystachyum densiflorum Keng	Js, Zj, Ah
†**Bretschneidera sinensis* Hemsley	Zj
Camellia chekiangoleosa Hu	Zj
Cardiandra moellendorffii Li	Zj, Ah
**Carpinus putoensis* Cheng	Zj
Carpinus tientaiensis Cheng	Zj
† *Carya cathayensis* Sarg.	Zj, Ah
Celtis chekiangensis Cheng	Zj
† *Cephalotaxus fortunei* Hook.f.	Zj, Ah
† *Cephalotaxus sinensis* Li	Js, Zj, Ah
†**Cercidiphyllum japonicum* Sieber & Zucc.	Zj, Ah
† *Cercis chingii* Chun	Ah
**Changium smyrnioides* Wolff	Js, Zj, Ah
**Changnienia amoena* Chien	Js, Zj
Chelonopsis chekiangensis C.Y. Wu	Js, Zj, Ah
† *Chimonanthus nitens* Oliver	Zj, Ah
Chimonobambusa quadrangularis Makino	Js, Zj
†**Cinnamomum japonicum* Sieber	Js, Zj, Ah
**Cinnamomum micranthum* Hayata	Zj
Conandron ramondioides Sieber & Zucc.	Ah
**Coptis chinensis* Franchet var. *brevisepala* W.T. Wang & Hsiao	Zj, Ah
**Croomia japonica* Miq.	Zj
† *Cyclocarya micro-paliurus* Iljinsk.	Ah
† *Cyclocarya paliurus* Iljinsk.	Zj, Ah
† *Decaisnea fargesii* Franchet	Zj, Ah
† *Deutzia glauca* Cheng	Ah
Didymocarpus cortusifolius Leveille	Zj
† *Diospyros glaucifolia* Metc.	Zj, Ah
Disanthus cercidifolius Maxim. var. *longipes* H.T. Chang	Zj
**Dysosma versipellis* (Hance) M. Cheng	Zj
†**Emmenopterys henryi* Oliver	Js, Zj, Ah
**Engelhardtia roxburghiana* Wallich	Zj
**Erythrophleum fordii* Oliver	Zj
**Euchresta trifoliolata* Merr.	Zj
†**Eucommia ulmoides* Oliver	Ah
†**Euptelea pleiospermum* Hook.f. & Thomson	Zj, Ah

Fagus engleriana Seem.	Zj, Ah
Fagus longipetiolata Seem.	Zj
†**Fokienia hodginsii* Henry	Zj
† *Fortunearia sinensis* Rehder & Wilson	Js, Ah
Fritillaria thunbergii Miq.	Js, Zj
†**Ginkgo biloba* L.	Zj
**Glehnia littoralis* F. Schmidt ex Miq.	Js, Zj
**Glycine soja* Sieber & Zucc.	Js, Zj, Ah
† *Gymnocladus chinensis* Baillon	Js, Zj, Ah
†**Halesia macgregorii* Chun	Zj
**Heptacodium jasminodes* Airy Shaw	Zj, Ah
† *Idesia polycarpa* Maxim.	Zj, Ah
Impatiens davidii Franchet	Ah
Indocalamus mijoi Keng f.	Js, Zj, Ah
**Isoetes sinensis* Palmer	Js, Zj
†**Keteleeria fortunei* Carr.	Zj
**Kirengoshema palmata* Yatabe	Zj, Ah
**Kolkwitzia amabilis* Graebner	Ah
† *Lilium lancifolium* Thunb.	Js, Zj, Ah
† *Lindera chienii* Cheng	Js, Zj, Ah
†**Liriodendron chinense* (Hemsley) Sarg.	Zj, Ah
†**Litsea auriculata* Chien & Cheng	Zj, Ah
† *Lonicera modesta* Rehder	Zj
Luisia hancockii Rolfe	Zj
Lychnis coronata Thunb.	Js, Ah
†**Magnolia amoena* Cheng	Js, Zj, Ah
Magnolia biondii Pampan.	Ah
†**Magnolia cylindrica* Wilson	Zj, Ah
†**Magnolia officinalis* Rehder & Wilson subsp. *biloba* Cheng & Law	Zj, Ah
†**Magnolia sieboldii* K. Koch	Zj, Ah
†**Magnolia zenii* Cheng	Js
† *Manglietia fordiana* Oliver	Zj, Ah
**Monimopetalum chinense* Rehder	Ah
**Neolitsea sericea* Koidz	Zj
Nepeta everardii S. Moore	Zj, Ah
† *Nyssa sinensis* Oliver	Zj, Ah
**Ophioglossum thermale* Kom.	Js
†**Ormosia hosiei* Hemsley & Wilson	Js, Zj, Ah
**Ostrya rehderiana* Chun	Zj
Paris polyphylla Sm.	Zj, Ah
Pentapanax henryi Harms var. *hwangshanensis* Cheng	Ah
†**Phoebe chekiangensis* C.B. Shang	Zj
† *Phoebe sheareri* Gamble	Js, Zj, Ah
Phyllostachys aurea Carr. ex A. & C. Riv.	Js
Phyllostachys pubescens Mazel ex Lehaie forma *gracilis* W.Y. Hsiung	Js
Phyllostachys pubescens Mazel ex Lehaie forma *granmica* W.Y. Hsiung	Zj
Phyllostachys purpurata McClure forma *solida* S.L. Chen	Js, Zj, Ah
Phyllostachys spectabilis C.D. Chu & C.S. Chao	Js
Phyllostachys viridi-glaucescens A. & C. Riv.	Js
Phyllostachys viridis (R.A. Young) McClure forma *houzeauana* C.D. Chu & C.S. Chao	Js
**Pinus dabeshanensis* Cheng & Law	Ah

*	*Platycrater arguta* Sieber & Zucc.	Zj
†	*Podocarpus naji* Zoll. & Mor.	Zj
†	*Poliothyrsis sinensis* Oliver	Js, Zj, Ah
†*	*Pseudolarix amabilis* Rehder	Js, Zj, Ah
†*	*Pseudotaxus chienii* Cheng	Zj
†*	*Pseudotaxus gaussenii* Flous	Zj, Ah
†*	*Pteroceltis tatarinowii* Maxim.	Zj, Ah
†	*Pterostyrax corymbosa* Sieber & Zucc.	Zj
†	*Quercus stewardii* Rehder	Ah
	Rhododendron anhweiense Wilson	Zj, Ah
†	*Rhododendron fortunei* Lindley	Zj
†	*Rhododendron shanii* Fang	Ah
†	*Rhodotypos scandens* Makino	Zj, Ah
†*	*Rosa odorata* Sweet	Js, Zj
*	*Sanicula tienmuensis* Shan & Constance	Zj
	Sasamorpha sinica Koidz.	Zj
†	*Sassafras tzumu* Hemsley	Js, Zj
	Saussurea hwangshanensis Ling	Ah
	Schizandra bicolor Cheng	Zj
	Sedum leptophyllum Frod.	Ah
	Shibataea chinensis Nakai	Js
†*	*Sinocalycanthus chinensis* Cheng	Zj
†*	*Sinojackia xylocarpa* Hu	Js
*	*Sorbus amabilis* Cheng & Yu	Zj, Ah
†*	*Stewartia sinensis* Rehder & Wilson	Zj, Ah
†	*Styrax dasyanthus* Perkins	Zj, Ah
	Sycopsis sinensis Oliver	Zj, Ah
†*	*Tapiscia sinensis* Oliver	Ah, Zj
†	*Taxus chinensis* Rehder var. *chinensis*	*Zj, Ah*
	Taxus chinensis Rehder var. *mairei* Cheng & L.K. Fu	Zj, Ah
†	*Tetrapanax papyriferus* K. Koch	Zj, Ah
	Tilia oblongifolia Rehder	Ah
	Tofieldia coccinea Richards	Ah
†	*Toona sureni* Merr.	Ah
†	*Torreya grandis* Fortune	Js, Ah, Zj
*	*Torreya jackii* Chun	Zj
*	*Trillium tschonoskii* Maxim.	Ah
*	*Tsuga chinensis* (Franchet) Pritzel var. *tchekiangensis* Cheng & L.K. Fu	Zj, Ah
†*	*Ulmus chenmoui* Cheng	Js, Ah
*	*Ulmus elongata* L.K. Fu & C.D. Ding	Zj
†*	*Ulmus gaussenii* Cheng	Ah
†	*Viburnum sargentii* Koehne	Zj, Ah
	Vitex trifolia L. var. *simplicifolia* Cham.	Js, Zj

5

Botanic Gardens — A World Network?

Cooperation in training and technical liaison between botanic gardens

J.B.E. SIMMONS and R.I. BEYER

Royal Botanic Gardens, Kew, U.K.

Summary

This paper explores the ways and means of encouraging greater cooperation in training and technical liaison between the botanic gardens of the world. Less than a quarter of gardens provide any training facilities, indicating the great need for training courses. The paper makes specific proposals about the training required at various levels. On technical liaison, the greatest need is for collaborative development on living collections. To achieve this, a draft multilateral agreement between gardens should be prepared, which could then be ratified by IUCN.

Resumen

Cooperación entre jardines botánicos en cuanto a formación y coordinación técnica

Esta comunicación examina los medios y la manera de estimular una mayor cooperación en formación y coordinación tecnica entre los jardines botánicos del mundo. Menos de una cuarta parte de los jardines presenta facilidades de formación, apuntandose la necesidad de cursos de formación. La comunicación propone proyectos específicos sobre la formación requerida en varios niveles. En cuanto a la coordinacion tecnica, sería mayormente necesaria para el desarrollo colaborativo de colecciones de colecciones de organismos vivos. Para su realización, se debería preparar un borrador con los acuerdos multilaterales entre jardines, que sería ratificado por la Union Internacional para la Conservación de la Naturaleza y de los Recursos Naturales (UICN).

Botanic Gardens and the
World Conservation Strategy

ISBN: 0-12-125462-3

Introduction

The fourth edition of the International Directory of Botanic Gardens contains nearly 800 entries spanning the globe and IUCN has identified many more botanic gardens through its questionnaire for the Botanic Gardens Conservation Strategy. While accepting that not all these are functioning as botanic gardens in the true sense, it does indicate that there is a tremendous pool of knowledge and expertise which, if harnessed, could play a significant rôle in the conservation of the world's diminishing plant resources. This paper explores the ways and means of encouraging greater cooperation in the areas of training and technical liaison in the hope that help can be offered to those botanic gardens who need it — particularly the tropical botanic gardens.

Since the pathfinding 1975 and 1978 Conservation Conferences at Kew, botanic gardens world-wide have significantly advanced their functions into *ex situ* and *in situ* conservation work. However, despite many achievements in this area, the requirements for plant and ecosystem conservation outstrip the resources available and increase the need for botanic gardens to pool their expertise. In the particular areas of training and liaison, experience over the last decade at Kew shows that the present situation is *ad hoc*. This can be illustrated by the occasional requests for formal collaborative agreements with tropical botanic gardens. Such an agreement is being developed with the new botanic gardens in Limbe, Cameroon, as an integrated project that uses the botanic garden for education and includes forest reserve management. More frequent requests for advisory visits to assist with the drafting of corporate plans for new botanic gardens are received by Kew and, where possible, help is given. In recent years Kew staff have advised botanic gardens in Mexico (Xalapa), Bangladesh, India, Malaya, Sabah and Venezuela. In most cases this has been followed by requests to provide training at Kew for technical managers so that they are more qualified to implement broad policies agreed during the advisory visit.

All these gardens have the potential to contribute significantly to conservation programmes, but need help through ongoing contact with established gardens who have the expertise they require.

Training

The questionnaire for the IUCN Botanic Gardens Conservation Strategy showed that less than a quarter of the gardens provide any form of formal training and that only 7% of the gardens that replied operate a diploma course. Of the latter, subjects covered relate mostly to public horticulture and dendrology, but none cover the subject areas of environmental protection or plant conservation. Although botanic gardens are not training centres *per se*, there is nevertheless a potential to develop training courses with a strong emphasis on conservation-related subjects at basic and advanced levels.

At the recent meeting of the European and Mediterranean Division of the International Association of Botanic Gardens (IABG) in Durham, a second questionnaire was circulated and completed by representatives of 25 gardens. All indicated that they would be willing to provide training opportunities, although only 40% had done so in the past.

Identifying the need — the new garden

Broad policies will have been formulated to justify a new garden obtaining funding from either institutional or federal bodies, but it is unlikely that these policies will be more than outline principles. An opportunity will always exist for guidance towards the local or regional needs that the garden should serve. Senior managers should be encouraged to carefully construct a definitive development programme to ensure that planning and implementation are dealt with logically and are fully understood at all management levels. Experienced assistance at this stage can be crucial in avoiding fundamental mistakes, and it is wise to seek the views of other scientific and technical managers on the broad perspectives of a successful botanic garden in order to turn policy into practical achievement. The technical staff who will be responsible for carrying out the planned activities often need a great deal of help as they are unlikely to have very much horticultural or management experience or expertise.

Each garden will be different in detail, but the basic problems and challenges are similar. From time to time Kew has been asked to provide a blueprint for an instant botanic garden: this is an impossible task. Brief visits by those involved, to or from an established temperate garden, will do no more than show what one garden has achieved and which, if copied, may have no relevance whatsoever to a new botanic garden. The need is to create a reference centre which can identify the requirements of a new botanic garden and indicate where help can best be obtained.

Again, the Durham questionnaire indicated that 88% of gardens were willing to take trainees for longer than a month and that all but 1 garden would be prepared to share the training of a particular individual with other gardens. It is encouraging to find that 36% were willing to seek sponsorship to pay towards living expenses, while 32% could give payment if the training included a significant amount of practical work. Living accommodation could be provided by 9 gardens.

Possible level and form of training

Senior Manager: study visits

From time to time Kew receives visits from senior managers who are building a completely new botanic garden or developing an existing one. As already indicated, Kew may not be the best place to seek advice, but it is usually included on the itinerary because it is well known. Before undertaking expensive study visits, senior managers should have access to a reliable source of information as to "where to go and what to see" — an international information bureau or publication which is well informed and broadly aware of what is available in the botanic garden world.

We recommend the updating and improvement of the published IABG Directory of botanic gardens, using the database now established by IUCN from the strategy questionnaire. This should include the training courses and other training opportunities offered by each garden.

Scientific Staff: postgraduate studies

These are likely to be more specific with the participant having some knowledge of his or her particular field of study and/or interest. A greater number of botanic gardens need to be encouraged to participate in such schemes and the existence of opportunities for postgraduate study should be more widely advertised. In the U.S.A., Longwood and Rancho Santa Ana offer facilities, but how many more do so, and what of Europe, Asia and Australasia?

Technical Manager: summer schools

Refining policies within the context of developing a botanic garden to meet specific objectives, discussing common problems and considering other management options are areas where an open forum learning programme would serve a very useful purpose. A possible solution would be to arrange summer schools based on centres which would be responsible for arranging a programme, and finding tutors to deal with particular topic areas. It is envisaged that these summer schools would not be attended by more than a small group of managers and, if possible, organised on a regional basis, making it possible to concentrate on problems common to the group.

We recommend that any host garden willing to establish a model conservation management course should be able to submit the prospectus to IUCN for approval and endorsement.

Technical Staff: internships

In many instances these will be at the first line management level where it is necessary to offer training in both horticultural and management skills. The most significant problem to be addressed is how to tailor the training to meet the requirements of individuals who will have come from differing backgrounds. Particular account has to be taken of ethnic origin, level of education and the need for an intern to adjust quickly to a different society. Is it fair, for instance, to immediately expose a trainee from the tropics to the European winter and to provide training in techniques which have little relevance back in the trainee's own botanic garden?

An increase in the number of botanic gardens willing to offer internships, similar to those offered by botanic gardens in North America, would make it easier to tailor the most appropriate training package to meet the needs of a particular individual. Sponsorship will be a limiting factor in providing money to support all the prospective interns wishing to be placed at any one time. However, if an international network of cooperating botanic gardens could be coordinated, it would be possible to ask for funds to support a programme rather than individuals.

The Durham questionnaire showed that about half the gardens provided "in house" training for their staff, but this was mostly "hands on" and not structured for career development. However, half intimated that they would be prepared to provide specific training programmes, with subject areas covering horticulture and botany. It was pleasing to note that all were prepared to arrange extension visits to other allied institutes, were willing to provide information on and to discuss financial budgeting and staffing policies.

We propose that:

(1) An IUCN agency is established as a clearing house linking students to scholarships, programmes and host gardens.
(2) Alternatively, a regional approach be adopted with centres in, say, Europe, America, Asia — perhaps through such as the American Association of Botanical Gardens and Arboreta (AABGA), IABG or the European Technischer Leiter Organisation — and that through the agency or groups international funding is sought.

Liaison

Views from the questionnaire

As indicated, international liaison between botanic gardens is currently *ad hoc* and minimal in relation to the need. The strategy survey indicates that the *greatest area for collaborative development occurs with the living collections*, with most being able to help and noting that this is also the most politically and economically viable area for developing accord. Joint scientific programmes were also considered as was exchange training, but with these, particularly the latter, problems such as cost, language, staff shortages and the administrative difficulties of exchanging paid personnel appear restrictive. However, the benefits of increasing liaison are manifold — the exchange of information, experience and plant material bring mutual understanding and move towards the sharing of resources and active collaboration on the conservation of germplasm.

Increasing the biological diversity and use of living collections in conservation work

Some 65% of those questioned had both the facility and wish to collaborate on improving the biological diversity of collections. The most vital areas for progress are inevitably in the development of *in situ* programmes. The full realisation of the potential of reserves attached to botanic gardens is illustrated by the Long Range Plan for the North Carolina Botanic Garden and the development of the Loder Valley Botanical Reserve for the Wealden Flora now established by Kew at Wakehurst Place in Sussex. Additionally, satellite areas can be established, such as those of the Jardín Botánico "Viera y Clavijo" in Las Palmas. From this, cooperation can extend to a regional or national level as shown by group botanic garden policies in India, Indonesia and U.S.S.R.

The *ex situ* work of botanic gardens is invariably oriented towards utility and education. At a basic level there is the need for the high quality curation that is able to guard against genetic erosion and to provide the technical facilities and skills that can adequately increase the progeny of endangered species. From this, there can be a move out to the *in situ* reintroducion of species. A sponsored programme at Kew for the symbiotic culture of terrestial orchids is attempting this with *Orchis morio*, by providing a large quantity of seedlings raised *in vitro* with their mycorrhizal fungus to the national Nature Conservancy Council, who will undertake the trial establishment and monitor the results.

The location, propagation and distribution of endangered species by botanic gardens under the Botanic Gardens Conservation Coordinating Body's monitoring scheme has amply demonstrated the potential of increasing liaison. The next step is shown by the development of agreements by botanic gardens to accept responsibility for groups of plants or regions, e.g. the Center for Plant Conservation's network in the United States or the similar, but less legally binding, schemes between gardens in Canada and in Holland.

Lessons from projects already undertaken

Recently the Living Collections Division (LCD) at Kew was coopted to help with a project set up by IUCN and the Fauna and Flora Preservation Society (FFPS) to propagate the 50 threatened endemic plant species of St Helena. An acceptable aid submission brought forth some national support that allowed an equipped Kew horticulturist (Simon Goodenough) to work on St Helena for 3 months, mapping the flora in collaboration with a taxonomist, and establishing an operational propagation unit on the island. Linked to this was an outreach commitment to the community which involved lectures to schools and other interested groups, as well as discussions with leading figures describing the value and interest of the island's unique floral heritage — largely unknown to the populace. Another member of the LCD staff (Chris Bailes) was also recently involved with a joint collecting visit to Nepal, where advice was also being sought on how to propagate orchids in cultivation to create local employment and reduce the pressure on the native flora. From both, the utility aspect of conservation emerged and the need for continuing liaison with such projects if they are not to falter.

To establish continuity and to sell conservation in a way that attracts aid support we must remember:

(1) to ensure outreach to schools and teachers;
(2) to support and train local and active conservation volunteers;
(3) to stress the broader economic advantages to a community of increased horticultural skills;
(4) to highlight the rôle of native flora in forming the most efficient watershed;
(5) to show the value of a unique flora in relation to tourism, and
(6) to screen native plants for utility and possible economic production.

Joint programmes can be seen to enrich collections and knowledge about individual taxa or habitats, for example, successful propagation by grafting the ebonies of St Helena. Joint expeditions, exhibitions and research projects can be used to increase the active use of specific collections and regional studies.

Suggested ways to progress

The initial suggestions of "adopt a garden" or "twinning" appear too restrictive (to some of those questioned) and may focus attention away from active conservation work. What is needed is a form of joint agreement between 2 or more gardens, conceivably regional, certainly multilateral, leading to a world network that promotes action on plant conservation activities by botanic gardens.

Conclusions

We recommend that IUCN be asked to form a model of a universally useable multilateral agreement of botanical conservation for use by 2 or more botanic gardens who wish to collaborate together.

This will allow twinning to occur where appropriate. It will also provide for multiple agreements between, say, 1 major garden and more than 1 developing garden, as well as encouraging the formation of regional groups. The aim is to increase the "north–south" transfer of technology. Such a scheme can provide a significant rationale to the work of gardens in temperate regions and should be so structured as to allow help at all levels.

It is further proposed that IUCN consider reforming and broadening the function of the "Body" to develop it into a Botanic Gardens Conservation Organisation. Its new remit would be as follows.

(1) To continue the Body's function of monitoring taxa *ex situ*.
(2) To ratify multilateral agreements on programmes for the shared development of biogenetically diverse collections and of *in situ* programmes.

This package should be made attractive to funding organisations so that members can more successfully apply for international resources. The umbrella organisations should also seek massive sponsorship. In authenticating and promoting such conservation work, botanic gardens could both improve their future security and the future of the flora with which they are involved.

Cultivation of rare cacti in Mexico

H. SÁNCHEZ-MEJORADA R.

Jardín Botánico, Instituto de Biología, Universidad Nacional Autónoma de Mexico, Mexico

Summary

In response to the sombre and grave threats that endanger the survival of many Mexican wild plants, especially many rare cacti, serious conservation efforts have been proposed and undertaken during the last few years by government agencies, scientific organisations, universities and amateur groups.

The most interesting and promising development is the growing concern to cultivate the rarest and most threatened species in botanic gardens, nurseries, private collections and even in the wild, in areas where they are native.

A major concern has been to encourage the establishment of commercial nurseries that could meet the demands from foreign trade and so decrease the strong pressure at present on wild populations.

Botanic gardens in Mexico are striving to cultivate and propagate the most threatened species, especially those native to the areas where they are located. The Jardín Botánico of the Instituto de Biología of the Universidad Nacional Autónoma de Mexico, besides currently trying to improve its cacti collection, has recently created a special research unit for the development of genetic resources. A section of this unit, devoted to tissue culture, is experimenting with useful and endangered species, including some cacti.

Plants confiscated by CITES authorities could be used to increase the botanic gardens' collections. Under special surveillance they could also be used by bona fide nurseries for commercial propagation. We should find a way to save these plants.

As part of the World Conservation Strategy, we strongly recommend

Botanic Gardens and the World Conservation Strategy

ISBN: 0-12-125462-3

that botanic gardens, in all continents, promote and increase the interchange of vulnerable, rare, threatened and endangered species, not only as seeds, but also as live specimens and tissue culture material, assuring, in this way, the survival of the disappearing cacti by cultivation in as many different places as possible.

Resumen

El cultivo de cactos poco comunes en México como un modelo para la colaboración internacional

En respuesta a las grandes amenazas que ponen en peligro la supervivencia de muchas plantas silvestres mexicanas, especialmente algunos cactos poco comunes, esfuerzos serios de conservación han sido propuestos y realizados durante estos últimos años por organismos gubernamentales, organizaciones científicas, universidades y grupos de aficionados.

El más interesante y prometedor desarrollo es el esfuerzo que concierne al cultivo de las especies más raras y amenazadas en los jardines botánicos, semilleros, colecciones privadas e igualmente en el campo, en áreas donde son nativas.

Un mayor interés tiene incitar el establecimiento de viveros comerciales que puedan hacer frente a las amenazas del comercio exterior y así disminuir la fuerte presión que se ejerce sobre las poblaciones silvestres.

Los Jardines Botánicos en México están esforzándose en cultivar y propager las especies más amenazadas, especialmente las endémicas de las áreas donde están localizadas. El Jardín Botánico del Instituto de Biología de la Universidad Nacional Autónoma de México, junto a su actual intento de incrementar las colecciones de cactos, ha creado recientemente una unidad de investigación especial para el desarrollo de recursos genéticos. Una sección de esta unidad, dedicada al cultivo de tejidos está experimentando con especies útiles y en peligro, incluyendo algunos cactos.

Las plantas confiscadas por las Autoridades CITES podrían ser utilizadas para incrementar las colecciones de los jardines botánicos. Bajo una especial vigilancia pueden también ser usadas de *bone-fide* por los viveros para propagación comercial. Deberíamos encontrar un camino para salvar estas plantas.

Como parte de la Estrategia Mundial para la Conservación, podemos recomendar enfáticamente que los jardines botánicos, en todos los continentes, promuevan o incrementen el intercambio de las especies vulnerables, raras, amenazadas y en peligro, no solamente como semillas, sino también como especímenes vivos y material para cultivo de tejidos, asegurando de esta manera, la supervivencia de los cactos que están desapareciendo, por medio de su cultivo en tantos sitios diferentes como sea posible.

All over the world, many plant species are endangered or threatened due to habitat destruction. However, some of them, because they have a high trade value, either as ornamentals or as the subject of collections, are also menaced by immoderate and exhaustive harvest from their habitats for commercial purposes. Among these, in Mexico, the most vulnerable are cacti, orchids, cycads and bromeliads.

In response to the grave and sombre threats that endanger the survival of many Mexican plants, during the last few years serious efforts for their conservation have been proposed and undertaken by government agencies, scientific organisations, universities and amateur groups.

A major concern of conservationists has been to encourage the propagation of rare species in botanic gardens, commercial nurseries, private collections and even in their natural habitats by the country people of the area. A priority is to promote the establishment of commercial nurseries that may meet the demands of foreign trade and so decrease the pressure now threatening the wild populations.

Botanic gardens in Mexico are striving to cultivate and propagate rare and endangered species, especially those plants that are part of the local flora.

The Jardín Botánico of the Instituto de Biología of the Universidad Nacional Autónoma de Mexico is situated at an altitude of 2,250 m above sea level, on an old lava field. Its main priority is the Mexican flora, especially desert plants. Its large and representative xerophytic collections are mostly in the open air where their upkeep is quite a challenge. This botanic garden is actively engaged in conservation, and is not only making efforts to spread the message to the public, especially to children, but also by taking specific action on both *in situ* and *ex situ* conservation.

The garden is currently trying to improve its collections of cacti and other succulent plants, both in quality and quantity. Special attention has been placed on rare, vulnerable, threatened and endangered species so as to create a reserve collection that may serve as a gene bank. The garden has recently created a special unit for the development of genetic resources, particularly potentially useful plants as well as threatened and endangered species. Part of this unit is devoted to tissue culture and is currently experimenting with useful plants that have potential in medicine or as crops. It has also experimented with endangered species, including orchids and cacti.

With regard to the orchids, it has been very successful in reproducing by tissue culture a terrestrial orchid *Bletia urbana*. This species is endemic to the lava flow where the garden is located and was on the verge of extinction due to the destruction and reduction of its habitat. It has now been reintroduced in its much reduced but now protected habitat.

With regard to cacti, 2 species also have been reproduced successfully by tissue culture. One of them, *Mammillaria sanangelensis*, is also endemic to the lava flow where *Bletia urbana* grows and is also on the verge of extinction. Only 6 individuals have been found recently. Several hundred plants, reproduced by tissue culture, will soon have attained a good transplanting size, and then can be reintroduced into their natural habitat, before the rainy season. The other, *Cephalocereus senilis*, a species that is now abundant but vulnerable, will be available for collectors and nurseries, so reducing the strong pressure on the natural populations.

We hope that by making available to the public the techniques of tissue culture, commercial nurseries will use them to meet the demands of the trade. We also hope that by cultivation in future of other rare, threatened and endangered species, we shall save many plants from extinction.

On a related topic, I would like to draw your attention to a grave problem faced by the authorities of the countries that are Parties to the Convention on International Trade in Endangered Species of Wild Fauna and Flora (CITES). This relates to confiscated shipments of plants listed in Appendix I of the treaty. Many hundreds of rare, threatened and endangered species, illegally introduced into countries that are Parties to the Convention, are currently being seized by the authorities. The problem is what to do with the plants and how to save them.

In some instances the plants are sent back to the country of origin, which may be costly. There, then, the problem persists. Three alternatives may be contemplated:

(1) To reintroduce the plants into their natural habitat
(2) To give the plants away to whoever wants them, including commercial dealers.
(3) To send them to botanic gardens.

These plants, after long journeys, are not always in good condition and each alternative provides problems. The first alternative requires that volunteers be found to work in the field, sometimes quite far from their homes. If the reintroduction is successful, the plants might be plundered again in a very short time. The second alternative is not very practical, since the commercial dealers will encourage the illegal trade in order to receive the confiscated plants.

The third alternative, sending them to botanic gardens, may be quite a problem in some cases, since not all the gardens have sufficient space or facilities to accommodate such large quantities of plants. In other cases, however, they may be willing to accept certain types of plants to enlarge their collections. Such is the case, as we have heard from our colleague Maurizio Sajevo, of the Palermo Botanic Garden, which would like to receive succulent plants confiscated in western Europe.

I am sure that among all the botanic gardens of the world there are many more like that of Palermo, which are willing and have the facilities to receive the impounded plants and thus save them, specially those that are rare or endangered. I urge my colleagues in botanic gardens, therefore, to contact the CITES authorities and to find out about seized shipments, and to receive some of these doomed plants for the benefit of future generations.

Finally, and returning to the main theme of this paper, we strongly recommend that, as part of the World Conservation Strategy, botanic gardens should promote and increase the interchange of rare, vulnerable and endangered species, not only as seeds, but also as live specimens and tissue culture, material, assuring, in this way, the survival of the disappearing cacti by cultivation in as many different places as possible.

The rôle of the International Association of Botanic Gardens (IABG) in conservation world-wide

K. LARSEN[1], B. MORLEY[2] and H. ERN[3]

[1] *Botanical Institute, University of Aarhus, Denmark, and President, IABG*
[2] *Botanic Gardens of Adelaide, Australia, and Secretary-General, IABG*
[3] *Botanical Garden and Museum, Berlin-Dahlem, F.R.G., and Chairman, European-Mediterranean Section, IABG*

Summary

The International Association of Botanic Gardens (IABG) is the official umbrella association for more than 800 botanic gardens and arboreta all over the world. In this paper we outline some of the conclusions to be drawn from the last edition of the IABG directory of botanic gardens and show how the activities of gardens can be better coordinated, especially between North and South. Various modern focusses for botanic gardens' work are outlined.

A new constitution for a more financially independent IABG will enable the organisation to obtain improved output, influence, and be of greater assistance to member gardens. The process of securing financial independence has, however, been slower than expected, and up to now funds are scarce.

A shift in attitude towards IABG is taking place. In particular regional groupings of IABG are being created, with European-Mediterranean and Australasian sections in place so far and a Latin American group being established. Closer and more active links have been sought with the American Association of Botanic Gardens and Arboreta (AABGA) and with IUCN. There is an urgent need to coordinate the work of conservation of endangered species in cultivation between IABG

Botanic Gardens and the World Conservation Strategy

ISBN: 0-12-125462-3

members, particularly those of more highly specialised botanic gardens in developed countries and the ways in which these organisations relate to botanic gardens in the tropics.

The responsibilities of the European-Mediterranean Regional Division of IABG are described in a separate section, by H. Ern.

Resumen

El papel de la Asociación Internacional de Jardines Botánicos (IABG) en la conservación a nivel mundial

La Asociación Internacional de Jardines Botánicos (IABG) es la asociación oficial de más de 800 jardines botánicos y arboretas de todo el mundo. En esta comunicación se subrayan algunas de las conclusiones de la edición mas reciente de la IABG y se muestra como las actividades de los jardines pueden ser mejor coordinadas, especialmente entre norte y sur. Se subrayan varios aspectos nuevos para el trabajo en jardines botánicos.

Una constitución nueva para la financiación de una organización más independiente puede facilitar a la IABG el obtener una mejora en el rendimiento, influencia, y tener una mayor asistencia de miembros a los jardines. El procedimiento de seguridad financiera independiente, sin embargo, ha sido más lento de los esperado, y hasta ahora el capital es escaso.

Un cambio de actitud hacia la IABG está teniendo lugar. La más significativa es la necesidad de crear grupos regionales de IABG, donde el grupo Europeo-Mediterráneo y el grupo Austaliano han sido ya creados. Un grupo latino-americano está en proceso de establecerse. Lazos más activos y cercanos han sido buscados con la Asociación Americana de Jardines Botánicos y Arboreta (AABGA) y con la UICN. Hay una urgente necesidad de coordinar el trabajo de conservación de las especies amenazados en cultivo entre los miembros de la IABG, particularmente en aquellos jardines botánicos más especializados de países desarrollados y las vías por las que éstas organizaciones se relacionan con los jardines botánicos en los trópicos.

Las responsabilidades del grupo Europeo-Mediterraneo de IABG han sido descritas en una sección aparte, par H. Ern.

The International Association of Botanic Gardens (IABG) is the official umbrella organisation for botanic gardens and arboreta all over the world. The IABG was established as a commission under IUBS in 1954.

In the last directory to be published approximately 800 gardens are enumerated, but we are aware that the actual number may be considerably higher. These gardens are very diverse in purpose and organisation, and the classic botanic garden with systematic and plant geographic sections and accompanying greenhouses comprise only one group.

A number of the larger gardens mentioned in the directory are nature reserves. On the other hand there are also several small private gardens with very limited staff and access; others are nurseries, agricultural or silvicultural experimental stations.

If at present we restrict ourselves to the classic botanic gardens, there is considerable diversity from small or medium-sized gardens with collections brought together from seeds or other material obtained from larger gardens, to institutions which actively gather new plant material from natural localities through collecting trips by staff members. There are also a number of gardens, such as we have had the opportunity of seeing, which specialise in the local flora, its endemics, its rare or endangered taxa, or others, as is the case with the Aarhus Botanic Garden, which take special interest in particular areas in the tropics.

There are gardens or stations developing gene banks for economically important species, or which maintain large collections of varieties of ornamental plants.

During the Presidency of the senior author, we have tried to draw conclusions from the last published directory of botanic gardens in order to obtain an overview of the potential for better coordination of activities, including conservation.

The total area of the 800 botanic gardens is *c.* 100,000 ha. Even if we deduct from this about 25,000 ha, representing those institutions that should rather be classified as nature reserves, it is still a considerable area for growing wild plants from the temperate and subtropical zones where by far the majority of these gardens are situated.

A survey of the greenhouse capacity was slightly more inaccurate as many gardens do not give data about the area under glass, but only cite the number of glasshouses. It is, however, possible to make an estimate concerning the 2,600 houses, which we have calculated to represent an area of 100 ha under glass for cultivating plant species from the warmer parts of the world. To this we add botanic gardens in the tropics, whose number is fortunately increasing. All in all a considerable reserve.

During the last 4 years as President of the IABG, the senior author has had the opportunity to visit a number of gardens in countries as different as Australia, the Soviet Union, eastern and western Europe and some Asian gardens. There are considerable differences in management, governmental influence and interest in the gardens and in nature conservancy strategies.

IABG has had sad letters from directors asking for help and have sent letters to ministers and local administrators in large western European countries trying to prevent functioning gardens from being closed, or from having their staff reduced to such a degree that many important activities have to be given up. But we have fortunately also seen several new gardens being established and old gardens receiving considerable funds for enlarging their activities, such as the construction of new and larger glasshouses in spite of the enormous heating expense involved.

It has become our firm belief that it is often in the power of the garden director to prevent disasters, or to create progress for their institution if the right steps are chosen. There are many ways to do this. IABG tries to help through an exchange of information; the importance of botanic gardens for nature conservation is only one point which can be emphasised, although a very important one, and it can be linked together with other activities of the classic botanic gardens.

1. Education of students

For this purpose most gardens maintain a standard collection of tropical and temperate species with emphasis on economically important genera. It may well be possible to substitute some of the old collections of, e.g. *Citrus, Dioscorea, Theobroma*, with a selection of wild species and rare cultivars showing not only the variation within a group, but at the same time serving as a conservation resource of important genes. We strongly recommend botanic gardens to make changes in their standard collections of economic plants.

2. Recreation and information for the public

The gardens' next important activity is as a recreation and information resource for the general public. Numerous gardens have been far too slow to realise that this service today must be more active, we would say aggressive, than 20 years ago. It is no longer enough to present the same nicely arranged flower-beds as 100 years ago. Special displays of economic plants, information about endangered species and the importance of saving them must be presented in pamphlets and displays. The importance of collaborating with other countries for our own and mutual benefit must be brought out to visitors through printed material, expositions, etc. Such activities cost money, but this is usually available through new thinking by directors. Also here the IABG tries to be helpful in suggesting new ideas and approaches.

3. Scientific activity

Fundamental to the gardens' work is scientific activity. Many gardens are traditionally part of universities or other centres of learning. The botanic garden is an excellent place in which to give taxpayers a tangible idea of the public use of their money. It is less obvious from expensive expeditions bringing back new plant species, or chromosome and pollen studies in the laboratory, and requires to be properly explained. It is the responsibility of the scientist to explain that these activities are vital to mankind and particularly necessary just now. It is not enough to complain that funding goes to high-technology instead of plant conservation and botanic gardens. We are responsible to the public to work through our gardens, and stress the necessity to maintain and develop them and their activities for the future.

4. Collaboration with industry

My fourth point about activities of botanic gardens is involvement and collaboration with industries working with plant products: this includes a large sector of industry as a whole. There is not much of a tradition for this and the 2 parties are in many cases hardly aware of each other. It is the responsibility of the scientist in this rapidly changing world, where plant resources are daily diminishing, to help exploit plant and other resources in a responsible way so that we leave a natural heritage to our children and grandchildren. Botanic gardens are placed in a key position to promote the conservation of plant resources.

The points chiefly refer to gardens in industrialised countries. The heaviest burden is, as already mentioned by several speakers, laid on the weakest shoulders: the gardens in the tropics. A closer link must be created now between North and South.

It was emphasised by the immediate past president of IABG, T.R.N. Lothian, during the Ninth General Meeting in Canberra in 1981, that a new constitution for a more financially independent organisation would enable IABG to obtain improved output, influence and be of greater assistance to member gardens. The process of securing financial independence has been slower than expected and up to now funds are scarce. Financing the activities of IABG is through the benevolence of a few institutions comprising less than 10% of botanic gardens in the world. Despite these difficulties, the IABG has taken part in a number of activities besides preparing the fourth edition of the directory and planning the fifth edition in collaboration with IUCN.

During the Canberra conference new activities were discussed; the most significant of these was the need to create regional groupings, an idea we have strongly encouraged. The establishment of regional divisions which send out newletters has occurred in the European-Mediterranean and the Australasian areas. A Latin American division is in the process of being established as well as a separate south Asian division. Closer and more active links have been achieved with the American Association of Botanic Gardens and Arboreta, and sought with the 134 Soviet gardens and with gardens of the Peoples' Republic of China.

There is an urgent need to coordinate the work of conservation of endangered species in cultivation between IABG members; the more highly specialised botanic gardens in developed countries have a responsibility towards botanic gardens in the tropics. The need for exchange and training of technical and scientific staff members between member-institutions is also receiving attention. Gardens are encouraged to establish educational and improved interpretative programmes for the general public. These and related issues are promoted through the regional newsletters financed by the gardens in each area, with further support from private funds.

The future role of IABG is first and foremost to represent all the botanic gardens of the world in international forums and to support individual gardens and help strengthen their position by mutual collaboration. It is also our opinion after the reorganisation of the secretariat last year that new commitments related to conservation should have the highest priority.

First of all collaboration with international organisations working for nature conservation should be established, such as WWF, FAO, IUCN and other bodies which are able to finance larger projects with private and public funds. In this connection the present conference has proved extremely useful and IABG is grateful to have been invited.

It is our hope that a firmer economic basis for IABG may be established with the new constitution prepared for the general assembly in Frankfurt-am-Main in August 1987 to which all botanic gardens have received an invitation.

We also point out that the regional IABG newsletters are open to contributions not only from members of IABG and AABGA, but from members of IUCN, WWF and others who in this way would be able to communicate with botanic gardens in a large part of the world. It is, furthermore, our hope that a general IABG newsletter can be issued in the near future.

Summarising our activities, we would like to emphasise that the regional meetings such as those in Nancy and Durham, the IABG board meetings and the regional newsletters, have brought botanic gardens closer together during the last 4 years. Our association will work continuously to promote a global network of gardens and nature parks and collaboration between them. Collaboration with conservation organisations also has a high priority to strengthen the role of botanic gardens in the World Conservation Strategy for this planet's natural resources.

Responsibilities of European-Mediterranean botanic gardens

H. ERN

The European and Mediterranean botanic gardens have been united into the first of the suggested regional divisions of IABG. A provisional board was set up in 1982 and immediately began work.

A modest Newsletter was started, of which Number 5 is now being prepared. Two international meetings were organised, thanks to the initiative and commitment of Monsieur Pierre Valck (Nancy) and Mr Peter Maudsley (then Durham). These meetings have each been attended by about 60 delegates from 20 different countries. The subjects dealt with at these meetings have been or will be published.

Within the region, we have several independent local associations, most of them in existence long before the foundation of the main division itself. The most recent of these is the Asociacion de Jardines Botanicos Ibero-Macaronesicos.

These are the results of 5 years of work. They have been possible because some individuals have taken a very considerable amount of work on their shoulders. We may say that the main responsibility of the European-Mediterranean botanic gardens, namely the formation of a working and communicating regional division of IABG, has thus been achieved. However, this is only the frame now to be filled with real activities. This will have to be as follows.

(1) Preparation of a constitution taking into account the future existence of several regional divisions of IABG (one of them, for the Australia–South Pacific Region, was founded in July 1985).

(2) Creation of a functioning infrastructure for the many administrative problems connected with the organisation of so many and so different botanic gardens. (At the board meeting of 1 September 1985, Mr Peter Maudsley has been proposed as Secretary.)

(3) Setting up of working groups: during the Durham meeting 2 working groups were set up to examine the policies of collections and seed distribution.

(4) Practical collaboration between botanic gardens of this huge region and those of other parts of the world, especially those of the tropics.

(5) Involvement in plant conservation projects within and outside the European-Mediterranean area.

Conservation of rare plant species and the threatened flora of the world is certainly a very prominent task for botanic gardens and its importance will grow. We must not forget, however, that botanic gardens also serve other purposes and their responsibilities go further than to plant conservation alone. By strengthening the organisational structures and the inter-relationships of the many botanic gardens of our area, our Regional Division tries to fulfil its purpose. Concerning the various aspects of conservation being expected to be resolved by our botanic gardens, we would be glad to get as much help and advise as possible from the big international nature conservation organisations.

Building a national *ex situ* conservation network — the U.S. Center for Plant Conservation

F.R. THIBODEAU and D.A. FALK

The Center for Plant Conservation, Jamaica Plain, Massachusetts, U.S.A.

Summary

In 1984 18 botanic gardens from throughout the United States joined with 7 national and international conservation organisations, including IUCN, to establish a U.S. national Center for Plant Conservation. The Center's primary goal is to create a complete programme of *ex situ* plant conservation, encompassing appropriate collection, storage, cultivation and research for the nearly 3,000 extremely rare, threatened or endangered native American taxa. The Center also encourages study and public enjoyment of the material in the collection, as well as assisting habitat conservation and management efforts. The Advisory Council of leading botanists and conservationists has been assembled to guide the technical aspects of the Center's programme.

At present, the Center's main activity is the development of the plant collection itself. To this end it maintains a database from which the highest priority species are chosen for accession by the gardens, in consultation with the Center's Advisory Council. This work is coordinated by the Center's National Office, which is also responsible for maintaining standards of collection, curation, and documentation.

In 1985, the Center's first year of programme implementation, protocols for operation were agreed upon, the first steps toward a

Botanic Gardens and the World Conservation Strategy

ISBN: 0-12-125462-3

secure financial base were taken, and most important, the first 40 taxa were brought into permanent collection. Professional and public reaction has been overwhelmingly positive.

Because the Center relies heavily on existing organisations, information and infrastructure, it provides a model applicable to other countries with significant institutional resources. We believe this model can be adapted to other countries' particular resources and conservation needs to develop coordinated and effective *ex situ* species protection programmes.

Resumen

Creando una red nacional de conservación ex situ; *el centro de los Estados Unidos para la conservación de plantas*

En 1984 se unieron 18 jardines botánicos de todos los Estados Unidos de América con 7 organizaciones de conservación nacional e internacional, incluida la UICN, para establecer un Centro nacional para la conservación de plantas. La meta primaria del centro consiste en crear un completo programa de conservación de plantas *ex situ*, acompañado de una apropiada recolección, almacenamiento, cultivo e investigación para todos los aproximadamente 3.000 taxones extremadamente raros, amenazados o en peligro que son de origen americano. El Centro también estimula el estudio y disfrute del material reunido así como ayuda a los esfuerzos por conservar el hábitat. El consejo consultivo de botánicos expertos y conservacionistas ha sido creado para aconsejar los aspectos tecnicos del programa del Centro.

Actualmente, la principal actividad del Centro es el desarrollo de la propia colección de plantas. A tal fin elabora una base de datos del que se eligen las especies altamente prioritarias para que sean adquiridas por los jardines, en consulta con el Consejo Consultivo del Centro, constituido por botánicos retirados del gobierno, organizaciones conservacionistas y académicas. Este trabajo está coordinado por la oficina nacional del Centro, la cual también es responsable por mantener unos estándares de colección, conservación y documentación.

En 1985, primer año del programa instrumental del Centro, se han elaborado protocolos para operaciones, se han dado los primeros pasos para asegurar una financiación segura y lo que es más importante, se han integrado en la colección permanente 40 taxones. La reacción profesional y pública ha sido abrumadoramente positiva.

Debido a que el Centro confía profundamente en las organizaciones, información e infraestructura existentes, proporciona un modelo fácilmente adaptable a otros países con significativos recursos institucionales, pero carentes de programas efectivos de conservatión de plantas *ex situ*.

The Center for Plant Conservation was established in April 1984 to develop a permanent, well-documented and accessible living collection of the 3,000 rare and endangered plants native to the United States. By the end of 1985, it had lived through 2 growing seasons — one for the organisation and one for its first taxa.* It should surprise no one who has tended either plants or organisations that the process of actually implementing a national programme is more complex in practice than in design. Nonetheless, the Center has demonstrated that programmes at a national scale can be conceived and carried out effectively.

The Center's metaphorical roots extend down to the IUCN conferences of the 1970s and early 1980s, when only a few of the major botanic gardens in the United States were playing an active rôle in the preservation of plants at risk of extinction. In 1978, the Smithsonian Institution, at the request of the U.S. Congress, published "Endangered and Threatened Plants of the United States" (Ayensu and DeFilipps 1978), listing 3,000 taxa at risk, some 10% of the American flora. They also noted that only about 200 of the 3,000 were known to be in cultivation at the time.

There were, however, extremely encouraging results from those gardens which were taking a responsible rôle in the preservation of their regional flora. For example, by the 1960s, *Agave arizonica* from the south-western United States had been reduced to only 3 known individuals. The staff of the Desert Botanical Garden in Phoenix, Arizona, began a programme that not only led to the recovery of the species, but also to its introduction into the horticultural trade. Today *Agave arizonica* is something of a symbol proclaiming the superiority of native flora in the ornamental horticulture of the south-western United States. Similarly, the Fairchild Tropical Garden in Miami, Florida, was cultivating the only *Amyris balsamifera* remaining in North America. It would be possible to draw examples from another 6 or 7 American gardens which were active at the time.

Not only was there botanical confidence that it would be feasible to establish an American collection, but there was institutional evidence as well. The botanic gardens of Arizona, Hawaii and a few other states were already working jointly to list rare and endangered plants they had in cultivation, and were taking responsibility for the protection of separate groups of their states' floras.

Viewed from a national perspective, however, the lack of coordination had extremely serious ill effects. Most sections of the country had no effective *ex situ* conservation programme; it was likely that plants without horticultural value would be overlooked; there was no way to determine

* Data are as of July 1986; by the end of 1986 the Center expects to have close to 200 taxa in the National Collection.

which taxa were in any particular collection without extensive research; and horticultural botanists often had no way of developing joint programmes with those managing plants *in situ*. Perhaps worst, there was no insurance that many garden collections had any meaningful degree of permanence. For instance, at roughly the same time that the last remaining *Pediocactus knowltonii* were being taken from its one remaining site in New Mexico, the only cultivated collection in a public garden was lost following the death of its curator, the single person at the garden who understood its significance.

This lack of a strategic response to plant endangerment did not go without note. The U.S. National Academy of Sciences (1978) found that "American botanical gardens and arboreta are the most logical agencies to take on the responsibility of protecting the thousand-odd plant species proposed for threatened or endangered status in the United States."

By the early 1980s, there was a clearly articulated rôle for gardens in plant conservation, primarily through the work of IUCN; some few gardens were demonstrating that they could carry out that rôle; and the larger scientific community was asking for an increased *ex situ* response. It was time for a comprehensive national programme. The Center for Plant Conservation thus cannot claim credit for the fundamental idea of building a U.S. National Collection of Endangered Plants, given the intellectual and practical work already being done, but it did catalyse the formation of one.

While all ideas begin with a spark at one point, how they grow is more important than how they begin. In the U.S., that spark was a series of conversations in 1982 between conservation botanists from the Boston area and the staffs of the Arnold Arboretum (Jamaica Plain, Massachusetts) and the New England Wild Flower Society (Framingham, Massachusetts). They defined a basic structure for cooperation among gardens that emphasised responsibility for the permanent curation of regional floras. By the following year, the Center also included the Desert Botanical Garden, Fairchild Tropical Garden, Missouri Botanical Garden, North Carolina Botanical Garden, Pacific Tropical Botanical Garden, Rancho Santa Ana Botanical Garden and Waimea Arboretum. Thus, there were gardens ready to begin a more extensive national programme with the capability to conserve plants from nearly half of the United States in a climate similar to that which the plants experience in the wild. This group then expanded to cover the remaining biogeographic regions of the U.S. through the inclusion of Berry Botanic Garden, Bok Tower Garden, Denver Botanic Garden, Holden Arboretum, Nebraska Statewide Arboretum, New York Botanical Garden and the San Antonio Botanical Center.

From the beginning, the Center was structured as an organisation with 4 operating bodies. First, there are the Participating Institutions which include the gardens and arboreta that collect, propagate and house material

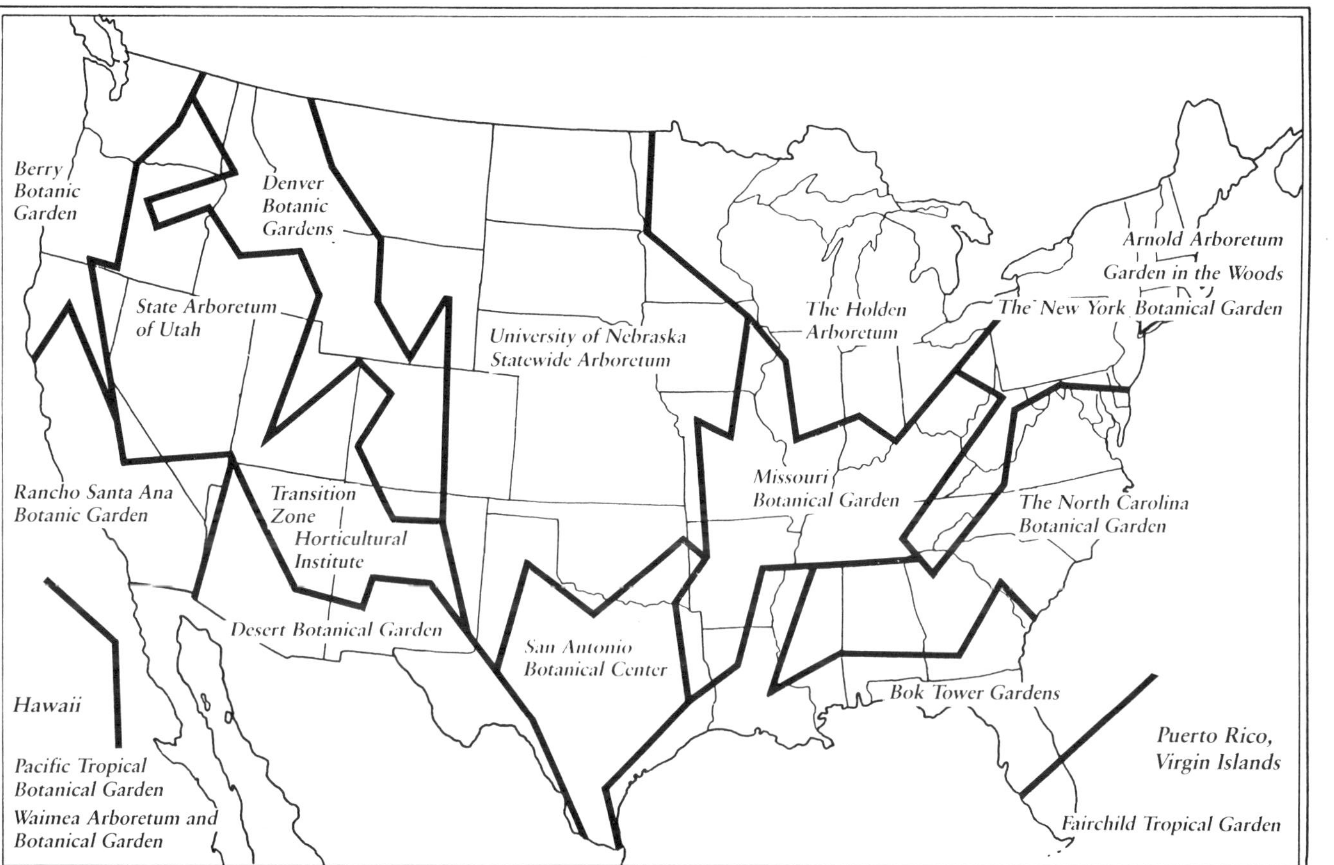

Fig. 1. U.S. biogeographical regions and the botanical institutions participating in conserving their flora.

for the National Collection, as well as the National Seed Storage Program of the U.S. Department of Agriculture, which provides long-term seed storage. Second, a scientific Advisory Council approves technical policy, including species selection and conservation plans for each taxon. Third, a Board of Trustees manages the Center as a corporation and links it to the world beyond plant conservation. Finally, there is a National Office — currently numbering 5 core professional staff — which coordinates activity, formulates policy for approval by the other groups, plans for the collections, and ensures that resources will be available for its development.

The Center's Advisory Council deserves special note. It is comprised of representatives of conservation organisations, the botanical branches of the federal government, and the academic community. While the Advisory Council was originally developed to ensure the quality of the botanical aspects of the programme, their rôle is already extending beyond internal overview in several important ways. For example, Dr Bruce MacBryde from the Scientific Authority of the U.S. Fish and Wildlife Service, one of the Center's advisors, is also a member of the Federal Working Group on the Biological Control of Weeds. While biocontrol is an important advance in ecological management, the Group is properly concerned that a parasite on a weedy species, for example, may also parasitise endangered ones. By linking the Center to the Working Group, 2 important conservation activities are advanced as one. The biocontrol researchers have ready access to plant materials otherwise unavailable, and the Center has an important client for its service. A similar rôle has been played by Council member Dr Robert Jenkins, Vice-President for Science of the Nature Conservancy (TNC). Through a cooperative agreement, TNC has supplied the Center with much of the computerised data on plants still needing protection. The Center, in turn, will supply the Conservancy both with propagules and, perhaps more importantly, basic biological information about species found on preserved land for which biological management information is lacking. The Advisory Council is presently chaired by Dr John Fay, Botanist with the U.S. Office of Endangered Species, Department of Interior, which is responsible for implementing provisions of the U.S. Endangered Species Act.

Central to the organisation's activity is a computerised data bank with 2 key sets of information. The first is the list of rare or declining taxa including key fields for federal classification of endangerment status, and states of occurrence, etc. This information enables the Center to identify those taxa which are most at risk and therefore in need of inclusion in the programme. The second component inventories plants included in the Center's collection, including information about successful horticultural techniques for their management. Almost without exception, this is information generated at the gardens individually, but never available before to the wider conservation community.

Data sharing is in many ways a key area for cooperation, because it underscores the essentially complementary nature of the *ex situ* and *in situ* approaches. *Ex situ* programmes are superb at providing two essential resources: information and live plants. Users who have indicated interest in research access include the U.S. Fish and Wildlife Service, the Nature Conservancy and several university departments of botany. An example of such cooperation is a preservation programme for *Hudsonia montana*, one population of which is found on land owned by the Nature Conservancy in Vermont. Despite its protected status, the population was nearly destroyed 2 years ago by off-road vehicles — a major problem in recreational areas in the United States. The Conservancy took cuttings of the remaining plants and contracted with a local greenhouse for propagation. The next year the rooted cuttings were reintroduced precisely to the original site, and the population is now recovering.

Another interesting example of cooperation with the habitat management community is the care of *Pediocactus knowltonii*, known from only one population in northern New Mexico. Populations of *P. knowltonii* had once numbered in the tens of thousands, but by the 1970s the species had been reduced to a single population, due to a combination of collecting, grazing and dam construction. The site was acquired and fenced by The Nature Conservancy, but the Conservancy also arranged with the Center to have plants established in cultivation in an Arizona botanic garden. In addition to growing a protected population of plants, the garden has carried out a research investigation into mycorrhizal associations in this and other species of *Pediocactus*. This research has identified a root symbiot which is critical to growing the plant in cultivation, and may be an essential piece of information for managing the wild population as well.

This theme of supplying plants and information is one that does not appear to have been fully appreciated in much of the literature on *ex situ* conservation. It has been the Center's experience that there are a large number of conservationists, horticulturists and botanical researchers — both pure and applied — whose work would be advanced significantly if a reliable and accessible source of research material was available. Although the collection *per se* is less than a year old, we are already receiving requests from biotechnologists for plant material that is otherwise unavailable. It is clear that the medicinal or agricultural potential of a rare plant is far less likely to be realised while it grows only at a remote, poorly-known location.

Another important service, seldom discussed in the literature, is the rôle of *ex situ* repositories as temporary refugia. In Puerto Rico several species are only found on land definitely scheduled for development this year.

These include *Zanthoxylum thomasianum* and *Banara vanderbiltii*, neither of which are likely to have any viable wild populations remaining by the end of 1986. The U.S. Fish and Wildlife Service has asked the Center to bring some of the remaining plants into cultivation at least long enough so that a preserve can be established and the plants reintroduced to undisturbed land in the same general area. While *ex situ* conservation does not substitute for habitat protection, together they are far more effective than either separately.

We have also found that there is danger, as well as opportunity, in such cooperations. *Exclusive* reliance on providing services to other programmes can easily lead to overlooking a plant under threat and already near extinction, but without a "sponsor". Thus, among the first procedures the Center established was that endangerment would be given highest consideration when planning new accessions to the collection. Potential utility would be considered, but only as a secondary factor.

The ranking of species for accession has been only one of the botanical issues faced by the Center. Among the first and most important questions to resolve were 2 that have never been addressed in detail. First, one must ask what constitutes a genetically representative sample of a rare taxon. This question, of some theoretical interest, becomes a pressing practical concern when it is necessary to decide how much and from which populations to sample. The best evidence now available suggests that a sample drawn from 50 unrelated individuals, taken from either one or several populations, will contain all but the most uncommon alleles of a taxon with 95% certainty (Chapman 1984). Unfortunately, much of the evidence one would need to study this problem thoroughly is unavailable or anecdotal; the most carefully surveyed taxa are small grain crops. While there is additional evidence that this recommendation is more broadly applicable, it must be tested with rare wild germplasm in particular. At present, there is scant empirical evidence. In fact, the Center may be able to contribute significantly to research on questions of genetic representation by providing facilities in which plants can be grown and studied.

It is also essential to address the question of how the living material is best maintained. Long-term seed storage, backed by growing collections for research, education and display, appears to be a secure and cost-effective method if the infrastructure is already available, but there remains the problem of taxa with recalcitrant seeds. For these plants, growing collections must do double duty for long-term conservation and active research use. Horticulturists often pale at the suggestion that several replicates of 50 growing individuals of each species may be thought essential, and that more are highly desirable. To address this question from a programmatic point of view, the Center (1986) has prepared

"Recommendations for the Collection and *Ex Situ* Management of Germplasm Resources from Rare Wild Plants", which have been approved by the Advisory Council. While collections of this size appear to be necessary, and are beyond the usual scope of garden activity, they also appear to be reliable, practical measures for long-term preservation.

One of the most striking aspects of programme development has been the enthusiasm with which both the more general public and the media have greeted the Center. This is due in great part to the excellence of the institutions which collectively constitute the Center as a whole. The Center conducts its public education programme largely by cooperating with member organisations such as the Garden Club of America and the American Horticultural Society. The Garden Club is making both its national membership and the general public aware of plant conservation issues, as well as the importance of the National Collection, through a jointly sponsored slide/tape programme entitled "The Garland of Generations". This 20-minute show will be presented to audiences throughout the country by specially trained members of the club during the coming 2 years.

It should be obvious from the foregoing discussion that the Center for Plant Conservation is not a single organisation in the traditional sense, but rather an organisation of organisations, all working together to build a National Collection. Such consortia take great advantage of skills and infrastructure that are already available and may be an important element in the development of other bio-regional *ex situ* plant conservation programmes.

While such programmes will draw heavily on the character of their member organisations, certain elements of the Center's design suggest themselves for replication. Foremost is an emphasis on tangible goals. The Center evaluates its progress in terms of the implementation of appropriate conservation strategies for individual taxa. Thus, progress can be measured in terms of plants added to the Collection, requests for research material, improvement in the status of wild populations based on information gained *ex situ*. Dr C. Ritchie Bell of the University of North Carolina put it most succinctly at the first meeting of the Participating Institutions when he said, "Plants first". Organisations must serve the plants, not the other way around.

In development of a *permanent* collection, it is important that each member of the organisation make a binding commitment to participate in an ongoing programme. The plants that are part of the collection, for example, are managed by the whole organisation and held in trust by the individual participating gardens. If a garden is unable to continue as a member, the National Collection remains intact. In the Center's case, each participating garden has signed an agreement affirming a committment to work according to common principles and policies set out in a series of uniform documents.

Botanic gardens, like the plants they conserve, grow and evolve. For many it has required a searching redefinition of their rôle to prepare them to enter the environmental fray. But organised on a national scale, they can be powerful allies in the common cause of biological conservation.

References

Ayensu, E. S. and DeFilipps, R. A. (1978). "Endangered and Threatened Plants of the United States". Smithsonian Institution and World Wildlife Fund, Washington, D.C.

Center for Plant Conservation (1986). "Recommendations for the Collection and Ex Situ Management of Germplasm Resources from Rare Wild Plants". Unpublished report.

Chapman, C. G. D. (1984). On the size of a gene bank and the variation it contains. *In* "Crop Genetic Resources: Conservation and Evaluation" (J. H. W. Holden and J. T. Williams, eds), pp. 102–119. Allen and Unwin, London.

U.S. National Academy of Sciences (1978). "Conservation of Germplasm Resources". N. A. S., Washington D.C.

A national plant conservation programme for Canadian botanic gardens

R.S. CURRAH, E.A. SMRECIU and P.N.D. SEYMOUR

Devonian Botanic Garden, University of Alberta, Edmonton, Canada

Summary

In 1984 the need for a national programme for *ex situ* conservation of native and non-native species in Canada was identified. This paper outlines subsequent efforts by staff of the Devonian Botanic Garden to develop such a programme and outlines its objectives: the programme will strengthen ties between Canadian gardens; ensure that a gene bank of horticultural material is maintained; create and maintain an information network of rare, endangered and threatened native plants in Canada; and provide a database which could lead to a Red Data Book for Canada. Two groups of plants are covered: rare wild plants of Canada in cultivation, and rare or uncommon cultivated species of either genetic or historical significance to Canada.

Resumen

Programa nacional de conservación vegetal para los jardines botánicos Canadienses

En 1984 se observó la necesidad de preparación de un programa nacional para la conservación *ex situ* de las especies autóctonas e introducidas en Canadá. Esta comunicación subraya los esfuerzos subsiguientemente realizados por los miembros del Jardín Botánico Devónico para desarrollar tal programa y sus objetivos son: estrechar los vínculos de unión entre los jardines canadienses; asegurar la

Botanic Gardens and the World Conservation Strategy

ISBN: 0-12-125462-3

mantención de un banco de genes de material de horticultura; crear y mantener una red de información de plantas nativas raras, en peligro y amenazadas; proveer un banco de datos que podría dar lugar al Libro de Datos Rojos (Libro de Especies en Peligro) de Canadá. El programa abarca a dos grupos de plantas: especies silvestres canadienses raras en cultivo y especies cultivadas raras o poco comunes de trascendencia histórica y genética para Canadá.

National activities in the field of plant conservation in Canada are currently centered on 2 main programmes. The Plant Gene Resources of Canada Programme under Agriculture Canada is concerned with crop plants of economic importance. The Committee on the Status of Endangered Canadian Wildlife (COSEWIC), under the Canadian Wildlife Service, is primarily concerned with habitat preservation. The National Museum of Natural Sciences, through its Rare Plant Atlas Project, has provided lists of rare native species for most Canadian provinces and territories.

The need for a national programme concerned with the conservation of cultivated collections of rare native and non-native species was identified during the annual meeting of the American Association of Botanic Gardens and Arboreta (AABGA) in Edmonton, June 1984. A discussion among directors and other representatives of Canadian botanic gardens resulted in a suggestion that the Canadian Plant Conservation Programme be founded.

During the autumn of 1984, the Devonian Botanic Garden assumed the responsibility of initiating a draft outlining some ideas for the organisation and definition of the Programme's rationale and objectives.

In the draft it was proposed that a Canadian Programme would fulfill 3 important requirements. Firstly, it would form an information network among Canadian botanic gardens. Prior to 1984, there was very little contact between gardens beyond that generated by attending the meetings of the AABGA.

Secondly, a Canadian programme would complement, and be complemented by, other plant conservation programmes such as the American Center for Plant Conservation, the National Council for the Conservation of Plants and Gardens, and the Botanic Gardens Conservation Coordinating Body of the IUCN Conservation Monitoring Centre.

Thirdly, a Canadian programme would fill an acute need for the provision of a national and professional avenue of communication among botanic garden staff and students about conservation programmes involving plants in cultivation, gardens, and plants and their natural habitats in Canada.

Seven specific objectives of the programme were outlined.

(1) To promote research into the propagation and maintenance in cultivation of genotypes considered rare, threatened or endangered.
(2) To organise a central secretariat (Canadian Centre for Plant Conservation — CPCP) which would coordinate a national inventory of specialised collections maintained at Canadian botanic gardens. This might also be extended to the identification and listing of rare and threatened plants in private gardens.
(3) To produce and circulate a newsletter to contributing institutions and individuals. Topics would include items of professional interest to conservationists and keen amateur plantsmen.
(4) To provide an advisory service to individuals (professional or amateur academic or commerical), that would locate materials (i.e. specific genotypes) for research, propagation, distribution or reintroduction.
(5) To assist in, and/or provide political and/or logistic support for, the conservation of valuable plant collections and/or species in danger of being destroyed.
(6) To sponsor a programme ensuring a uniform signage system to identify, to the public, participating gardens and their specific major collections. (This would inform the public of the programme and help them understand its objectives.)
(7) To encourage systematic maintenance of various plant groups at their most appropriate geographic centres.

Finally, the brief included a proposal for the establishment of a national botanic garden database using the Stanford Public Information and Retrieval System (SPIRES). The SPIRES program offers "user-friendly" access, update, and output protocols, and is already in use at 2 major universities associated with botanic gardens in Canada. Comments on the draft were positive. Based on respondents' comments, it was proposed that the CPCP focus efforts on cultivated plants of two categories:

Category 1: Rare wild plants of Canada in cultivation, and
Category 2: Rare or uncommon cultivated species of either genetic or historic significance to Canada.

In October 1985 an informative package was made up and dispatched to 27 gardens and 72 interested individuals. The package contained the following materials.

(1) A short review of the proposal for creating a national plant conservation programme for Canadian botanic gardens.
(2) Copies of a master list of some 2,500 plants considered rare in most of the Canadian Provinces.
(3) Data forms for identifying rare native plants in cultivation in the garden.
(4) A list of objectives concerning preservation of horticultural plants and a questionnaire on special horticultural collections.

Gardens were requested, (a) to identify rare native plants in their gardens; (b) by using the IUCN Red Data Book categories, to indicate their opinion of the status of plants on the list; and (c) to identify rare horticultural plants.

The objectives of the Horticultural Plant Project are as follows:

(1) To identify Canadian botanic gardens holding specific collections of horticultural plants both for ornamental and genepool purposes, as a first step in ensuring ornamental horticultural genepool maintenance.
(2) To identify Canadian botanic gardens which hold collections of plants of known date of introduction to Canada; to determine which are rarest and to urge botanic gardens to propagate and maintain these plants.
(3) To cooperate with amateur horticultural societies to seek out garden varieties which are of genepool merit and to obtain material from these plants for botanic gardens. (Many old horticultural varieties are held almost solely in amateur hands and are no longer available commercially.)
(4) To survey commercial nurseries in Canada in order to ascertain their holdings. It is apparent that many varieties are available in Canadian nurseries that are no longer offered for sale in other parts of the world.
(5) To coordinate the holdings of the various plant societies and to hold such information for retrieval upon request. In Canada we have many specialist societies (e.g. rose, orchid, gesneriad). Their holdings are often overlooked.
(6) To encourage the upkeep and expansion of horticultural herbaria which are necessary to serve as references for taxonomy and identification of cultivated ornamental plants.
(7) To encourage botanic gardens to conduct a survey of garden plants in their areas and to rescue and propagate old garden varieties.

Finally gardens were asked to provide comments on the following questions. Does your Garden have a collections policy for: (i) rare native plants? (ii) cultivated plants? Do you feel there is a need for a national collections policy? Replies are now coming in. We hope to tabulate these and distribute the data in 1986. A national meeting to review the programme will also be called in the autumn of 1986.

A word about seeking funding. It is our intention to keep the endangered native species and horticultural programmes completely and utterly separate.

To sum up, the objectives of the entire programme are as follows:

(1) To strengthen ties among Canadian gardens.
(2) To ensure that a gene bank of horticultural material be maintained.
(3) To create and maintain an information network of rare, endangered and threatened native plants in Canada.
(4) To provide a database which could lead to a Red Data Book for Canada.

Principles and standards for the computerisation of garden record schemes, as applied to conservation, with proposals for an International Transfer Format

J. CULLEN

Royal Botanic Garden, Edinburgh, U.K.

M. LEAR

IUCN Consultant

D.C. MACKINDER and H. SYNGE

IUCN Conservation Monitoring Centre, 53 The Green, Kew, Richmond, Surrey, U.K.

Executive Summary

Botanic gardens have traditionally kept records of their accessions, often on card indexes. An IUCN survey has shown that many have now put theirs on computer and that more plan to do so. The advent of the low-cost personal computer (PC), with disk storage of 10 mb or more, means that a computerised system is within the reach of virtually all gardens.

Botanic Gardens and the World Conservation Strategy

ISBN: 0-12-125462-3

The principal advantage in computerising garden records is to enable the garden to manage its collections better. In itself this is important for conservation. But computerisation also opens the door to a far more effective and powerful means of communication between gardens; if botanic gardens are to be effective as a world network for *ex situ* conservation of threatened plants, it is essential that excessive duplication is avoided and all the threatened species are covered. To do this, gardens must not only be able to share records on their collections with each other, but with IUCN or any other central agency so as to receive assessment of their holdings in comparison with those of others

At present about 250 botanic gardens share data with IUCN on their holdings of threatened plants. This is done manually at present and is both time-consuming for the garden and not up to date. The work could be greatly eased and improved if garden collections were on computer and each garden permitted IUCN access to its database, say on tape or disk once a year. This idea was the spur for the present paper.

To promote computerisation and to encourage a necessary degree of standardisation and harmony between record schemes, IUCN is:

(1) Promoting the creation of a model system, to run initially on an IBM PC, which will be available for any garden which does not wish to have professional programmers or contract out the software to software houses.
(2) Defining the conceptual and logical basis for an optimal scheme, as an aid to programmers, based on the best advice available and with an analysis of the objectives of garden record schemes.
(3) Proposing the definition of an International Transfer Format, which will provide a means by which any garden can communicate with any other and can exchange data with the IUCN database.

This paper is the first contribution to provide gardens with at least part of (2) and a proposal for (3). It covers only those elements of the data that were thought most necessary for conservation purposes. These are: Accession Data, Plant Name, Verification, Donor, Provenance and Material, and Survival Status. Others may be just as important to gardens but are not needed for international conservation. The intention is not just to define an optimal and minimal format for the transfer, but to gather together all the relevant facts on each data element that a computer programmer may need when writing software for a garden record scheme. In this way we hope to assist those gardens starting to computerise and to encourage gardens to gather and store the data in ways that are both botanically sound and are compatible from garden to garden. It has been a great pleasure to find that after discussion on each item, there does seem to be a most concise and elegant way of recording the data, so providing a good incentive for standardisation.

Introduction

One of the main features distinguishing botanic gardens from other kinds of garden is that a botanic garden will maintain some kind of record of its collections: what plants are grown, where they came from, where in the garden they are situated, etc. Such a record is essentially a management tool for the staff of the garden, allowing for the planning of the collection in a rational way in accordance with an agreed policy. It also allows for interchange of information on collections between different gardens, and between gardens and the scientific community and, indeed, the public generally.

The type of record system in use in any particular botanic garden will depend on the precise purposes of the garden (as perceived and expressed in its overall policy) and the resources available to it. Systems may range from the simple, manual card index containing a basic minimum of information, to very sophisticated computers offering access to the record from various starting points.

Whatever record system is in place, however, it is necessary that it be maintained. Changes in the collection must, fairly rapidly, be reflected in the record. If this is not done conscientiously and accurately, the record quickly becomes a fiction, either of little use as a source of information, or even a hindrance to the acquisition and dissemination of information. Updating and maintaining records requires staff time and enthusiasm — characteristics greatly in demand in other areas of the garden's work. The amount of time and effort required to maintain a good and accurate record system should not be underestimated.

The information recorded in a botanic garden system can vary enormously, depending on the factors mentioned above: the precise purposes of the garden and the resources (including staff time) available. The simplest case is probably a record containing the names of the plants, where they came from and where in the garden they are planted. These 3 areas of information are probably basic to all systems. On top of these one can elaborate *ad infinitum*. Taxonomic information may be included (plant family, synonyms, authorities for Latin names, details of verification, natural distribution, etc.), as may horticultural information (condition and numbers of plants, details of propagation, cultivation treatments, performance in terms of growth, flowering and fruiting, etc.), information on usage of the plants (which have been used for research projects, etc.) and on distribution of material to other gardens, and so on. There is really no limit to the degree of elaboration that can be built up in any particular situation. The dangers of over-elaboration are obvious, and what is required of the system should always be realistically assessed before new complications are added.

The increasing realisation, over recent decades, that wild plants are under threats of varying extents in all parts of the world, and the consequent understanding of the need for conservation, has given an increased importance to plants in cultivation and the records which relate to them. Persistence in a garden may be the last resort for some species when their native habitats have been destroyed or significantly altered; plant material for research may be more suitably provided from cultivated stocks than from further raids on diminishing wild populations; and cultivated plants may provide a reservoir of material for the re-stocking of wild areas.

The importance of botanic gardens collections in this respect was quickly recognised by IUCN, and the staffs of many gardens will be familiar with the work of the Conservation Monitoring Centre (CMC) at Kew in coordinating and distributing lists of threatened plants in cultivation. This represents an effective beginning, but leads on to the consideration of wider issues. Conservation is a world-wide problem, and some means of coordinating the efforts of botanic gardens is clearly necessary. In order to develop policies for effective conservation of plant stocks it is necessary that botanic gardens communicate with each other about their collections. Of course, they do this to some extent now: by means of seed-lists and catalogues and by *ad hoc* requests. But this communication is random: it must be developed and extended so that clear and coordinated policies can be agreed. Botanic gardens must be able to transmit accurate information to each other and to the CMC in an easily understood and adaptable form. Computerisation provides the key as far as this is concerned.

Garden record systems are ideal candidates for computerisation, and many gardens have taken this step or are thinking of doing so. Because of the variation in systems outlined above, and because of the need to improve communication, this seems an appropriate time to consider the future and to attempt to set some minimum standards for computerised botanic garden record systems, so that the interchange of information between gardens and between gardens and the CMC is facilitated. Some gardens have already begun this process. For instance, the Dutch and Belgian gardens have agreed among themselves a computerised record system to which they all contribute and which they all use (see Aleva *et al.* 1984). Similar developments are also proposed for the U.S.A. and Canada where the American Association of Botanical Gardens and Arboreta is taking the lead.

During early 1985, the CMC undertook a survey of computerisation of botanic garden record systems. The results are available as an appendix to this paper, and only the main conclusions are important here. Of the gardens which contributed to the survey, about a quarter are already computerised, and about half were thinking of computerisation, some of them well into the planning stage. These gardens cover a wide spectrum of size (from 200 to over 100,000 accessions) and degree of specialisation. Many of the gardens expressed an interest in a collaborative project to design compatibility standards, or, at least, in the results of such a project. All of this indicates that there is a will in the botanic gardens community, first of all to consider computerisation, secondly to attempt to prepare and use standards; this paper contains the first proposals for such standards.

It is necessary, as a first step, to consider what can be standardised. It is very clear from the CMC survey that it is not possible to standardise on equipment (hardware). Too many different computers are already in use, or planned for. Most of these are micros, but some gardens, especially those run by Universities, tend to have their records on the parent organisation's mainframes. Of the micros in use or planned for, the IBM PC is the most popular (8 gardens), but, even so, there is clearly no possibility of standardising on this. Similarly, there are several operating systems already in use, so, again, it is not possible to standardise on any one of these.

This leaves, essentially, the applications software and the data themselves; and it seems possible to standardise only on the latter. In terms of software, IUCN is promoting the creation and development of a model system to run initially on an IBM PC, which will be available for any garden which does not wish to have professional programmers or to contract out the software to software houses. This system, when developed, will incorporate such standards as have been agreed, and will, in itself, set new standards. However, as these developments are for the future, this project is not relevant to present purposes, and the rest of this paper will be concentrated on data standards.

As outlined above, there is a huge mass of potential data that can be incorporated into a garden record system. The first task, then, must be to sift through this mass of potential data to extract a minimum necessary set which will fulfil the purposes of better communications between gardens and the CMC (in terms of threatened plants in cultivation). At present, the amount of communication is limited; but we have to look to the future, and the pace of technological development is very rapid. At the moment it is quite difficult to exchange floppy disks between systems. This, however, will undoubtedly become easier and developments in the use of telephone lines and even communications satellites will eventually have profound effects.

In July 1985, a small group of people interested in such matters met at the CMC, at the invitation of Hugh Synge, for a workshop on the possibilities for the future and the minimum data-set necessary for effective communication. The rest of this paper is essentially the outcome from that workshop and the discussions which followed from it. It is not possible to impose standards, but it is hoped that the combination of the need for effective communication and the simplicity of the standards suggested will guarantee their acceptance. In the first instance, it is recommended that those gardens currently planning computerisation will incorporate these standards into their system from the beginning, and that those already computerised do so when they upgrade their current systems.

The standards proposed here deal only with the information required for the international exchange of data for conservation purposes. Individual gardens will be able to add whatever other information they choose to their own systems. It is not the intention to limit anyone's freedom of action; only to indicate that, within any system which is prepared to participate fully in the international exchange of information about cultivated plants, collection planning and the management of plants of conservation interest, certain items of data should be present and dealt with in a particular way.

The areas of information that are the basic minimum for all systems are:

Accession data
Plant name data
Taxonomic verification data
Source and origin data

These are all of cardinal importance to the CMC and are likely to be included in any form of garden record system. Later in this paper we present proposals for the formats in which these data should be included; together these formats can be assembled into an *International Transfer Format* (ITF).

It is recommended that gardens consider these formats when designing their own systems. The ITF, when agreed, should be followed exactly for communication purposes. Those who have computer schemes and those planning such schemes are respectfully asked to look at the ITF and consider: (a) whether it answers their needs; and (b) whether a computer program to import and export records in ITF from their own data file is likely to be straightforward or not.

In the first instance, the priority will be to produce data files in ITF, i.e. each record system should be able to produce output (either on paper or on disk) in the ITF; this does not mean that gardens necessarily have to hold their records in this form. Building data from ITF into a garden's own file will come later. This will allow for rapid updating of records of conservation interest, as individual gardens will be able to send their ITF file to the CMC, who will be able to compare it rapidly with their master files, extract the necessary information and update the garden's record, at least as far as threatened plants are concerned. Individual gardens will be able to compare collections more easily, leading to rationalisation of collections throughout the world. Eventually it should be possible to combine all these records so that a master catalogue of plants in cultivation can be produced, either as an accessible database or as printed copy. A work such as this would be of enormous value, not only to botanists, horticulturists and conservationists, but also to nurserymen, landscape architects, agronomists, plant breeders and, indeed, any researchers who need plant material for their work.

If any of this is to come about, we need active agreement about the ITF. The proposals presented here have been seriously considered and are now available for discussion and comment. Clearly, this paper represents only a first stage in the development of ITF; experience with it will no doubt suggest alterations. The proposals are flexible and should be able to accommodate future developments. The next stage, following agreement on the ITF, will be the production of a "User's Manual" which will give more details than is appropriate here.

Potential data elements

From the IUCN survey it is clear that different gardens use different sets of data in their computerised record systems. The following list includes all the areas of data covered in the survey; those marked with an asterisk are included in the ITF; suggestions for dealing with some of the others are presented on pp. 330–332.

Accession number*
Plant name*
Plant family
Location in garden
Provenance type*
Material type*
Collector and number*
Collecting data (especially locality and altitude)*
Donor*
Wild distribution of the taxon
Country of origin*
Synonymy
Taxonomic verification*
Date of planting
IUCN conservation category
Vernacular names

Codings for label-making
Authorities for Latin names
References to literature
Whether or not the plant is a type specimen
Dated checks on the persistence of the plant(s)
Cytological information
Existence of a herbarium specimen of the accession
Existence of a photograph or illustration of the accession
Presence of the accession in seed, pollen or gene bank
Presence in *Index Seminum*
Other plant-exchange data
Evaluation of function in garden
Horticultural assessment
Plant habit
Management information (health, treatment, etc.)
Seasonality
Miscellaneous characteristics (plant poisonous, etc.)
Propagation methods
Number of plants in mass planting

The above list indicates the variety of data-types that can be included in a record system. The ITF will include only a few of these types, but this list includes most of those likely to be considered for inclusion. Individual gardens are free to decide which (or, indeed, types not included in the list) they wish to have in their own systems additional to the ITF information.

The International Transfer Format: Description of data elements

From the outset it was agreed that the format should be made up of fields of fixed lengths, rather than fields of undefined length separated by break characters. Fixed length fields are preferable for the following reasons: (1) they allow for faster processing; (2) it is easier to prepare complex output programmes using them; (3) they are easier for "computer amateurs" to use; and, (4) they are a requirement for programmes written in dBase, a widely used language that was frequently mentioned in the CMC survey.

The data to be included can be grouped under 6 broad headings:

(1) Internal management information.
(2) Accession data.
(3) Plant name.
(4) Plant verification and identification.

(5) Source and origin information.
(6) IUCN Conservation Categories.

Internal management information

In order to make the ITF flexible and capable of further development and growth, it is necessary to include some information which allows for internal computer management. This information can be accommodated in 2 single-character fields (Fields 1–2).

Field 1 is labelled "Record Type", and will contain the single digit "1", which will indicate that the record that follows is in ITF. Later developments may lead to somewhat different formats becoming compatible with the ITF, and these, when and if they are developed, will be indicated by the figures 2–9 in this field. Initially, however, all records will have a 1 in Field 1.

Field 2 is labelled "Record Status", and, again, is a single character field, used for indicating the status of the record with respect to the CMC. The field is made blank if the record has been transmitted to the CMC; it will contain "A" (meaning addition) if the record has been added to the system since the last time the record as a whole was made available to the CMC, "M" if the record has been modified since that time, or "D" if the record has been deleted since that time. Initially, therefore, every record will contain A in this field. Later, as the system develops, gardens will send to the CMC only those records which have A or M in this field, as records already scanned by the CMC computer will have had this field made blank.

"Record type 0" will appear once only at the beginning of the file, and will contain merely a description of the fields and their lengths.

Problems can arise when a plant dies; what should be done with its record? In some systems the record is completely deleted, in others it is marked in some way but retained in the same file as those for living plants, and in yet others it is transferred wholly or in part to another file (the Dead file). For the purposes of ITF it is recommended that the whole record, or at least its ITF component, be transferred to a separate Dead file. Such a Dead file can be retained for any convenient length of time; if it becomes too large, it can be printed out and then cleared for future use.

Accession data

Accession number

The IUCN survey found that all gardens that had computerised record systems (and most others) had their own individual accession number system. The accession number is often the *only* unique item of information for a particular plant, and is therefore of great relevance, both to the individual garden and to the ITF. It is a principle of most garden record schemes and of the ITF that each record in the file has a unique accession number. That single record may, of course, refer to one plant, a clump of plants or to a tray of seedlings.

The accession number is usually given to the plant material when it arrives at a garden, and remains with it throughout the life of the plant and its record; it is the only item of the record that is not subject to change (apart from the addition of prefixes or suffixes in some systems). In most gardens a unique accession number is given to the arriving material of each taxon; this is the system recommended here, as it prevents confusion of numbers later on. Some gardens give all the taxa in the one incoming batch the same accession number (i.e. the accession number is really a batch number); this system is not recommended as it produces duplicate numbers. The solution is to add an incrementing number on the end of the existing number within the 10-character field (see below).

Individual gardens use varying forms of number. The simplest is a running (serial) number, but this is not widely used. More common is a system incorporating the year of accession as the first 2 digits, followed by a 4- or 5-character running number, e.g. 852102 — the 2102th accession during the year 1985. The system is simple to use, but in older gardens has the problem that there is no distinction between 1885 and 1985. At Edinburgh, the surviving accessions from the 19th century will be re-accessed as necessary. Other gardens use more complicated systems involving up to 10 characters including letters and signs (hyphens, slashes) as well as numbers, e.g. 210 85 00198 (batch number, year, serial number) or 82 BG 24031 (year, origin of material, serial number).

Where the accession number includes punctuation, e.g. the fourth character is a hyphen, each garden should decide whether or not to include the punctuation in the ITF version of the number. Consistency within each garden is vital!

For the ITF a 10-character field (Field 3, labelled "Accession number") will be used. This field *must* contain an entry for the record to be accepted as being in the ITF system.

Garden code

In order fully to ensure that each garden's accession system is distinct from others (and to prevent the remote chance of duplication of numbers for the same taxon) the ITF should contain a field for a code for each garden.

The best approach to this is not to invent a new set of codes, but to use those well-known codes designed by the International Association of Plant Taxonomists (IAPT) for herbaria, as outlined in *Index Herbariorum* (Holmgren *et al.* 1981). These are mnemonic codes, e.g. G for Geneva, AD for Adelaide, SBBG for Santa Barbara, DAKAR for Dakkar, LMU for Lourenco Marques, COL for Bogotá in Colombia. Where a garden does not have a herbarium, a new code can be devised; IUCN has been doing this for some years in collaboration with IAPT, so all 250 or so botanic gardens that have contributed records to the CMC have a mnemonic code in harmony with their IAPT code. It is recommended that the next edition of the "Directory of Botanic Gardens" include codes in this system for *all* botanic gardens (the programme booklet for this conference contains a list of several hundred gardens and their codes).

The inclusion of the garden code, though repetitious, has another advantage in the long term: when one garden gives a plant to another garden, the donor garden should also provide the full data relating to the plant from its record file; using the ITF, the receiving garden could even include a copy of the record from the donor garden in its own file, linked to its own new record. The garden code could be used to distinguish a garden's own records from those of other gardens for plants it had received from them. Of course, this would require a rather expanded version of the ITF from the one proposed here for conservation purposes.

In the ITF the garden code will form Field 4 (labelled "Garden Code") which will be 5 characters long. This field *must* be filled.

Plant name

The names of the plants in a collection are, of course, a basic part of any record system; so much will be generally agreed. But what does the plant name actually consist of? Does it necessarily include, for instance, the name of the plant family (or higher group) to which the plant belongs?; authorities for the various Latin names? etc. These are points which require some consideration.

For the purposes of the ITF, the plant name is taken to be the Latin name as generally understood, i.e. Genus name, specific epithet, infraspecific epithet(s) if any, Cultivar name if any, together with any other signs (e.g. "×" to show hybridity or "+" to indicate a graft-chimaera) generally associated with these terms. The names of families and higher groups to which the plants belong are not required for the ITF; nor are the names of the authorities which conventionally follow the specific and infraspecific epithets. The reasons for these omissions, and some suggestions as to how these elements can be dealt with in a computerised system are given on pp. 330–332.

Generic, specific and infraspecific names

The "name" of a wild plant that is not a hybrid consists of the genus name, the species name and perhaps 1 or more infraspecific epithets. Under article 11.3 of the "International Code of Botanical Nomenclature" (Voss *et al.* 1983):

> Every name at the level of genus or above is unique;
> Every binomial of genus and species is unique;
> Below the species level, every trinomial is unique.

This means that every wild plant can be uniquely described by a three-part Latin name. Hence, *Rhododendron arboreum* subsp. *delavayi* var. *peramoenum* can be known uniquely as *Rhododendron arboreum* var. *peramoenum*. Therefore, in the ITF, the name will consist of 3 main fields: genus, species name and *lowest* infraspecific taxon name. This infraspecific taxon name must be qualified by its rank, which may be:

Subspecies (subsp. or ssp.) — S;
Varietas (var.) — V;
Forma (form, f.) — F.

The following ranks are omitted from the ITF: Subgenus, Section, Subsection, subvarietas and subforma. Of these, the first 3 are not strictly part of the plant name and so do not require to be included; the final 2 are very rarely used, and, for all practical purposes can be treated as equivalent to Forma.

It is necessary that the various components of the name be stored in the computer as separate fields, e.g.:

Genus name: *Rhododendron*
Species name: *maddenii*
Subspecies (infraspecific) name: *crassum*

rather than as a single field, e.g.

Name: *Rhododendron maddenii crassum.*

The *disadvantages* of holding the name as a single field are: (1) the greater likelihood of mistypings; (2) the difficulty of validating those parts of the name that can be validated, e.g. that the rank of the lowest taxon must be "subsp.", "var." or "forma"; (3) the difficulty of asking questions which depend on rank, e.g. which Chrysanthemum cultivars are grown? What percentage of the collection is hybrid?

Of course, output programs can be written which translate the first format into the second, but translation in the other direction is much more difficult (though possible). IUCN did this for Kew in 1983–4, using a program written in COBOL, and would be happy to share the program with other institutions; the file included many troublesome orchid hybrids, conifer cultivars, and so on.

The genus name can be accommodated in a field of 20 characters in length. In those very rare cases where this length is overstepped (e.g. *Pseudoblepharispermum),* the name must be truncated (i.e. entered as *"Pseudoblepharispermu")* rather than abbreviated, to avoid mismatches due to different systems of abbreviation. The specific and infraspecific name fields will each need to be 30 characters in length, which should be sufficient to prevent any need for truncation or abbreviation. IUCN is presently undertaking a survey to compare frequency with length for very long names of genera and species; in the meantime, taxonomists and others describing new genera and species are earnestly requested to control their ingenuity and to avoid names of more than 12 letters.

The genus name should have the initial letter in upper case, the remainder of the word in lower case letters. Under the rules of the Code of botanical nomenclature, the genus name must be one word, of characters a–z and may include 1 or 2 hyphens (no more) and no other characters. The specific and infraspecific epithets should be entirely of lower case letters, with up to 2 hyphens; the code recommends, but does not rule, that specific and infraspecific epithets should be less than 7 syllables long.

Many plants in garden collections are not fully identified, so any or all of these 3 name fields may be left blank. If the genus of the plant is unknown, but the family has been determined, it is permissible in the ITF to enter, in the Genus name field, the name of the family, followed by "sp.", e.g. "Acanthaceae sp.". In the rare cases where a new species or infraspecific taxon is entered into a record system before it has been formally described, "sp. nova" or "subsp. nova", "var. nova" or "f. nova" should be entered into the appropriate name field.

Cultivars and hybrids

Cultivars

A cultivar can be defined as "a segment of the variation of cultivated plants being clearly distinguished by any characters which are retained when the cultivar is reproduced (sexually or asexually)". Although some cultivars have originated in the wild, the majority have arisen in cultivation. The important point, however, is that all cultivars are *maintained* by cultivation.

The naming of cultivars in governed by the rules of "The International Code of Nomenclature for Cultivated Plants" (Brickell *et al.* 1980). This code rules that on or after 1 January 1959, cultivar names must be in a modern language, may be of up to 3 words, and may include hyphens, numbers, etc Cultivars named before this date, or cultivars subsequently formed by changes in rank from specific or infraspecific taxa named under the botanical code, may keep their Latin form, e.g. *Malus sikkimensis* cv. Rockii, which is based on *Malus rockii*.

The first character of the cultivar name is upper case, as are the first characters of other words in the name, except when linguistic usage determines otherwise, e.g. *Ilex aquifolium* cv. J.C. van Tol. In outputs, cultivar names are not underlined (printed in italic type) and are *either* placed between single quotation marks or preceded by "cv.", e.g. *Acer palmatum* 'Elegans' or *Acer palmatum* cv. Elegans.

For the ITF a single field of 30 characters has been reserved for the cultivar name; therefore the cv. notation does not have to be entered; the quotation marks would occupy needless space in the computer, and therefore should also be omitted.

There are cases where the cultivar name follows the generic name directly, for example when the parentage of the cultivar is unknown, e.g. *Betula* cv. Jermyns. In the ITF only the genus and cultivar name fields are completed in such cases, the other name fields being left blank.

Cultivar Group names are used to designate assemblages of similar cultivars within a species or hybrid, e.g. *Lolium perenne* Early Group. Such names are often associated with a cultivar name, when the group name is entered in parentheses, e.g. *Lolium perenne* (Early Group) cv. Devon Eaver. It is proposed that these group names can be accommodated in the ITF by entering them in the field for the infraspecific taxon, and entering "U" rather than "S", "V" or "F" in the field which indicates the rank of the infraspecific taxon (the Infraspecific type flag field, see below); the parentheses are not required in the ITF.

Hybrids

Hybrids produced by sexual crossing can be named in one of three ways (A, B and C below); A and B are governed by the botanical code, C by the horticultural code. Hybrids cause great problems for any schemes of this kind, as they imply a doubling of the name-fields. Several ideas have been proposed for the treatment of these, and a simplified scheme is proposed here. For the purposes of exchange with the CMC, hybrid names are not, for the present, very important.

A. Hybrid formula

This method consists of the names of the parent taxa connected by the multiplication sign (letter x on many typewriters, computers and word-processors, usually a separate sign in printing fonts), e.g. *Acer davidii* × *rufinerve; Polypodium vulgare* subsp. *prionodes* × subsp. *vulgare; Magnolia campbellii* subsp. *mollicomata* × *M. sprengeri* var. *elongata.* Where the individual sexes of the parents are known, the names may be given in alphabetical order with the sexes indicated by (M) or (F), or by the special signs conventionally used, or the female parent is listed first. For many garden plants, however, the direction of the cross is *not* known, and these refinements cannot be incorporated in the ITF.

As the above examples show, formulated hybrids can be very complex. This naturally presents problems for computerised records in terms of the space needed as they could require double the amount of space normally allocated.

B. Latin collective names

These are the most commonly encountered hybrid names. In the case of an intergeneric hybrid an upper case multiplication sign (or capital X) is placed before the name of the hybrid genus, e.g. × *Cupressocyparis* (the generic name for all crosses between species of *Cupressus* and *Chamaecyparis*, or, in the case of an interspecific hybrid, a lower case multiplication sign or x is placed between the genus name and the species-equivalent name, e.g. *Lonicera* x *tellmanniana*. Hybrids between two infraspecific taxa of the same species are more rarely encountered, but are treated in the same way, e.g. *Rhododendron cinnabarinum* subsp. *cinnabarinum* × subsp. *xanthocodon*.

C. Non-Latin collective names

As the above heading suggests, these are modern language equivalents of Latin collective names and similarly cover all the progeny of any particular hybrid combination. The name includes both directions of the cross, for example the taxon that is the pollen parent on one occasion can become the seed parent on another.

The name must consist of not more than 3 words. Often they include the terms Hybrid, Hybrids, Cross, Crosses, Grex (abbreviated g.), for example: *Lilium* Bellingham Hybrids (the collective name for interspecific hybrids between *L. humboldtii* and *L. pardalinum*). The usage of the term grex has become institutionalised and widely used, particularly with Rhododendrons and Orchids, e.g. *Rhododendron* Fabia Grex (which covers all the progeny of *R. dichroanthum* × *R. griersonianum).*

The above example is a straight cross between 2 species; however, there are many collective names which have more complex parentage, involving hybrids or infraspecific taxa, e.g. *Rhododendron* Jalisco Grex, which is a compound of 4 species. If the collective epithet is followed by a cultivar name, the collective name is usually placed in parentheses, e.g. *Lilium* (Bellingham Hybrids) cv. Shuksan. However, the code for cultivated plants rules that the parentheses may be omitted, and, where the context is clear, the word grex or its abbreviation may also be omitted, e.g. *Rhododendron* Fabia cv. Minterne Apricot. It should be noted that the collective (grex) name can also be used in a clonal sense, e.g. *Rhododendron* Fabia cv. Fabia, where this is the clone descended from the original cross between the 2 species. It is not surprising that non-Latin collective names can cause problems for record systems.

It is proposed that for the ITF the non-Latin collective name can be entered in the field for the infraspecific taxon, with "G" rather than "S", "V" or "F" in the infraspecific type flag field. Any parentheses should be omitted. If the cultivar name follows, this can be placed in the normal way in the Cultivar name field.

So far, only hybrids produced by sexual crossing have been considered. However, hybrids can also be produced by grafting 2 different taxa together to produce a graft-hybrid or graft-chimaera. Such plants are named in exactly the same way as sexual hybrids except that the plus sign (+) replaces the multiplication sign in the Intergeneric Hybrid/Graft Chimaera flag field e.g. + *Laburnocytisus adamii*, which is the graft-chimaera between *Cytisus purpureus* and *Laburnum anagyroides*. When the components of a graft-chimaera belong to the same genus, the name consists of the generic name followed by a "+" and an epithet, e.g. *Syringa* + *correlata*, the graft-chimaera between S. × *chinensis* and *S. vulgaris*. In all cases, the name of a graft-chimaera should not be the same as that of a sexual hybrid of the same parentage.

Conclusions, format and examples

The discussion of plant names shows how complex this area of nomenclature is, particularly with hybrids and cultivars. The ITF must cater for both straightforward species as well as all possible methods of naming hybrids, and this is done by the scheme outlined below.

The plant name area of the format will consist of 7 fields:

Field 5. Intergeneric hybrid or graft-chimaera flag. A single-character field which may be blank or may contain X (for an intergeneric hybrid) or + (for an intergeneric graft-chimaera).

Field 6. Genus Name. A 20-character field for the genus name.

Field 7. Inter-specific name type flag. A single character field, the contents of which qualify the species name (field 8), if any. The field may be blank, or may contain x (to indicate a Latin collective name for an inter-specific hybrid) or H (to indicate a hybrid formula) or + (to indicate an inter-specific graft-chimaera).

Field 8. Species name. A 30-character field to contain the species name or its equivalent, i.e. the species name itself or a Latin collective name for an inter-specific hybrid, or, in the case of a formulated hybrid, the two species names connected by x, e.g. *davidii* x *rufinerve.*

Field 9. Infraspecific type flag. A single character field (qualifying the names in the next two fields) which may be blank or which may contain S (subspecies), V (varietas), F (Forma), G (non Latin collective name), B (hybrid between 2 subspecies), R (hybrid between 2 varieties), I (hybrid between 2 infraspecific ranks which are different) or U (to indicate a cultivar group which is not a grex).

Field 10. Infraspecific name. A 30-character field for the infraspecific name or its equivalent — a grex or cultivar group name or 2 names connected by x, e.g. *mollicomata* x *elongata.*

Field 11. Cultivar name. A 30-character field for the cultivar name (neither cv. nor single inverted commas are required).

Field 12. Though not strictly part of the name, this is a single character field used to denote the sex of the plants cultivated. This is particularly important with dioecious species, especially with regard to those of conservation interest, where both sexes should be in cultivation if possible. The field may contain F (female), M (male), B (both sexes), ? (for a plant known to belong to a dioecious species whose sex has not yet been determined), or may contain spaces when the field is not applicable (i.e. for bisexual or monoecious species).

Identification qualifiers and verification

Identification qualifiers

Botanists sometimes add various terms, such as "cf." or "aff" to the name of a plant to indicate a degree of uncertainty of identification. This is often the case with garden plants which are of uncertain origin and do not produce flowers or fruits. The terms used can be placed in front of any element of the name, e.g. cf. *Rhododendron*; *Rhododendron* aff. *zaleucum*; *Rhododendron zaleucum* cf. subsp. *zaleucum*, etc. The terms in general use are:

aff. (Latin *affinis*): akin to or bordering on;
cf. or cfr. (Latin *confer*): compare with;
forsan (Latin): perhaps;
near or nr.: close to.
(Meanings from Stearn 1973).

A question mark may also be similarly used, e.g. *Acer ? davidii.* These various terms, though originally subtly different in meaning, are now used more or less interchangeably, and can be accommodated in a single-character field in the ITF (Field 13) which can contain A, C, F or N (for the terms above), or ? or space. Following this is Field 14, labelled "Rank Qualified", which contains appropriate letters, G (genus), S (species), I (infraspecific rank) or C (cultivar) to indicate the rank whose identification is doubtful.

Similarly, the term "agg.", meaning aggregate species, is used when botanists are prepared to make an identification to an aggregate species but not to any of the segregate microspecies. A separate single-character field, (Field 15, labelled "Aggregate Flag") is proposed for this, which may contain A or space.

Verification

The name of the plant is important from every point of view. However, it is necessary to indicate what degree of confidence can be placed in the name; is it the name the plant arrived with?, or has the plant be re-identified, and if so, by whom and when?

It is obvious that there are several levels of possible taxonomic verification. The simplest scheme, and that recommended for the ITF, is to use a single character field (labelled "Verification Level", Field 16) which may contain the numbers 0–3:

0 indicates that the name has not been checked by any authority.
1 indicates that the plant was named in a garden by comparison with other, named, living plants.
2 indicates that the plant has been named by a taxonomist or other competent person using the facilities of a library and herbarium, or other documented living material.
3 indicates that the plant has been named by a taxonomist who is currently or has been recently involved in a revision of the family or genus.

These 4 levels are fairly easy to apply and show clearly what degree of confidence can be placed in the name as given; the levels apply to the whole name as recorded: to qualify every level would be unnecessarily repetitious.

Field 17. (labelled "Verifier's Name") is a 20-character field for the name of the person making the verification. This should be entered in the form surname, comma, initials of given names, e.g. Smith, J.P. or Sosnowsky, D.F.

Field 18 (labelled "Verification Date") contains 6 characters. The date should be given in the form YYMMDD, e.g. 851109 for the ninth of November 1985.

Source data (donor, provenance and material)

This is a very complex area of information and one likely to lead to confusion if not properly understood and applied.

Donor

Plant materials may come to a garden from various sources and in various ways. Common to all of these is, however, the name of a donor. Six broad categories occur:

(1) Plants derived directly from the wild, received usually from the collector.
(2) From various horticultural or specialist plant societies.
(3) From parks and other institutional but non-botanic gardens (e.g. in the U.K., The National Trust).
(4) From nurseries or garden centres or any commercial plant suppliers.
(5) From private individuals.
(6) From other botanic gardens.

The information under these heads is very various, and (apart from (6) cannot be coded or fully standardised. It is proposed that the information be stored in a single, 20-character field (Field 19, labelled "Donor's Name"), using the following guidelines.

(1) For materials from expeditions, the name of the collector (surname, comma, initials) should be given first, followed by the name of the country or area to which the expedition was made; or, if the expedition has its own title, e.g. S.B.E.C. (Sino-British Expedition to China), this should be given first, followed by the name of the country or area. Abbreviations or truncations should be used as necessary.
(2) For societies, the name of the society (abbreviated or truncated as necessary) should be entered, followed, if space permits, by the name of the country in which the society is based.
(3) Parks and gardens. The name of the institution (abbreviated or truncated as necessary) should be entered, followed by the name of the country in which it is situated.
(4) Nurseries and garden centres: as for (3).

(5) Private individuals. The minimum information should be name (surname, comma, initials) and country; more detail, e.g. town or province, can be included if space permits.
(6) Botanic gardens. Instead of the name, the standard IUCN/IAPT abbreviations as used in field 4 should be entered.

If the donor is unknown, the word "unknown" should be entered.

When the donor is another botanic garden, it is likely that the donor garden's accession number will be given as part of the documentation accompanying the plant. This should be entered in Field 20 ("Donor's Accession Number"), which is subject to the same rules as is Field 3 pp. 315–316).

Provenance and material type

Most gardens use a code system to distinguish between the different provenances of plants. The IUCN method of coding, used by the Botanic Gardens Conservation Co-ordinating Body, is clearly inadequate. What is needed is a coding system which covers both the provenance (origin) and the material type.

Plant material may derive from 3 types of provenance:

(1) Directly from the wild, in which case there should be information concerning its origin (place, date, altitude, ecological notes and the collector's name and number).
(2) From a cultivated plant whose origin was originally of type 1, when there should also be collecting details available.
(3) From a cultivated plant *not* of known wild origin. This includes plants which are without the necessary documentation to confirm wild origin. Also included are plants of known or unknown cultivated origin.

Plant material can be of many types, seed, unrooted cuttings, layers, grafted plants, bulbs, etc. Clearly it is important for individual gardens to record this information, and again, most do so by using a code system. However, it is possible to group these various forms under 2 heads.

(1) Material received vegetatively. This can be considered as representing the same genetic stock as the original plant; hence, any collecting information with the original plant should be passed on to the receiver garden.
(2) Material received as seed is more complex: if the seed derives directly from the wild, then it can be considered as an (admittedly small) sample of the genetic stock of the wild population, and

collecting details should be retained and transferred with the seed. If the seed is derived from a plant of known wild origin currently cultivated in a garden, 2 possibilities occur. If the seed is produced by open pollination, then the resultant plants can *not* be considered as of known wild origin, and collecting details should not be transferred with the seed; or, if the seed is produced by selfing, which could either be due to controlled pollination or because there was certainly no other related plant flowering at the same time, or the parent plant is known to be apomictic, then the resultant material can be considered as representing the original collection, and collecting details should be transferred with the seed.

In the ITF a system of coding is proposed as outlined in Table 1. However, it should be pointed out that this scheme is orientated towards wild source plants in that it attempts to tackle the problems of genetic integrity. For purely cultivated plants (i.e. plants of known garden origin, whether cultivars or not), many of these details are not relevant. Other information about such hybrids, selection procedures, etc., is best accommodated in a free-text area of the record.

These various codes are entered in 2 single character fields. Field 21 ("Provenance Type") may contain the letters G, W or Z. Field 22 ("Material Type") may contain the letters O, S or V.

Geography and collecting details

Garden records mostly have 2 sections dealing with geography:

(1) one for the range of the taxon (its wild distribution), which is often used on garden labels; and
(2) one for the locality where the actual plant was collected; this is often supported by other information.

Gardens handle information on these sections in a mixture of complex codes (especially for range) and in free text. It is proposed that taxon range (i.e. wild distribution) should not be included in the ITF: accurate information is not easily obtained, and is more a matter for taxonomic databases than horticultural ones. Whether gardens include this information in their file, and, if so, how they deal with it, are primarily internal matters for each garden to determine for itself.

The location where the original material was collected, however, is vital both for the garden concerned and for the ITF. It is proposed that the ITF include the location and altitude as far as possible, but not details of ecology, vegetation type, etc. If standard ways of handling such data

Table 1

		MATERIAL TYPE		
		Seed (Code S or O)		*Vegetative Material (Code V)*
PROVENANCE TYPE	*Direct from the wild (Code W)*	Location data, etc. from collector WS		Location data, etc. from collector WV
	From a cultivated plant of known wild origin *(Code Z)*	*Selfed seed (including apomicts)* Location data etc. from donor ZS	*Seed from open pollination* No location data ZO	Location data, etc. from donor ZV
	From a cultivated plant not of known wild origin *(Code G)*	No location data GS	GO	No location data GV

Notes to Table 1. Three grades of integrity are formed by the combinations of codes, as follows:

Grade 1 — WS, WV, ZV (unshaded area of table)
Grade 2 — ZS (pale shading)
Grade 3 — ZO, GS, GO, GV (darker shading).

Grade 1 plants are guaranteed of known wild origin. Grade 2 plants involve a *slight* chance of hybridisation, though this will vary with the type of fertilisation of the "selfed" plant:

(1) Controlled pollination is most certain because of the total exclusion of foreign pollen.
(2) Apomictic species are the next safest, although there are 2 problems here: some apomictic species occasionally outbreed, and there is no definitive list of apomictic species.
(3) Selfing due to isolation from potential foreign pollen sources is likely to be the poorest area of quality control because of the lack of controlled conditions. If there are any doubts as to whether a plant has been selfed, code ZO should be used instead of code ZS.

Grade 2 plants ("selfed seed") are problematic, but they are included because there are many known wild source plants in cultivation that are very difficult to propagate vegetatively, e.g. species of *Castanopsis* and *Lithocarpus*.

Grade 3 plants have either no documentation to prove that they are wild, or they are clearly plants that have arisen in cultivation. However, it is important to remember that these plants may include some that are very important, e.g. old cultivars that may be of value in plant breeding. If necessary, an additional code could be devised by the sponsors of the ITF to cover such cases.

Both GS and GO are permitted codings. GO plants are more likely to be of hybrid origin than those coded GS. Only WS (not WO) is permitted in the section for seeds from the wild.

An unresolved area is connected with the term *known* wild origin. For example, what provenance category should be used for a plant where there is circumstantial evidence that it is of wild origin? This problem may require consideration case by case; the possibility of using a free-text area of the record for information of this kind should also be considered.

eventually are agreed, they could be incorporated into later versions of the ITF.

The geographical system recommended here is a provisional version of that being developed by the Taxonomic Databases Working Group, which met for the first time in Geneva in September 1985. This system is designed to cope with the complex problems of coding geographical data, and is by no means finalised. The explanation and justification for the system will be published elsewhere in the near future.

Geography and collecting details are handled in the ITF in 8 fields:

Field 23. Country of origin. This is the name of the country (or important island) where the original material was collected. The field is of 2 characters to accommodate the ISO 2-letter codes for these names (International Standards Organisation 1981).

Field 24. Sub-country unit. This is a 20-character field for major political divisions (states, provinces, counties, etc.) of larger countries, e.g. Yunnan, Seine-et-Oise, Tadzhikistan, etc. It is hoped that it will be possible to produce a standard list of these before too long. If there is doubt in the case of any particular record as to what should be entered here, the field should be left blank.

Field 25. Location. A 30-character field to contain the actual collecting location, e.g. Mt Emei.

Field 26. Altitude of collection. This is information of particular value to temperate gardens as an indicator of potential hardiness. It is accommodated in a 9-character field to allow a range of altitude to be given, e.g. 1800–2400. Altitude should always be given in metres.

Field 27 and *Field 28.* Latitude and longitude of collection. These are 2 6-character fields which can be filled in if the relevant information is available (3 digits for degrees, 2 for minutes, 1 for N, E, S or W, e.g. 02430S 04715W).

Field 29. Collector's name. A 30-character field. Generally the name of the collecter will be entered in full, as name, comma, initials, e.g. Forrest, G. When there is more than one collector, space may become a problem, in which case initials should not be entered, and some or all of the names truncated as necessary.

Field 30. Collector's number. A 10-character field for the collection number (if any) for the particular plant.

IUCN Conservation Category

Many gardens will wish to record data on the conservation status of the plants they grow, whether it is rare, threatened, etc. The standard system to record degree of threat is the set of Red Data Book categories designed by IUCN and used virtually universally. These codes are Ex (Extinct), E (Endangered), V (Vulnerable), R (Rare), I (Indeterminate), K (Insufficiently known) and nt (neither rare nor threatened). These are defined, with examples, in a booklet available on request from the CMC. In furture K is to be subdivided into several categories. Some combinations of these codes are also allowed; these are: Ex/E, E/V, V/R, E/R, V/nt and R/nt.

These categories can of course be used at any geographic level, e.g. country, district, region, world. It is expected that most gardens will only want access to the world categories. Accordingly a 5-character field is included in the ITF (*Field 31*, labelled "Conservation Category"). However, it is expected that this will be filled *by IUCN* when returning records from reading, giving gardens the latest status of their plants from IUCN. It is *not* expected that the ITF will be a mechanism whereby institutions inform IUCN about the status of species in the wild; other mechanisms exist for this. It is important to remember that the categories refer to the status of the plant in the wild and do *not* take account of the abundance of the plant in cultivation.

Suggestions for computerisation of some data elements not required for the ITF

Plant family (and higher groups)

The family to which an individual plant belongs is of importance for the individual garden containing it. Many gardens have been asked such questions as "What Liliaceae do you grow?" by researchers, teachers and others. However, the matter, though apparently simple, is beset with problems. There is no general agreement among taxonomists, even with the Angiosperms, as to what families there are or which genera belong to which families. There are numerous systems available (e.g. Bentham & Hooker, Engler & Prantl, Cronquist, Thorne, Takhtadjan, etc.) and some gardens use one, others another. Thus, Saxifragaceae in the Edinburgh system (which follows Engler & Prantl as modified by Melchior) will not be quite the same as Saxifragaceae in the Kew system (which follows a modified Bentham & Hooker system). Here, clearly, there are problems which are even more complex if higher groups such as Orders are considered. Because the family is not part of the name, it is not required in the ITF; this leaves gardens free to deal with the family as they choose. Several systems of coding are already in use in various gardens.

Authorities for Latin names

These are generally considered to be part of the name by taxonomists (at least as far as the species level and below that) and are often included in record systems. The reason for using them is generally stated to be the need to prevent confusion between homonyms (the same name having been used for 2 different taxa; one or more of these names will be technically illegitimate). For example, the name *Juniperus excelsa,* standing alone, may mean either *J. excelsa* Bieberstein, a legitimate name for a widely recognised species, or *J. excelsa* Brandis, an illegitimate name and a synonym of the species known usually as *J. recurva* Buchanan-Hamilton. While this may be important to technical taxonomy, its effects on garden record systems are small and usually obvious. Authorities (even if, as they usually are, annoyingly abbreviated) can take up a vast amount of space in the computer, while adding little or nothing of substance.

It is unlikely, for instance, that it would ever be necessary to search a file for all species with authority (Humboldt, Bonpland & Kunth) Chamisso & Schlechtendahl, to take a fairly ridiculous example. Accordingly, the authority is not needed in the ITF. Once again, individual gardens may make up their own minds. If authorities are included, however, we recommend that they should be dealt with in separate fields from those containing Latin names.

It may be of interest to note that experience with building the IUCN database on threatened plants has shown that: (a) there is relatively little variation in the spelling of botanical names; and (b) there is great variation in the authorities used, both in the abbreviations made and the conventions adopted.

Synonyms

Individual gardens will want to treat synonyms in different ways, depending on their purposes, but, in terms of transmission of information, there are many problems connected with synonyms, not all of which are easily soluble.

In the first case, there may be ''synonyms'' within the overall record as held by the CMC. Garden A may well call a plant by the name *Planta vulgaris*, while garden B may well call the same taxon *Planta officinalis*, even though these 2 names may not necessarily be synonyms in the taxonomic sense (i.e. one or both names may be wrong). Thus, 2 different names for the same plant will be presented to the CMC, who will not necessarily know that they do refer to the same plant. Only careful use of names and scrutiny of the combined records can overcome problems of this type.

Secondly there are botanical synonyms, which all professional gardeners must have experienced. There are basically 2 kinds; the nomenclatural synonym, where 2 to several names are based on the same type specimen, and the taxonomic synonym which arises when 2 species (or other taxa), previously thought to be distinct, are combined into one, or when an earlier name is discovered for a species generally known under a later name. Nomenclatural synonyms do not usually create any problems, as the priority principle gives clear guidance on which name should be used. Taxonomic synonyms are more difficult, because of the part played by taxonomic opinion. Further consideration of this matter is necessary, perhaps in consultation with those involved in the production of taxonomic databases, who will have met the problem in a more acute form.

Because of the problems, synonyms are not included in the ITF. For gardens' own use, it is recommended that synonyms are stored in a separate file, containing old name and new name, and possibly a pointer to the first occurrence of the correct name in the master file.

Acknowledgements

Many botanists, gardeners and computer experts have contributed ideas which have been incorporated in this paper, though its content is completely the responsibility of the authors. Especial thanks are due to D. Clayton, C. Jeffrey and N.P. Taylor (Kew) and to E.H. Hamlet and R.L. Shaw (Edinburgh), and well as to all who took part in the 1985 workshop: R. Allkin, Southampton; K. Bowen-Nellthorp, Reading; S. Coom, Kew; C. Erskine, Kew; W. Loader, Kew; S. Luckcock, Kew; G.L. Murray, Kirstenbosch; N. Phillips, CMC; G. Van Vliet, Leiden; K. Walter, Arnold Arboretum. The costs of the workshop and of a consultancy for Michael Lear were covered by a grant for the Canaries Conference from WWF to whom we are most grateful. Other costs were covered within the CMC budget from IUCN, WWF and UNEP, to whom we also record our thanks.

References

Aleva, J.F. *et al.* (1984). "Computerisation in the Botanic Gardens of the Netherlands and Belgium".

Brickell, C.D. *et al.* (1980). "International Code of Nomenclature for Cultivated Plants" Regnum Vegetabile vol. 104.

Holmgren, P. *et al.* (1981). "Index Herbariorum", edition 7. Regnum Vegetabile vol. 106.

ISO (International Standards Organisation) 3166 (1981). "Specification for Codes for the Representation of Names of Countries". (British Standard 5374:1981).

Stearn, W.T. (1973). "Botanical Latin", edition 2.

Voss, E. *et al.* (1983). "International Code of Botanical Nomenclature". Regnum Vegetabile vol. 111.

Appendix 1

Results of the IUCN questionnaire

In early 1985, IUCN circulated a questionnaire to the 138 botanic gardens that were subscribers to the Botanic Gardens Conservation Co-ordinating Body. The questionnaire had been carefully designed by the Computer Services Unit of IUCN's Conservation Monitoring Centre to find out the extent to which gardens had computerised their records. For those that had computerised or that had made detailed plans to do so, we asked for details of hardware and software, with emphasis on the ability to communicate and exchange data; for these that had not, we asked about their future plans.

By November 1985 79 Botanic Gardens had replied. Although this is less than had been hoped, most of the replies give precisely the data we were seeking. The attached Tables summarise the main points, omitting the technical information on types of floppy disks, of magnetic tape drives, etc.

Of the 79 gardens:

23 (29%) are on computer so far
of the remaining 56 (71%):
  42 (53% of total) intend to computerise
  13 (16%) do not
  1 was imprecise

A similar survey by the AABGA of North American gardens gives similar results (Murbach, D. (1984), "Directory of Computer Use in Plant Record-Keeping", AABGA, Swarthmore, P.A.). This recorded that 42 gardens (29%) are using computers to assist them in their record-keeping; of these 31 (22%) keep their plant records on computer. A further 20 (14%) planning to computerise in 1985, and another 62 (36%) planning to computerise within the next 5 years.

A copy of the IUCN questionnaire is reproduced (Fig. 1) on pp. 340–343. The tables analysing its results follow.

Our tentative conclusions are as follows.

(1) The gardens that have computerised appear at first sight to have similar data requirements but have chosen many different types of hardware and software. Fourteen use mainframes or minis, 15 micros.

(2) A larger number of gardens would like to computerise and plan to do so, with plans at many stages of development. Of these, 11 plan to use mainframes or minis, 18 micros, showing a shift to micros. Many had clearly looked into computerisation in some detail.

(3) Of the hardware, among the 47 gardens in (1) and (2) above who listed their hardware, no machine appeared more than once except:

Onyx C-8001 & C-8002	1 of each
Amdahl V7/8 & 470 V/8	1 of each
Burroughs B7700 & B4900	1 of each
Cyber 180/855 & 170/825	1 of each
ACT Sirius	2
IBM PC	8

Table 2
Summary of garden replies — their collections and extent of computerisation

The replies to questions 1, 2, 3 and 5 of the questionnaire

Codes 1, 2 and 3, below, refer to the following in question 3:

(1) Are your garden records stored on a computer? (Y, N)
(2) If "No", do you intend to store them on a computer? (Y, N)
(3) If so how detailed are your plans? N: No formal plan yet
O: Outline
F: Fully developed

The final column shows whether or not the garden has provided data on its planned or actual computer system.

Name	1.Accessions	2.Cvs	Spp	Spp and/or Lower taxa	3.Status 1	2	3	5.	Computer Data
Aarhus	7,600		6,000		N	Y	O	Y	Y
Adelaide	20,000	1,500	12,500	14,000*	N	Y	O	Y	Y
Amsterdam (Univ.)	10,000	500	6,000	6,250	N	Y	F	Y	Y
Arnold Arbor.	15,000	2,443	2,525	1,224	Y			Y	Y
Assoc. Orquid. Mex.	8,000		2,000		N	Y	O	Y	Y
Bergen	10,000				N	Y	O		Y
Berlin	–		18,000		N	Y	O	Y	
Berne	200–250	30		200	N	Y	N	Y	Y
Berry, Portland	2,800	100	1,950	750	N	Y	O	Y	Y
Blanes	12,000	500	2,500	300	N	N		Y	
Bochum		100	15,000	500	N	Y	O	N	Y
Brest	1,100	50	900	150	N	Y	N	Y	
Bristol	16,000				N	Y	O	Y	Y
Buckout, Penn.	700	140	560		N	N		N	
Cambridge, U.K.				8,000*	N	N			
Canberra	150,000	400		15,000	Y/N			Y	Y
Chicago	18,000				N	Y	F	Y	Y
Coffs Harbour, NSW	398	64	304	20	N	N		Y	
Coimbatore	350	68	350		N	Y	N	Y	Y
Copenhagen	25,000	500	15,000	10,000	Y			Y	Y
Cordoba, Spain				1,200	N	Y	O	Y	Y
Desert BG, Phoenix	17,350		3,000		Y			Y	Y
Dunedin					N	Y	N	Y	
Dusseldorf					N	N			
Edinburgh	29,960		12,500		Y			Y	Y
Edmonton	5,261	1,300	2,000		N	Y	N	Y	Y
Elsloo	2,400	750	1,500		N	N		Y	
Fairchild	6,500				Y			Y	Y
Frankfurt (Palm.)	18,000		10,000		N	Y	O	Y	Y
Fullerton Arb.	3,000			2,000	N	Y	O	Y	Y
Glasgow	12,000				N	Y	O	Y	Y
Glasnevin	20,000				N	Y	N	Y	

Continued

Name	1.Accessions	2.Cvs	Spp	Spp and/or Lower taxa	3.Status 1	2	3	5.	Computer Data
Goteborg	30,000	1,000		12,000	N	Y	N	Y	
Gottingen (Neuer)					N	Y	O/F	Y	Y
Guelph	6,000	1,000	4,000	1,000	Y			Y	Y
Hamilton, Ontario	8,000				Y			Y	Y
Harrogate					N	Y	N	Y	Y
Helsinki	9,000	6,000			N	Y	N		
Hull					N	Y	N	Y	Y
Huntington					N	Y	N		Y
Irvine	3,500	500	2,750	250	Y			Y	Y
Kew	85,000	6,763	61,000	8,700	Y			Y	Y
Kiel			8,000		N	N			
Kirstenbosch	70,000	800	5,500		Y			Y	Y
Kowloon (letter)					N	N		Y	
Lae	101,334	510	11,500		N	Y	O	Y	Y
Leiden	9,000	500		8,500	Y			Y	Y
Lisbon (letter)					N				
Liverpool (Cald.)	3,000	800	2,100	100	N	N		Y	
Los Angeles (Arb.)	7,200	2,400		4,800	Y			Y	Y
Marburg					N	Y	N	Y	
Meise	50,000		20,000		N	Y	O	Y	Y
Melbourne	35,000	2,000	10,000		Y			Y	Y
Missouri	15,000	6,000			Y/N			Y	Y
Monaco					N	Y	N	Y	Y
Moscow (Main)		11,100	10,200		Y			Y	Y
Nancy			10,000		N	Y	O	Y	Y
New England WF	1,500		1,100	400	N	Y	N	Y	Y
Nijmegen	3,000	30	2,500		Y			Y	Y
Pavia					N	N			
Porquerolles	3,000		2,000	1,000	Y			Y	Y
Pretoria (Univ.)	3,000	100	500		N	Y	N	N	Y
Reading	3,000–3,500	2–3,000	2,500		N	Y	O	Y	Y
Rotterdam	3,000	2,000	1,000		N	N		N	
Stockholm					N	Y	N	Y	Y
Tenerife	3,500				N	Y	N	Y	
Timaru	1,400	900	500		N	Y	N	Y	Y
Trinity, Dublin	5,000	300	4,000	300	N	Y	N	Y	Y
Turku				2,200	Y				Y
Utrecht	14,198		7,969	3,436*	Y			Y	Y
Vancouver	14,000	500		3,000	Y				Y
Wageningen	12,000	500	10,000		Y			Y	Y
Waimea	5,000				N	Y	O		
Warsaw		300		4,500	N	N		Y	
West Perth	4,000		3,000		N	Y	O	Y	Y
Westonbirt	14,000		2,835		Y			Y	Y
Wisley (RHS)					N	Y	O	Y	Y
Xalapa					N	Y	O	Y	Y
Zurich (succ.)				8,000	N	N		Y	

Notes. * Including cultivars.
N.B. Many of the figures for number of accessions, species, etc, are approximate.

Table 3
As table 2 but sorted in approximate order of taxa held

Name	1.Accessions	2.Cvs	Spp	Spp and/or Lower taxa	3.Status 1	2	3	5.	Computer Data
Kew	85,000	6,763	61,000	8,700	Y			Y	Y
Meise	50,000		20,000		N	Y	O	Y	Y
Berlin			18,000		N	Y	O	Y	
Bochum		100	15,000	500	N	Y	O	N	Y
Canberra	150,000	400		15,000	Y/N			Y	Y
Copenhagen	25,000	500	15,000	10,000	Y			Y	Y
Adelaide	20,000	1,500	12,500	14,000*	N	Y	O	Y	Y
Edinburgh	29,960		12,500		Y			Y	Y
Goteborg	30,000	1,000		12,000	N	Y	N	Y	
Lae	101,334	510	11,500		N	Y	O	Y	Y
Moscow (Main)		11,100	10,200		Y			Y	Y
Frankfurt (Palm.)	18,000		10,000		N	Y	O	Y	Y
Melbourne	35,000	2,000	10,000		Y			Y	Y
Nancy			10,000		N	Y	O	Y	Y
Wageningen	12,000		10,000		Y			Y	Y
Leiden	9,000	500		8,500	Y			Y	Y
Kiel			8,000		N	N			
Zurich (succ.)				8,000	N	N		Y	
Cambridge, U.K.				8,000*	N	N			
Utrecht	14,198		7,969	3,436*	Y			Y	Y
Amsterdam (Univ.)	10,000	500	6,000	6,250	N	Y	F	Y	Y
Missouri	15,000	6,000			Y/N			Y	Y
Aarhus	7,600		6,000		N	Y	O	Y	Y
Helsinki	9,000	6,000			N	Y	N		
Kirstenbosch	70,000	800	5,500		Y			Y	Y
Los Angeles (Arb.)	7,200	2,400		4,800	Y			Y	Y
Warsaw		300		4,500	N	N		Y	
Guelph	6,000	1,000	4,000	1,000	Y			Y	Y
Trinity, Dublin	5,000	300	4,000	300	N	Y	N	Y	Y
Desert BG, Phoenix	17,350		3,000		Y			Y	Y
West Perth	4,000		3,000		N	Y	O	Y	Y
Vancouver	14,000	500		3,000	Y				Y
Westonbirt	14,000		2,835		Y			Y	Y
Irvine	3,500	500	2,750	250	Y			Y	Y
Arnold Arbor.	15,000	2,443	2,525	1,224	Y			Y	Y
Blanes	12,000	500	2,500	300	N	N		Y	
Nijmegen	3,000	30	2,500		Y			Y	Y
Reading	3,000–3,500	2–3,000	2,500		N	Y	O	Y	Y
Turku				2,200	Y				Y
Liverpool (Cald.)	3,000	800	2,100	100	N	N		Y	
Fullerton Arb.	3,000			2,000	N	Y	O	Y	Y
Assoc. Orquid. Mex.	8,000		2,000		N	Y	O	Y	Y
Edmonton	5,261	1,300	2,000		N	Y	N	Y	Y
Porquerolles	3,000		2,000	1,000	Y			Y	Y
Berry, Portland	2,800	100	1,950	750	N	Y	O	Y	Y
Elsloo	2,400	750	1,500		N	N		Y	

continued

Name	1.Accessions	2.Cvs	Spp	Spp and/or Lower taxa	3.Status 1	2	3	5.	Computer Data
Cordoba, Spain				1,200	N	Y	O	Y	Y
New England WF	1,500		1,100	400	N	Y	N	Y	Y
Rotterdam	3,000	2,000	1,000		N	N		N	
Brest	1,100	50	900	150	N	Y	N	Y	
Buckout, Penn.	700	140	560		N	N		N	
Pretoria (Univ.)	3,000	100	500		N	Y	N	N	Y
Timaru	1,400	900	500		N	Y	N	Y	Y
Coimbatore	350	68	350		N	Y	N	Y	Y
Coffs Harbour, NSW	398	64	304	20	N	N		Y	
Berne	200–250	30		200	N	Y	N	Y	Y
Bergen	10,000				N	Y	O		Y
Bristol	16,000				N	Y	O	Y	Y
Chicago	18,000				N	Y	F	Y	Y
Dunedin					N	Y	N	Y	
Dusseldorf					N	N			
Fairchild	6,500				Y			Y	Y
Glasgow	12,000				N	Y	O	Y	Y
Glasnevin	20,000				N	Y	N	Y	
Gottingen (Neuer)					N	Y	O/F	Y	Y
Hamilton, Ontario	8,000				Y			Y	Y
Harrogate					N	Y	N	Y	Y
Hull					N	Y	N	Y	Y
Huntington					N	Y	N		Y
Kowloon (letter)					N	N		Y	
Lisbon (letter)					N				
Marburg					N	Y	N	Y	
Monaco					N	Y	N	Y	Y
Pavia					N	N			
Stockholm					N	Y	N	Y	Y
Tenerife	3,500				N	Y	N	Y	
Waimea	5,000				N	Y	O		Y
Wisley (RHS)					N	Y	O	Y	Y
Xalapa					N	Y	O	Y	Y

Table 4
Gardens that do not intend to computerise (with reasoning where given)

Blanes	Do not yet have an effective card system
Buckout, Penn.	Collection too small
Cambridge, U.K.	Investigated in 1974 when it was found that likely gains were not commensurate with effort required; may be reconsidered with computerisation of university libraries
Coffs Habour, NSW	No funds likely in foreseeable future
Dusseldorf	Would like to, but staff too few
Elsloo	Financial reasons
Kiel	
Kowloon	Unlikely in near future due to financial constraints
Liverpool (Cald.)	No computer available.
Pavia	
Rotterdam	Present card system, maps and catalogue work easily and very well
Warsaw	Not possible at present; hope it will be possible in future
Zurich (succ.)	(a) finance, (b) not connected to an institute or university

(4) The gardens that do not intend to computerise do not for the most part, close the door to computerisation; rather, resources are not available at present.

(5) An overwhelming majority of gardens sampled — 64 out of 79 — expressed interest in collaborative projects to design compatibility standards and/or garden records software. Among those that were not interested, several said this was because they had little to contribute rather than that they opposed the idea.

Table 2 provides a summary of garden replies, showing the extent of their collections and the extent of computerisation. These are the results of page 2 of the Questionnaire.

Table 3 reproduces the same data but in descending order (approximate) of number of taxa.

Table 4 lists those few gardens that do *not* intend to computerise and gives their reasoning.

The AABGA report (cited above) provides data on many other gardens that have computerised their records. Among the most prominent are:

Name	*No. Plant Records*
Brookside Gardens	10,000
Central Park Conservancy, NY	24,000
Denver Botanic Gardens	14,500
The Dow Gardens, MI	10,000
Holden Arboretum	6,000
Kentucky Botanical Garden	2,200
Longwood Gardens	12,000
Mitchell Park Conservatory, WI	2,000
N. Carolina Botanical Garden	4,000
N. Carolina State University NCSU Arboretum	3,000
Northwest Plant Germplasm Repository, OR	4,500
Pacific Tropical Botanical Garden	13,000
Rhododendron Species Foundation, WA	3,144
San Diego Wild Animal Park	3,500
Santa Barbara Botanic Garden	2,193
Marie Selby Botanical Garden	10,000
Smithsonian Institution (Horticulture)	21,000
USDA Subtropical Hortic. Research Unit, Miami	30,623
UCLA Mildred E. Mathias Botanical Garden	4,000
UC Riverside	3,000
University of Michigan Matthaei Botanical Garden	6,400
University of Minnesota Landscape Arboretum	6,000
San Diego Zoo	10,000

Postscript (July 1987)

The ITF User Manual is virtually complete and is available in draft form. Please do not try to implement the ITF without it. Copies are available from Botanic Gardens Conservation Secretariat, 53 The Green, Kew, Richmond, Surrey TW93AA, UK.

Fig. 1

INTERNATIONAL UNION FOR CONSERVATION OF NATURE AND NATURAL RESOURCES

CONSERVATION MONITORING CENTRE

BOTANIC GARDENS CONSERVATION CO-ORDINATING BODY

Questionnaire on computerized garden record schemes

Name of garden:............................

Completed by:........................

Date:................

If your garden records are on computer, please accompany the questionnaire with as many as possible of the following items:

1. Papers or booklets explaining the system;
2. Sample print-outs;
3. Most useful of all, a printout of 50 or so records from the data file, annotated to show the meaning of each field.

Please return to:

IUCN Conservation Monitoring Centre,
c/o The Herbarium,
Royal Botanic Gardens,
Kew,
Richmond, TW9 3AE,
Surrey, U.K.

It would be appreciated if you could return the questionnaire by 1 May 1985.

Section A: GENERAL QUESTIONS

1. How many Accessions do you have? [ ]

2. How many taxa do you believe they represent?
   [ ] Cultivars
   [ ] Species
   [ ] Species &/or lower taxa

3. Are your garden records stored on a computer? [Yes/No]
   If "No", do you intend to store them on a computer? [Yes/No]
   If so, how detailed are your plans?
   No formal plan yet [ ]
   Outline [ ]
   Fully developed [ ]

4. Briefly, can you outline the factors which led your garden to adopt the course of action indicated by your answers to Question 3.

5. The development of compatible record systems will simplify the exchange of data with IUCN and between individual gardens. Would you be interested in collaborative projects to design compatibility standards, and/or garden records software? [Yes/No]

If your garden records are stored on a computer, or you have plans to computerise, please complete Sections B and C as best you can. Don't worry if you cannot answer all of the more technical questions.

Section B: QUESTIONS CONCERNING YOUR COMPUTER SYSTEM (Planned or actual)

Please tick as many of the following boxes as are applicable. Section C asks more detailed supplementary questions.

6. How many simultaneous users will your system support? [ ]

7. Do/Will you have your own programming staff? [Yes/No]
   If "Yes", how many? [ ]

8. Do/Will you use
   Your own computer? [ ]
   A computer belonging to another institution? [ ]
   A bureau service? [ ]

   If you use computer facilities that are not your own, please provide the name(s) of the institution(s) or bureau(x) concerned.

9. Type of hardware used/to be used:
   Mainframe or mini-computer(s) [ ]
   Micro-computer(s) [ ]

Section C: DETAILED INFORMATION ABOUT YOUR COMPUTER SYSTEM

Hardware and Operating Systems

10. Please give make, model and operating systems of computers used/to be used:

Applications Software

11. Do/Will you use:
    Purpose-written software
    written in-house [ ]
    written commercially [ ]
    written by staff of "host" institution [ ]
    Please indicate the languages used:

    Commercial software packages
    supplied with your system [ ]
    purchased for the purpose [ ]
    Please indicate the package(s) used:

Communications

12. Does/Will your system have the ability to communicate with other computers through the public telephone network? [Yes/No]

If so, please complete one row in the following table for each type of telecommunication your computer(s) will support:

Baud rate(s)	Synchronous/ Asynchronous	Protocol eg IBM2780, TTY	Modem modulation type eg CCITT V26

13. Are you/Will you be linked to a data network? [Yes/No]
If so, which?

Data Transfer Media

14. Do/Will you have floppy disk drive(s)? [Yes/No]

If so, please complete one row in the following table for each type of diskette that your computer(s) is/are capable of reading or writing. If you cannot complete all columns don't worry - just complete as many as possible.

Physical Size	Sides	Sectors	Tracks	Density	Capacity	Format(s)	Capabilities
3.5, 5.25, 8"	Single/ Double	No. Type Soft/ Hard	No.	Single/ Double/ Quad	k bytes	eg CP/N, MS-DOS	Read/Write

15. Do/Will you have magnetic tape drive(s)? [Yes/No]

If so, please complete one row in the following table for each type of tape your computer(s) is/are capable of reading or writing

Tracks 7 or 9-track	Density & recording format eg 800bpi NRZ1, 1600bpi PE, 6250 bpi

16. Do/Will you have any other means of transferring data electronically (eg cartridge tapes, telex)? [Yes/No]

If so, which?

Thank you

6

Short Communications

A rare plant programme for southern Ontario

JOHN D. AMBROSE

University of Guelph Arboretum, Guelph, Ontario, Canada

The University of Guelph Arboretum has been compiling information on and studying the rare trees and shrubs of southern Ontario for the past 6 years. More recently, detailed studies on potentially endangered species have been conducted, both for the preparation of status reports for the Committee on the Status of Endangered Wildlife in Canada (COSEWIC) and to provide basic information on their biology. From the latter, effective strategies can be developed for their conservation management. In addition, a gene bank, or conservation stand, is being developed at the arboretum for many of these rare or endangered species, with propagules taken from a broad sampling of the remaining natural stands in Ontario.

Regional rarity, is it important?

All of the rare species of southern Ontario have a broad distribution, most through eastern or central North America. Many of them become relatively common to the south. Why then should we concern ourselves with these species?

A species exhibits variation in its genetic make-up, which typically reflects the pattern of variation in its environment. At the limits of its distribution, unique gene complexes, representing adaptations to the local environment, are likely to be present, especially in isolated populations. The

Botanic Gardens and the World Conservation Strategy

ISBN: 0-12-125462-3

Carolinian flora of extreme southern Ontario exemplifies such a situation, where many species of eastern North America reach their northern limits of distribution and free genetic exchange to the south is restricted by the Great Lakes. The loss of these populations could result in losses of genetic diversity of these species, especially genotypes adapted to Ontario's environment. In species rare throughout their range, such regional losses could further threaten their long-term survival.

Approaches to rare plant conservation

Habitat (*in situ*) conservation is considered the most effective and important means to ensure the long-term survival of species. Greater genetic diversity can more readily be maintained in natural habitats than under cultivation, and natural populations continue to respond to the selective forces of their environment, as well as to support associated species, including pollinators and dispersal agents, which may themselves be threatened by the loss of a particular plant species.

Identifying the species that are potentially endangered is the first step to initiate action by appropriate conservation agencies. In Ontario, both legislation and guidelines call for protective action, once the official status of a species has been determined. For the most significant habitats, nature reserves are being established, by the province, by naturalists' clubs, and by conservation bodies. Promoting stewardship by private landowners, through contact and discussion, is another means of protecting natural areas that has proven to be successful. Further studies on the species' biology can provide the necessary background for a management strategy.

To complement habitat conservation, consider the establishment of off-site (*ex situ)* conservation stands. They can provide material for experimentation or evaluation, without threatening natural populations. When natural populations are highly depleted, or species have been locally extirpated, these stands also can provide propagation material for controlled reintroduction or population enhancement. It is important to consider the sampling of genetic diversity that such a stand represents. A large stand based on a single collection, vegetatively propagated, could give a false sense of security. To maximise genetic diversity, it is important to collect seeds (or other propagules) from a large percentage of the remaining populations. If only a few populations remain, then collect from various individuals across each population, and from different microhabitats within each site

Rôle of botanic gardens and arboreta

We tend to think that the only rôle such institutions can play with regard to rare plants is in acquiring and growing them, and possibly conducting propagation trials. Our experience has been that we have been successful in contributing to both *in situ* and *ex situ* conservation through the following aspects of our programme.

(1) Regional coordination of data collection on rare woody species and acting as an information centre.
(2) Liaison with government agencies and conservation bodies — regional, national and international.
(3) Field studies on the rare and endangered species of southern Ontario, leading to the official status of several trees and shrubs, and subsequent studies on their biology to provide information for conservation management.
(4) Establishment of conservation stands of regionally rare species based on broad sampling procedure.

Acknowledgements

The various aspects of this programme have been supported by the Ontario Ministry of Natural Resources, the World Wildlife Fund (Canada), and the Natural Sciences and Engineering Research Council of Canada. Field assistance and information was provided by numerous associates across Ontario and research cooperation was provided by faculty associates at the University of Guelph.

Rimba Ilmu Universiti Malaya — preserving the natural plant heritage of the world's oldest rain forest

DAVID T. JONES

Department of Botany, University of Malaya, Kuala Lumpur, Malaysia

The Botanic Garden of the University of Malaya, or Rimba Ilmu, was established in 1974 as a botanical facility of the university for research and education. Only recently has it become more involved in conservation projects. Specifically, the garden is devoted to the study and conservation of the immense diversity of the tropical rain forest flora and to analyse its floristic components employing a kind of "rescue and curation" operation. It is one of the garden's interests to try and preserve through *ex situ* conservation at least a representative cross-section of the variation from within the Malaysian lowland rain forests. In addition to fulfilling these scientific objectives, the garden also works toward increasing awareness of the natural environment by encouraging student and public utilisation.

Today, the garden's 1,200 living taxa are divided between a general collection and several "speciality" collections, which reflect the specific research interests of the Department of Botany. These collections include medicinal plants, orchids, tropical fruit trees, citroids, palms and pandans. Most of these accessions are of local (Malaysian) origin.

A number of significant and noteworthy research projects have been or are being carried out by staff and students utilising the resources of the garden. Some examples are phytochemical surveys of medicinal plants; phenological observations on selected taxa; germination trials; detailed soil

Botanic Gardens and the World Conservation Strategy

ISBN: 0-12-125462-3

surveys; litter decomposition studies and the effects of acid rain; phycological and limnological surveys of the aquatic habitats; and reproductive biology studies of native fruit trees.

More recently, Rimba Ilmu has taken measures to augment its collections of select genetic "germplasm" collections. Examples of these include not only specific taxonomic groups of plants (i.e. citroids, wild ginger) but also taxa occupying threatened or vulnerable habitats such as limestone outcrops.

We look forward to seeing a concern for the problems of valuable and potentially valuable species conservation reflected in a diverse collection. The provision of a broad genetic basis, by means of accumulating related species, is a very suitable programme for any botanic garden.

The Mexican Association of Botanic Gardens

GRAHAM PATTISON

National Council for the Conservation of Plants and Gardens, Wisley, U.K.

Back in 1980 the eminent Mexican botanist Dr Arturo Gómez-Pompa mentioned at the first Course for Technical Staff of Botanic Gardens the importance of cooperation and unity between botanic gardens in Mexico. This lay dormant until 1983, when an inaugural meeting was held by some 13 Mexican institutes and universities. After cooperation from the Instituto Nacional de Investigación sobre Recursos Bióticos (INIREB), the Universidad National Autónoma de México (UNAM), the Secretariá de Desarrollo Urbano y Ecología (SEDUE) and other institutes/universities, the National Association of Botanic Gardens was formed.

The objectives of the Association are as follows.

(1) To establish contact with other botanic gardens in the country and to exchange ideas and information;
- (i) by publishing a newsletter to keep other members informed;
- (ii) by organising conferences, symposia, etc;
- (iii) by forming a bank of botanical and bibliographical information.

(2) To promote the establishment of other botanic gardens within the country to form a representative collection of the local flora;
- (i) by studies of different ecosystems;
- (ii) by giving suggestions as to possible site for botanic gardens.

Botanic Gardens and the World Conservation Strategy

ISBN: 0-12-125462-3

(3) To contribute to prevent the extinction of native species and promote the use of native flora.
(4) To promote finance for research, specifically for:
  (i) courses in horticulture, taxonomy and botanic garden administration;
  (ii) buying materials in bulk.
(5) To promote the botanic gardens and an understanding of their importance.

The Association had its first National Conference at UNAM in May 1985. This was attended by 125 delegates. The conference took the form of presentations on the various projects within the country. The 2 principal speakers were Dr Thomas Elias, Director of Rancho Santa Anna Botanic Garden, California, and Biol. Wilfredo Contreras, Director of Flora and Fauna in SEDUE.

The 1986 Conference was held in the north of Mexico at the Agricultural University ''Antonio Narro'' in Saltillo. The most important part of this is the National Inventory of Botanic Gardens in Mexico, to outline where they are and what plants they have. It will publish an inventory of all living collections which will be available towards the end of 1987 or early 1988.*

It is hoped that this Association will go from strength to strength and result in the growing of rationalised collections.

* Both the 1985 and 1986 Memorias can be obtained from A. Vovides at INIREB, Apdo Postal 63, Xalapa, Veracruz, México, or Edelmira Linares, Jardín Botánico, Ciudad Universitaria, Apto Postal 70614, México D.F. 04510 México.

Moroccan gardens: past, present and future

MOH. REJDALI

Institut Agronomique et Veterinaire Hassan II, Rabat, Morocco

Through their changing history, Moroccan gardens have remained as areas of public recreation and enjoyment, serving mainly usefulness and aesthetics. No attempt was made to develop their considerable potential in the functions of systematic plant resources.

The early gardens, "Riad" as they are called, were inspired from the Moorish gardens. They consisted of internal halls equipped with natural elements of water, light, trees and flowers. They were well known at Fes, Meknes and other towns. By the time of the Protectorate, urban public recreational gardens had developed and became widespread all over the country, in every large town. Examples of such gardens are: Jnane S'bil (Fes), Triangle de Vue (Rabat), Jardin les Majorelles (Marrakesh). Some of these gardens are equipped with areas for children and games. At the same time some private or public gardens were created to act as recreational public gardens, and also some information and education centres were created. Such gardens are usually equipped with small nurseries and limited areas for research activities. Examples of these gardens are:

> *Les Jardins Exotiques:* Located 13 km west of Rabat, Sale, this garden contains an enormous collection of exotic plants from the tropics and other regions, but unfortunately most of the plants are unlabelled and no records have been kept of their origin.

Botanic Gardens and the World Conservation Strategy

ISBN: 0-12-125462-3

Le Jardin d'Essais (Rabat): This garden, which belongs to the Institut National Agronomique, covers 6 ha, but has a less diverse collection of plants. In this garden experiments are carried out, mainly on fruit trees.

Nowadays there is an increasing awareness among Moroccan scientists and politicians of the rôle that a botanic garden can play in various scientific fields such as the promotion of research, and the conservation of threatened and endangered plants, besides its use for the education of the local community.

However, no botanic garden, in the normal sense of the word, can be said to exist up to now in Morocco, although both the gardens mentioned above could form the basis of one.

Taking into account the great diversity and richness of the Moroccan flora (around 4,000 species of which 19% are endemic), the wide range of vegetation types from desert to alpine, the diversity of the Moroccan climate (all the Mediterranean subdivisions are represented) and finally because of the disastrous threat to which these resources are exposed, there is a genuine need for at least one botanic garden, if not a network of small botanic gardens in the major cities, as well as other conservation measures to save this wealth. A national conservation strategy would be an ideal goal to achieve this.

Protection of endangered succulent plants at the Botanic Garden of Palermo

MAURIZIO SAJEVA

Dipartimento di Scienze Botaniche dell' Universita di Palermo, Italy

Overcollecting of wild plants for commercial purposes is one of the major threats to many succulents, and propagation of documented material could take the pressure off wild populations threatened by over-collecting. The conservation strategy at the Botanic Garden of Palermo (Laboratorio di Fisiologia Vegetale, Gruppo Biologia delle Piante Succulente) is based on a programme to propagate and distribute endangered succulent species. Financial help from the Regional Sicilian Government has enabled us to acquire the necessary equipment to begin this programme, which is undoubtedly helped by the favourable climate of our location. We have been able to fit out a small laboratory to be used for micropropagation, with the aim of increasing reproduction of documented plant material.

The collection of Cactaceae has been revised as it was found to be the best represented group among the succulents in the garden, with the aim of using it for propagation. During the revision it was noticed how often the seeds produced in botanic gardens are hybrids, and are consequently useless for research and conservation. We discussed the possibility of creating a nursery, where documented succulent plants could be grown under strict pollination control. Before setting up of this type of nursery, we contacted the International Organisation for Succulent Plant Study (IOS) as it is the best authority in the field of succulents. The proposal was presented to the first IOS intercongress seminar meeting (Zurich, 20–24 July 1985) as "Proposal of setting up an IOS nursery at the Botanical Garden of

Botanic Gardens and the World Conservation Strategy

ISBN: 0-12-125462-3

Palermo" and was approved by the participants. During the discussion it was thought possible to use plants confiscated under the CITES convention. There could be 2 advantages in using these plants; on the one hand the plants would be better looked after than left in the custom sheds and on the other hand they would give us good material for study and propagation. Cultivated plants would remain the property of the supplying governments, and any use of the plants would be under IOS supervision. We were assured by the Italian Ministry of Agriculture that licences will be issued for importation of confiscated plants in consideration of scientific and conservation purposes.

Nurserymen producing plants grown from seed are of great help for conservation; IOS Reference Collections are a very important source of documented material, very useful for research and conservation. The nursery we are setting up should be complementary to the above-mentioned ones; without commercial or display purposes much care could be devoted to reproduction and distribution.

We believe that this type of nursery used could be set up in other parts of the world with suitable climate or qualified help, for other groups of plants in danger of extinction. This would be a step in the right direction for helping conservation. We would like, however, to emphasise that any contribution that a botanic garden can give to conservation is limited without the protection of the natural habitat of endangered species. Protection of the wild remains the basic goal.

Rare plant propagation in Mauritius

P. WYSE JACKSON,[1] Q. CRONK,[2] J. PARNELL[3] and W. STRAHM[4]

[1] *Botanic Gardens Conservation Secretariat, Kew, U.K.*
[2] *Department of Botany, University of Cambridge, U.K.*
[3] *School of Botany, Trinity College, Dublin, Ireland*
[4] *IUCN/WWF Project Officer, Mauritius*

Introduction

The island of Mauritius lies in the western Indian Ocean some 900 km east of Madagascar. Its closest neighbours are the islands of Réunion and Rodrigues; the three forming the Mascarene Island group.

Mauritius was first settled by the Dutch at the beginning of the 17th century. They quickly began to exploit the ebony forests for timber, which rapidly resulted in the extinction of many plant and animal species, including the dodo. The expanding population of Mauritius (now *c.* 1 million) has made heavy demands on the limited land area available. Most native forests and marshes have been cleared for sugar-cane production (the major industry) and in the plateau region, large areas of introduced species such as *Eucalyptus* and pines have been planted on the denuded landscape. Other introduced species such as "Chinese" Guava (*Psidium cattleyanum* Sabine), Privet and Traveller's Palm, have spread to encroach onto the remaining small patches of native woodland, bringing the regeneration of native species to a standstill in most cases. Most Mauritian households use wood for cooking, having no other natural domestic fuel source; thus the remaining native forests continue to be under considerable pressure, also from illegal wood-cutting.

Botanic Gardens and the
World Conservation Strategy

ISBN: 0-12-125462-3

Many endangered endemics

Mauritius has a large number of endemic plant species, many of which are extremely rare and endangered. Several are reduced to a single known example and others exist as small vulnerable populations in badly degraded native habitats. For several years IUCN/WWF have sponsored Wendy Strahm's work in Mauritius towards the production of a Red Data Book for the island. Strahm aims to find out which patches of native vegetation remain, identify examples for protection as nature reserves and assess the size of remaining populations of each endemic species and the threats they face. To complement her work it was deemed essential to begin propagation of many of the most threatened species so that wild populations could be supplemented with new, young cultivated examples both as a Mauritian *in situ* collection and *ex situ* in botanic gardens. For this purpose an expedition was organised for June 1985, consisting of a small group from Dublin and Cambridge Universities.

The Dublin/Cambridge Expedition

The aim of the expedition was to propagate many of the most severely threatened species, to build a propagation facility for rare native plants in Mauritius and to establish the beginnings of a rare plant nursery for the Mauritian Forestry Service. A shade house and mist bench were constructed on arrival, into which cuttings of some 30 species were placed. At the time of writing, 3 months after the commencement of the project, successful propagation by means of cuttings has been achieved in more than 20 species, most of which have never been propagated or cultivated before. Most of the resultant plants have been retained by the Mauritian Forestry Service and, in due course, will be planted out in managed nature reserves.

A few of the most notable species whose propagation has been achieved are outlined below.

Elaeocarpus bojeri R.E. Vaughan — Elaeocarpaceae

Only 5 individuals of this species are known. They all grow in one extremely small and vulnerable locality, at Grand Bassin on the southern plateau. All are old and decrepit and so far the limited amount of seed produced has proved impossible to germinate and there appears to be no natural regeneration. We have successfully propagated it and several young plants have been established in cultivation.

Claoxylon linostachys Baillon — Euphorbiaceae

Three subspecies of this species are known. All are rare and indeed one may be extinct. The other 2 subspecies are known only at one locality each. Of subsp. *linostachys*, only 2 individuals are known and of subsp. *brachyphyllum*, less than 10 plants can be found. It appears to be most intolerant of the invasion of its habitat by exotic weeds. Few of the remaining known plants of either subspecies are vigorous and so few cuttings were obtained. However, several individuals of each have been propagated by cuttings and are now in cultivation.

Tetrataxis salicifolia Baker — Lythraceae

This is a monotypic genus of trees, confined to badly degraded forest at one locality in the upland forest of the southern plateau. Less than 20 specimens remain at this site and little regeneration takes place. Three more individuals have recently been found by Strahm at a new locality in the Black River Gorges in south-west Mauritius, but these too are threatened. The expedition successfully propagated this species by cuttings.

The Declaration of Gran Canaria

For centuries, botanic gardens have been major centres for the scientific study of plant diversity, providing a mechanism for introduction and assessment of plants for agriculture, horticulture, forestry and medicine.

They attract more than 100 million visitors a year, affording havens of beauty and tranquillity for an increasing urban society, and a spiritual link with the plant world on which we all depend.

They inform and educate; they are showcases for the living world, places where science and people meet.

For historical reasons, most botanic gardens are in the cooler, more industrialised countries of the world, but two thirds of all plant species occur in the tropics and subtropics. More than 60,000 species risk extinction within our lifetimes because of the destruction and degradation of the earth's vegetation, which is the basis of human survival. Recently, many of the world's botanic gardens have mobilised their resources for conservation action to avert this threat. They are conserving plants in the wild, cultivating them in the gardens themselves, and preserving them in gene banks.

Recognising that they can only succeed in achieving these objectives if they work together, botanic gardens throughout the world are uniting to apply the *World Conservation Strategy* to the special predicament of plants. Basing their efforts on this global plan for sustainable development and conservation of living resources, they will produce, adopt and implement a *Botanic Gardens Conservation Strategy*.

This declaration is the result of the 1985 Las Palmas Conference on Botanic Gardens and the World Conservation Strategy, involving more than 200 leading specialists from countries throughout the world.

They as a body assert their determination to work together to defend plant life for the benefit of all people now and in the future. They call upon Governments to provide the necessary support and resources, in accordance with their responsibilities.

Recommendations

Recommendation 1

The Botanic Gardens Conservation Congress

Aware of the value of the deliberations of this Conference;

Appreciative of the most welcome support for this Conference from sponsors;

The International Conference on Botanic Gardens and the World Conservation Strategy:

Proposes that a Botanic Gardens Conservation Congress meet every 3 years;

Recommends participating gardens and institutions to support the IUCN Botanic Gardens Conservation Co-ordinating Body and to provide the necessary financial support for the Body, and that this Body provides the Secretariat for the Congress;

Invites international and national agencies to help sponsor the Congress.

Recommendation 2

The Botanic Gardens Conservation Strategy

Aware of the fundamental need that utilisation of species and ecosystems be sustainable;

Appreciating that botanic gardens have a rôle in the conservation of living resources as a basis for sustainable development;

Recognising that botanic gardens need to identify and agree on the objectives and priority tasks they can undertake in implementing the World Conservation Strategy;

The International Conference on Botanic Gardens and the World Conservation Strategy:

Considers that a Botanic Gardens Conservation Strategy is essential;

Calls on IUCN to complete the draft of the Botanic Gardens Conservation Strategy in the light of the deliberations of the Conference, to publish it and to make it as widely available as possible;

Invites the individual botanic gardens of the world to pledge their support for the Botanic Gardens Conservation Strategy, to identify their missions and to define the tasks that they will undertake for its implementation;

Invites IUCN to assist botanic gardens in setting up a Secretariat to co-ordinate and promote the implementation of the Strategy, incorporating the activities of the IUCN Botanic Gardens Conservation Co-ordinating Body, with the aim of making this Secretariat self-supporting.

Recommendation 3

Worldwide network of botanic gardens

Acknowledging that botanic gardens and arboreta are essential instruments in the conservation of plant genetic resources throughout the world;

Noting the serious imbalance between the location of botanic gardens and the floristic richness of the areas in which they occur;

Further noting the small number of active botanic gardens in tropical countries compared with the great richness of tropical floras;

Welcoming the initiative of IUCN and IABG to collaborate on the next International Directory of Botanic Gardens;

The International Conference on Botanic Gardens and the World Conservation Strategy:

Calls on those responsible for existing botanic gardens to acknowledge their gardens' actual or potential rôle in genetic resource conservation, to strengthen their existing conservation activities and to take steps to guarantee the long-term survival of their gardens.

Recommends governments and states, in association with international agencies, to review the distribution of botanic gardens and to recommend, where appropriate, the creation and funding of new gardens with a major conservation mission, the emphasis being on regions with rich endemic floras, such as tropical forests and islands;

Urges the botanic gardens of the world to unite forces and organise themselves in such a way as to provide effective channels of communication and means of cooperation at national, regional and international levels by creating a botanic gardens worldwide network, in association with the International Association of Botanic Gardens (IABG).

Recommendation 4

Ex situ conservation

Recognising that no single approach to the conservation of endangered species can be relied upon;

Appreciating that *ex situ* conservation is a necessary adjunct to *in situ* conservation;

Acknowledging the importance of seed banks in the long-term conservation of genetic resources;

The International Conference on Botanic Gardens and the World Conservation Strategy:

Urges botanic gardens to recognise their responsibility to maintain, propagate and make available stock of critically threatened species for scientific and horticultural research, for reintroduction (where appropriate) and to provide suitable stock for horticulture;

Recommends that exploration and collection of species be based on concepts of infra-specific diversity so that ecogeographical diversity and diversity between and within populations be sampled in such a way that a maximum of genetic diversity be captured and stored;

Urges botanic gardens to become involved with seed conservation and recommends that the International Board for Plant Genetic Resources (IBPGR) be approached so as to establish closer collaboration on the conservation of threatened species;

Recommends that every effort be made to maintain minimal international standards for seed storage and rejuvenation;

Recommends IUCN to continue and expand the monitoring and co-ordination of *ex situ* conservation, presently carried out by the Botanic Gardens Conservation Co-ordinating Body, as an integral part of the implementation of the Botanic Gardens Conservation Strategy;

Recommends botanic gardens and other relevant institutions to support this essential work and to provide the necessary finance for it.

Note: The term 'seed' is used here to include spores of pteridophytes and lower plants.

Recommendation 5

Botanic gardens in the tropics and subtropics

Recognising that two thirds of the world flora (some 170,000 species) are in the tropics and the subtropics;

Realising that recent calculations show that at least 60,000 plant species are at risk of extinction in the next 30–40 years;

Recognising that these constitute nearly a quarter of the world's total plant species diversity;

Acknowledging that botanic gardens in the tropics and subtropics can help study and provide some of the vital solutions to this horrifying threat;

Convinced that botanic gardens can play a very important part in the conservation of the flora and the vegetation of those regions and are essential in furthering the maintenance of national plant genetic resources, both *ex situ* and *in situ*;

Being aware that many tropical and subtropical countries have botanic gardens that are moribund or barely function;

Realising that a further range of tropical and subtropical countries do not have a single botanic garden;

The International Conference on Botanic Gardens and the World Conservation Strategy:

Urges institutions of higher learning, Governments, aid agencies and financial institutions to put more resources into existing botanic gardens and where necessary to establish new botanic gardens in the tropics and the subtropics;

Invites botanic gardens both in these countries and elsewhere to provide all possible training facilities and support in kind, particularly by enabling their staff to work with colleagues in the tropics and subtropics, and vice versa;

Calls upon educational and scientific institutions, Governments, bilateral and multilateral aid agencies to provide sufficient funding to facilitate this interchange of staff and technology.

Recommendation 6

Protected areas

Aware that deforestation and destruction of ecosystems adversely affect the lives of millions of people by causing periodic flooding, scarcity of fuelwood, degradation of soil and water and reduction of agricultural productivity;

Recognising the importance of protecting a worldwide network of representative ecosystems for maintaining biological stability and diversity;

Recognising that the benefits of protecting representative ecosystems will accrue to all mankind and not just to those nations in which the ecosystems occur;

Noting that many of the botanic gardens of the world are caretakers of nature reserves to conserve local floras *in situ*, individual species or particularly important plants;

Realising that global frameworks exist in the form of Unesco's Action Plan for Biosphere Reserves and IUCN's Bali Action Plan, and a regional framework in IUCN's Corbett Action Plan for Protected Areas of the Indomalayan Realm;

The International Conference on Botanic Gardens and the World Conservation Strategy;

Recommends botanic gardens to take active steps to cooperate with national and international agencies responsible for protected areas so as to ensure the conservation of plants in the wild by the establishment of networks of reserves within and across national boundaries;

Urges the botanic gardens of the world to take an active part in aiding the conservation *in situ* of the floras and plants of the regions in which they are situated;

Calls upon the botanic gardens to cooperate with land-use agencies to provide data and to train staff in those aspects of the conservation and monitoring of flora that will aid the management of plants in protected areas.

Recommendation 7

Critically threatened environments

Recognising the threat confronting many fragile ecosystems consisting of highly specialised and endemic floras;

The International Conference on Botanic Gardens and the World Conservation Strategy:

Calls upon Governments, conservation organisations and aid agencies to support botanic gardens and related institutions involved with the most critically threatened plant ecosystems, particularly rain forests, wetlands, Mediterranean-type vegetation such as Cape Fynbos, and island ecosystems.

Recommendation 8

Plant records

Acknowledging that botanic gardens have kept plant records in various forms for centuries;

Being aware that the needs of conservation have placed even greater emphasis on full and accurate records, especially on the origins of wild source plants;

Recognising that computerisation can bring benefits of up-to-date retrieval and data exchange;

Recognising the need for the international exchange of data between gardens and between gardens and IUCN;

Appreciative of the experience of those gardens which have computerised their plant records, for example the botanic garden network in the Netherlands;

Realising that many more gardens are about to computerise their records;

Considering that now is an ideal time to agree minimal data standards for the storage and exchange of plant data;

The International Conference on Botanic Gardens and the World Conservation Strategy:

> Recommends botanic gardens to computerise their plant records so as to aid management of their collections and to permit data exchange both between gardens and with international organisations;
>
> Requests the IUCN Conservation Monitoring Centre to continue its work on the conceptual basis for computer record schemes and on the International Transfer Format (ITF) by further collaboration with botanic gardens.

Recommendation 9

Availability of germ plasm

Aware that the world community has a responsibility to conserve wild plants and their genetic diversity;

Recognising that a very wide spectrum of plant resources is necessary to combat desertification, stop erosion, restore the vegetation of degraded lands and to provide the basic materials necessary to ensure present and future supplies of food, timber, fuelwood, fibre, medicines as well as a host of other products;

Realising that plants do not recognise political frontiers and that many plants are extremely well suited for cultivation in places far from their native localities;

Certain that the success of conservation efforts to preserve useful or potentially useful plants, especially those threatened by the destruction of ecosystems, may depend upon the free exchange of germ plasm;

The International Conference on Botanic Gardens and the World Conservation Strategy;

> Recommends Governments and international organisations to facilitate the full availability of plant material of threatened species, under adequate quarantine and related measures.

Recommendation 10

Conservation research

Recognising that a sound information base, well managed and accessible, is of critical importance for effective conservation;

Further recognising the lack of available information about plants and the central rôle of taxonomy in organising this information and making it available, and that insufficient is known about the chemistry, cytology, physiology (including seed physiology), autecology, reproductive biology, methods of cultivation and propagation, which are essential for plant conservation both *in situ* and *ex situ*.

The International Conference on Botanic Gardens and the World Conservation Strategy:

> Calls upon botanic gardens to incorporate conservation-oriented research into their programmes as a matter of urgency, including taxonomy, chemistry, cytology, physiology (with seed physiology), autecology, reproductive biology and methods of cultivation and propagation;
>
> Strongly endorses and reiterates to Governments, aid agencies and non-governmental organisations the calls for increased support taxonomic and systematic research in the tropics and subtropics;
>
> Recommends that as a matter of urgency relevant taxonomic and associated data should be made available for botanic gardens and other botanic institutions, using computerised data-bases wherever possible;
>
> Urges that botanic gardens and local herbaria should forge effective working links where these do not already exist.

Recommendation 11

Education and public awareness

Recognising the vital importance of community understanding and awareness in achieving conservation of biological resources;

Acknowledging that botanic gardens of the world are visited by over 100 million people each year;

Aware of the increasing inadequacy of courses in many aspects of whole plant biology and systematics in schools and universities;

Accepting that botanic gardens are a unique resource for increasing the understanding of plants and plant conservation;

The International Conference on Botanic Gardens and the World Conservation Strategy:

Calls on Governments, conservation organisations, schools and colleges, industry and concerned people to support educational programmes in botanic gardens by funding, moral support and direct involvement;

Urges botanic gardens to:

develop education programmes for people of different ages, backgrounds and interests;

engage people with professional training in education and botany to develop and implement programmes;

promote conservation education both within gardens and in the broader community;

engage in cooperative education programmes with conservation agencies, zoos, museums, societies and clubs;

Also calls upon botanic gardens and all media agencies, particularly television, to work together to generate a fuller understanding of how the plant kingdom is essential for the survival and well-being of all people on earth.

Recommendation 12

Liaison and training

Aware of the urgent need and desire of botanic gardens to develop long-term liaison through national, regional and twinning relationships leading to the North-South and South-South transfer of botanical science and technology;

Recognising the existence of training programmes offered by botanic gardens and their willingness to provide more opportunities for training;

The International Conference on Botanic Gardens and the World Conservation Strategy:

Recommends IUCN to validate formal agreements on long-term conservation action between two or more botanic gardens;

Recommends botanic gardens to provide suitable biological and horticultural training for staff responsible for critically threatened species;

Requests botanic gardens and other organisations to make known the existence of current training programmes and to encourage the provision of IUCN-validated conservation training.

Recommendation 13

The members of the International Conference on Botanic Gardens and the World Conservation Strategy:

Wish to express their profound gratitude to the Excmo. Gobierno de Canarias and to the Excmo. Cabildo Insular de Gran Canaria for the invitation to hold their meeting in Las Palmas de Gran Canaria, famed throughout the world for the beauty and rarity of its extraordinary plant life;

Congratulates the Excmo. Cabildo Insular on its foresight and commitment in establishing and maintaining the Jardín Canario Viera y Clavijo which is widely admired as a model, and in implementing their campaign to "Make Gran Canaria Green Again";

Welcomes the Declaration of Gran Canaria, which will be seen as a major contribution to the world conservation movement;

Wishes to thank all the sponsoring and supporting institutions for their generous financial support and all the individuals, especially the Director and Staff of the Jardín Canario Viera y Clavijo, for their work in making this conference possible.